EXTREME FINANCIAL RISKS AND ASSET ALLOCATION

Series in Quantitative Finance

ISSN: 1756-1604

Series Editor: Ralf Korn *(University of Kaiserslautern, Germany)*

Editorial Members: Tang Shanjian *(Fudan University, China)*
Kwok Yue Kuen *(Hong Kong University of Science and Technology, China)*

Published

Vol. 1   An Introduction to Computational Finance
         by Ömür Uğur

Vol. 2   Advanced Asset Pricing Theory
         by Chenghu Ma

Vol. 3   Option Pricing in Incomplete Markets:
         Modeling Based on Geometric Lévy Processes and
         Minimal Entropy Martingale Measures
         by Yoshio Miyahara

Vol. 4   Simulating Copulas:
         Stochastic Models, Sampling Algorithms, and Applications
         by Jan-Frederik Mai and Matthias Scherer

Vol. 5   Extreme Financial Risks and Asset Allocation
         by Olivier Le Courtois and Christian Walter

Series in Quantitative Finance – Vol. 5

EXTREME FINANCIAL RISKS AND ASSET ALLOCATION

Olivier Le Courtois
EM Lyon Business School, France

Christian Walter
Fondation Maison des Sciences de l'Homme, France

Imperial College Press

Published by

Imperial College Press
57 Shelton Street
Covent Garden
London WC2H 9HE

Distributed by

World Scientific Publishing Co. Pte. Ltd.
5 Toh Tuck Link, Singapore 596224
USA office: 27 Warren Street, Suite 401-402, Hackensack, NJ 07601
UK office: 57 Shelton Street, Covent Garden, London WC2H 9HE

British Library Cataloguing-in-Publication Data
A catalogue record for this book is available from the British Library.

Series in Quantitative Finance — Vol. 5
EXTREME FINANCIAL RISKS AND ASSET ALLOCATION

Copyright © 2014 by Imperial College Press

All rights reserved. This book, or parts thereof, may not be reproduced in any form or by any means, electronic or mechanical, including photocopying, recording or any information storage and retrieval system now known or to be invented, without written permission from the Publisher.

For photocopying of material in this volume, please pay a copying fee through the Copyright Clearance Center, Inc., 222 Rosewood Drive, Danvers, MA 01923, USA. In this case permission to photocopy is not required from the publisher.

ISBN 978-1-78326-308-0

Printed in Singapore

Foreword

Each financial crisis calls, by its novelty and the mechanisms it shares with preceding crises, for the appropriate means to analyze financial risks. In *Extreme Financial Risks and Asset Allocation*, Olivier Le Courtois and Christian Walter present in an accessible and timely manner the concepts, methods, and techniques that are essential for the understanding of these risks in an environment where asset prices are subject to sudden, rough, and unpredictable changes. These phenomena, mathematically known as "jumps", play an important role in practice. Their quantitative treatment is generally tricky and is sparsely tackled in similar books. One of the main appeals of this book by Olivier Le Courtois and Christian Walter rests in their approachable and concise presentation of the *ad hoc* mathematical tools without sacrificing the necessary rigor and precision.

Before concentrating on the measurement of risks and the optimal allocation of assets, it is necessary to construct models that allow the representation of price evolutions. In order to be realistic, these models must include the impact of potential crises and therefore the possibility of jumps. The need to analyze economic and financial time series is at the heart of the use of new stochastic processes. Brownian motion, introduced by Louis Bachelier in 1900, is the most classic example. However, Brownian motion does not permit the incorporation of jumps in the modeling. Nowadays, financial theories use a large panel of models that include in particular Lévy processes. These processes display jumps and constitute a natural generalization of Brownian motion.

The Gaussian hypothesis associated with Brownian motion, and constant volatility and drift terms, is often not very realistic from a finance viewpoint. However, this hypothesis leads to explicit results such as the Black–Scholes formula for option prices. Theoretical developments making use of hypotheses that are more general than the Gaussian one use the same dynamic replication and absence of arbitrage concepts as those impregnating the analysis by Merton of the Black–Scholes model. These developments are not limited to the Brownian/Gaussian case. Olivier Le Courtois and Christian Walter's book allows the application of these developments to a Lévy process framework; for instance, a stable Paretian marginal distribution or any other representation related to a Lévy process.

The first six chapters of the book are dedicated to the detailed study of the non-Gaussian laws and the Lévy processes that are relevant for the analysis of risks and the modeling of the impact of crises on asset prices. Chapters 7 and 8 provide the risk management tools, and Chapters 10 and 11 develop the portfolio theory of optimal asset allocation. Chapter 9 models the psychology of risk.

This book offers an excellent synthesis of the academic literature in a clear, ordered, and intuitive way. The authors do not just stick together the main contributions and results of the field: they order them based on an intuitive reasoning. The continuous-time theory of the choice of portfolio is exposed with a particular care when asset dynamics are modeled with processes having a jump component. This is a technically difficult topic that is tackled here with a lot of clarity. The authors also propose original approaches. The reader can for instance benefit from the relations that are constructed between the theory of extreme values and the study of the maxima of Lévy processes, or from the computation of Value-at-Risk (VaR) and Conditional Tail Expectation (CTE) using Fast Fourier Transforms (FFTs). The reader can also discover an interesting presentation of static portfolio choice with moments of order three and four and interesting economic interpretations of the asymmetry and kurtosis coefficients.

The synthesis and all the contents comprising this book allow the reader to save precious time by facilitating the access to the most recent techniques. Further, the authors show that the theoretical developments related to Lévy processes are not only interesting from a mathematical viewpoint, but also bring solutions to concrete management problems. Numerous illustrations confirm this point.

This book comprises theories and methods that can usually be found in highly technical books destined for mathematicians or in scattered, and most often very recent, research articles. The authors have created a remarkable pedagogical work that makes these difficult results accessible to a large readership. Master's and Ph.D. students, but also financial engineers, will find this book useful. And no doubt, the book will also reside in the personal libraries of the researchers in the field.

Yacine Aït-Sahalia
Otto Hack 1903 Professor of Finance and Economics
Director, Bendheim Center for Finance
Princeton University

Preface

This book is the result of a long thought process undertaken by the authors on the use of Lévy dynamics in finance. From a practical standpoint, the book reflects a work program carried out in collaboration with the insurance group SMABTP. The goal of this program was to think of potential ways of improving the theoretical representations of finance used by practitioners in order to better take their concerns into account. More precisely, the scientific project was divided into three steps. The first step was to find a scientific approach that could explain the empirical management practices not currently addressed by mainstream finance theory. The second step was to create new possibilities for the application of risk computations and portfolio management methods. The final step was to promote the results obtained, which yielded a new approach to uncertainty in daily management, to regulators and supervisory authorities.

The origin of this program can be traced back to the following problem. The desired proximity of the models to the empirical needs of practitioners cannot be perfectly achieved within the paradigm of mainstream finance theory. This theory is based on a principle of continuity that dates back to Leibniz and that has long been abandoned by the physical sciences. However, neoclassical economics, and now modern finance, have maintained the principle despite a large body of empirical evidence to the contrary. Indeed, financial phenomena are by nature discontinuous. Our goal is simple: we wish to take into account in an appropriate way, by use of updated models, the discontinuity that lies at the heart of all scales of financial phenomena.

We first started by exploring the most visible group of discontinuities: those that occur over large scales, as examined by extreme value theory. However, the analysis of the financial phenomena rapidly showed that they should not be cut into two parts: smooth market periods alternating with turbulent ones. The inability to construct a unified version of market dynamics can only make investors helpless when extreme events, unpredictable by definition, occur. Therefore, it became clear that a change in the methods and techniques of finance corresponded to a change in the theoretical representation of the nature of financial risk itself.

The exploration of the modeling possibilities offered by Lévy processes in their modern (non-fractal) version was then conducted in a systematic way. The understanding of the precise articulation of the non-fractal Lévy processes and the first (non-Gaussian) law of Laplace allowed us to further revise the models and to adapt them to the needs of practitioners. The reconstruction of Lévy processes by using convolutions of Laplace distributions yielded very encouraging practical results in the context of the computation of risk budgets under the Solvency II Directive and in the context of the design of portfolio allocation strategies without benchmarks. Ultimately, the 2007–2008 crisis highlighted the relevance of our approach by casting doubt on the extant theoretical representations and the ensuing professional practices.

The collaboration with SMABTP was conducted under the leadership of Hubert Rodarie, Deputy CEO of the company. This work was made possible thanks to a close partnership led with this company's teams in direct link with Philippe Desurmont and Mohamed Majri, Managing Director of SMAGestion and Head of Financial Modeling, respectively. This interaction, and the very useful engineering contribution of Anthony Floryszczak, enabled in-depth programming and statistical work to be performed. It was therefore possible to move from an initial mathematical concept (Lévy processes resulting from a mix of Laplace laws) to a substantial practical improvement in professional finance, both in terms of portfolio and risk management. The models presented in this book have all been implemented and integrated in the thought toolbox. Some of these models enable the production of regular VaR reports that are included in the Group's reporting documents and that satisfy its regulatory obligations in terms of risk.

The studies described in this book gave rise to presentations at the conferences of the French Finance Association (AFFI); the European Financial Management Association (EFMA); the Actuarial Approach for Financial Risk (AFIR); the Australasian Finance & Banking Conference (AFBC); the Campus for Finance; the French Society of Statistics (SSF); and the Insurance: Mathematics and Economics (IME) association. Additionally, the studies related to the computation of risk budgets were presented in two research seminars at the Universities of Lille and Hitotsubashi and at a workshop organized by RiskLab-Madrid. These studies received the prize for the best article presented at the 2010 International Congress of Actuaries (ICA). We wish to thank the participants of these seminars and conferences for the fruitful debates that subsequently allowed us to improve our work.

Another source of inspiration led to a closer historical analysis of the probabilistic paradigm of the finance theory. The discussions that took place on the seminar on the history of probability and statistics, hosted by the Centre d'Analyse et de Mathématiques Sociales (CAMS) at the École de Hautes Études et Sciences Sociales (EHESS) and during the History and Epistemology of Finance Research program, of the Fondation de la Maison des Science de l'Homme (FMSH), were extremely beneficial. We express our greatest thanks to all of the colleagues who participated in these discussions.

We wish to thank Carole Bernard, Abdou Kélani, François Quittard-Pinon, Rivo Randrianarivony, Aurélie Sannajust, Lorenz Schneider, and Jacques Lévy-Véhel for their detailed reading of the initial version of this text. Their comments allowed us to greatly improve the quality of our book. Further, this book benefited from numerous exchanges with several colleagues, including Michel Armatte, Marc Barbut, Jean-Philippe Bouchaud, Éric Brian, David Crainich, Amy Dahan, Louis Eeckhoudt, Monique Jeanblanc, Benoit Mandelbrot, and Daniel Zajdenweber.

Despite several rereadings, we are solely responsible for any errors or mistakes that may remain in the text.

Contents

Foreword		v
Preface		ix
1. Introduction		1
2. Market Framework		9
2.1 Studied Quantities		10
2.1.1 Financial Assets		10
2.1.2 Portfolios		12
2.1.3 Distribution Parameters		19
2.2 The Question of Time		22
2.2.1 Choosing the Measure of Time		22
2.2.2 Choosing the Scale of Time		25
3. Statistical Description of Markets		31
3.1 Construction of a Representation		32
3.1.1 Role of the Statistical Description		32
3.1.2 Continuous or Discontinuous Representations		32
3.2 Normality Tests		34
3.2.1 The Pearson–Fisher Coefficients		35
3.2.2 Kolmogorov Test		37
3.3 Discontinuity Test		39
3.3.1 Definition of Estimators		39
3.3.2 Confidence Intervals		41
3.4 Continuity Test		45
3.4.1 Definition of the Estimators		45

		3.4.2	Confidence Interval	46
	3.5	Testing the Finiteness of the Activity		47
		3.5.1	Construction of the Tests	48
		3.5.2	Illustration	50
4.	Lévy Processes			53
	4.1	Definitions and Construction		54
		4.1.1	The Characteristic Exponent	54
		4.1.2	Infinitely Divisible Distributions	54
		4.1.3	A Construction with Poisson Processes	55
	4.2	The Lévy–Khintchine Formula		60
		4.2.1	Form of the Characteristic Exponent	60
		4.2.2	The Lévy Measure	62
	4.3	The Moments of Lévy Processes of Finite Variation		67
		4.3.1	Existence of the Moments	68
		4.3.2	Calculating the Moments	69
5.	Stable Distributions and Processes			77
	5.1	Definitions and Properties		78
		5.1.1	Definitions	78
		5.1.2	Characteristic Function and Lévy Measure	81
		5.1.3	Some Special Cases of Stable Distributions	90
		5.1.4	Simulating Paths of Stable Processes	94
	5.2	Stable Financial Models		100
		5.2.1	With Pure Stable Distributions	100
		5.2.2	With Tempered Stable Distributions	101
6.	Laplace Distributions and Processes			105
	6.1	The First Laplace Distribution		106
		6.1.1	The Intuitive Approach	106
		6.1.2	Representations of the Laplace Distribution	108
		6.1.3	Laplace Motion	117
	6.2	The Asymmetrization of the Laplace Distribution		129
		6.2.1	Construction of the Asymmetrization	129
		6.2.2	Laplace Processes	134
	6.3	The Laplace Distribution as the Limit of Hyperbolic Distributions		136
		6.3.1	Motivation for Hyperbolic Distributions	138

| | | 6.3.2 | Construction of Hyperbolic Distributions | 139 |
| | | 6.3.3 | Hyperbolic Distributions as Mixture Distributions | 143 |

7. The Time Change Framework — 147

- 7.1 Time Changes — 148
  - 7.1.1 Historical Survey — 148
  - 7.1.2 A First Modeling Example — 149
- 7.2 Subordinated Brownian Motions — 155
  - 7.2.1 The Mechanics of Subordination — 155
  - 7.2.2 Construction of a Time Change — 158
  - 7.2.3 Brownian Motion in Gamma Time — 165
- 7.3 Time-Changed Laplace Process — 173
  - 7.3.1 Mean-Reverting Clock — 174
  - 7.3.2 The Laplace Process in ICIR Time — 178

8. Tail Distributions — 181

- 8.1 Largest Values Approach — 181
  - 8.1.1 The Laws of Maxima — 182
  - 8.1.2 The Maxima of Lévy Processes — 190
- 8.2 Threshold Approach — 194
  - 8.2.1 The Law of Threshold Exceedances — 194
  - 8.2.2 Linearity of Means beyond Thresholds — 198
- 8.3 Statistical Phenomenon Approach — 202
  - 8.3.1 Concentration of Results — 202
  - 8.3.2 Hierarchy of Large Values — 216
- 8.4 Estimation of the Shape Parameter — 220
  - 8.4.1 A New Algorithm — 221
  - 8.4.2 Examples of Results — 224

9. Risk Budgets — 227

- 9.1 Risk Measures — 228
  - 9.1.1 Main Issues — 228
  - 9.1.2 Definition of the Main Risk Measures — 230
  - 9.1.3 VaR, TCE, and the Laws of Maximum — 233
  - 9.1.4 Notion of Model Risk — 235
- 9.2 Computation of Risk Budgets — 242
  - 9.2.1 Numerical Method — 242
  - 9.2.2 Semi-Heavy Distribution Tails — 247

| | | 9.2.3 | Heavy Distribution Tails | 250 |

10. The Psychology of Risk — 253

- 10.1 Basic Principles of the Psychology of Risk … 254
  - 10.1.1 The Notion of Psychological Value … 254
  - 10.1.2 The Notion of Certainty Equivalent … 255
- 10.2 The Measurement of Risk Aversion … 256
  - 10.2.1 Definitions of the Risk Premium … 256
  - 10.2.2 Decomposition of the Risk Premium … 258
  - 10.2.3 Illustration … 264
- 10.3 Typology of Risk Aversion … 267
  - 10.3.1 Attitude with Respect to Financial Risk … 268
  - 10.3.2 The Family of HARA Functions … 269

11. Monoperiodic Portfolio Choice — 275

- 11.1 The Optimization Program … 277
- 11.2 Optimizing with Two Moments … 279
  - 11.2.1 One Risky Asset … 280
  - 11.2.2 Several Risky Assets … 282
- 11.3 Optimizing with Three Moments … 284
  - 11.3.1 One Risky Asset … 284
  - 11.3.2 Several Risky Assets … 288
- 11.4 Optimizing with Four Moments … 289
  - 11.4.1 One Risky Asset … 289
  - 11.4.2 Several Risky Assets … 292
- 11.5 Other Problems … 294
  - 11.5.1 Giving up Comoments … 294
  - 11.5.2 Perturbative Approach and Normalized Moments … 296
- Appendix: Dealing with Uncertainty … 297

12. Dynamic Portfolio Choice — 303

- 12.1 The Optimization Program … 304
  - 12.1.1 The Objective Function … 304
  - 12.1.2 Modeling Stock Fluctuations … 308
- 12.2 Classic Approach … 315
- 12.3 Optimization in the Presence of Jumps … 319
  - 12.3.1 Presentation of the Model … 319
  - 12.3.2 Illustration … 322

Appendix: Dealing with Uncertainty 325

13. Conclusion 331

Appendix A Concentration vs Diversification 333

Bibliography 341

Index 349

Chapter 1

Introduction

In physics, the principle of continuity states that change is continuous rather than discrete. Leibnitz and Newton, inventors of differential calculus, said "Natura non facit saltus" (nature does not make jumps). This same principle underpinned the thoughts of Linné on the classification of species and later Charles Darwin's theory of evolution (1859). In 1890, Alfred Marshall's *Principles of Economics* assumed the principle of continuity, allowing the use of differential calculus in economics and the subsequent development of neoclassical economic theory. Modern financial theory grew out of neoclassical economics and naturally assumes the same principle of continuity.

One of the great success stories of modern financial theory was the valuation of derivatives. Examples include the formulas of Fisher Black, Myron Scholes, and Robert Merton (1973) for valuing options, and the subsequent fundamental theorem of asset valuation that emerged from the work of Michael Harrison, Daniel Kreps, and Stanley Pliska between 1979 and 1981.

In the 20th century, both physics and genetics abrogated the principle of continuity. Quantum mechanics postulated discrete energy levels while genetics took discontinuities into account. But economics – including modern financial theory – stood back from this intellectual revolution. An early attempt by Benoit Mandelbrot in 1962 to take explicit account of discontinuities on all scales in stock market prices led to a debate in the profession. By the 1980s the consensus reaffirmed the principle of continuity – and this despite the repeated financial crises following the 1987 stock market crash. Many popular financial techniques, such as portfolio insurance or the calculation of capital requirements in the insurance industry assume that (economic) nature does not make jumps. Most statistical descriptions of time series in finance assume continuity. The predominant example of this is Brownian motion.

One consequence of the assumed continuity is the truncation of financial time series into "normal" periods and periods of "insanity" where markets are "irrational". This dichotomy leaves the profession unable to explain the transition from one period to another. For example, in an editorial for the Financial Times (16.3.08) Alan Greenspan commented on the financial crisis of 2007–2008, "We can never anticipate all discontinuities in financial markets." For Greenspan, (financial) nature does not make jumps. This demonstrates the limits of traditional risk management.

The object of this book is to provide a remedy. We will take explicit account of discontinuities using non-Brownian probability distributions. We will apply the resulting models both to risk management and to portfolio construction. We believe that the difficulties of traditional risk management techniques are not the result of discontinuities *per se* but rather of the failure of traditional models to encompass them. We propose a more general approach putting discontinuities at the heart of the model. The traditional continuous model is merely a special case of the above.

We propose a more general representation using Lévy processes. Lévy processes are probabilistic descriptions of random dynamics with jumps, whose marginal probability distributions are non-normal. By contrast, Brownian motion is a particularly simple form of Lévy process: a continuous process whose underlying probability distribution is normal. The discontinuous nature of Lévy processes makes them particularly useful for measuring and controlling risk in finance, particularly for banks and insurance companies that are faced with frequent jumps in asset prices (e.g. exogenous shocks, micro crises in liquidity, etc). Lévy processes allow us to model these discontinuities at all scales.

Lévy processes allow us to use the same techniques for large and small discontinuities. Large discontinuities, or extreme financial risks, are the subject of "Extreme Value Theory". Examples include the absence of a functioning market, liquidity crises, or large jumps in equity prices that can be observed from time to time. Prudential regulation (Basel III, Solvency II, UCITS IV) seeks to protect institutions against such events, but considers these risks as a separate category.

We believe that understanding small-scale discontinuities reveals the intrinsic fragility of a market and thus its exposure to large-scale discontinuities, rather like geologists who examine small-scale faults to determine the probability of large-scale events. Therefore the modeling of these small-scale discontinuities (such as temporary shortages of liquidity) improves the identification and management of large-scale risks. Lévy processes – which operate at all scales – allow us to understand extreme financial risks with the same tools that are used for small changes in asset prices. To summarize, discontinuities are a general feature of the paths of asset prices, whatever the scale.

This proposition is in the spirit of Mandelbrot but without the scale invariance of the models of the 1960s. The early models of Mandelbrot used Lévy processes that were stable in addition (scale invariant). The Lévy processes we shall use do not have this property, unless they contain an *ad hoc* additional specification. Such an *ad hoc* specification is too strong an assumption if we are to represent actual market movements.

How shall we name markets that are "continually discontinuous"? Mandelbrot felt that the name should reflect the fractured nature of the market. He coined the term "fractal" (from the Latin *fractus*, meaning fractured) to indicate the discontinuities at all scales. The way in which we describe markets in this book leads us to call them "fractal markets" as opposed to "regular markets". However, the term "fractal" also connotes scale invariability or long-memory processes, which we do NOT assume in general.

On another linguistic level, Mandelbrot proposes two measures of variability: smooth ("normal" markets) and granular (markets with discontinuities). The Lévy–Khintchine formula incorporates this distinction. The "smooth" component of asset price variations is measured by a diffusion process while the granular component is measured by a pure-jump Lévy process. If we use that distinction, volatility is a risk parameter describing the scale of variation while granularity describes its discontinuities. The choice of vocabulary remains open.

This book contains 11 chapters. Chapter 2 introduces basic concepts and defines the debate regarding the appropriate measure of time. The first part of the chapter introduces general notation common to the different models used in the research literature. The second part introduces two possible measures of time: calendar time (corresponding to physical time and measured by a clock) and trading time (corresponding to market quotations and measured by a stochastic clock). These two measures of time yield different analytical and modeling frameworks. The issues here are the choice of stochastic clock for measuring trading time and the choice of resolution scale for measuring calendar time. The rules for changing time scales are presented in a summary table. From this it can be seen that the "square root of time" rule underlying the regulatory requirements

(Basel III, Solvency II, UCITS IV) for calculating minimum capital is a very narrow subset of scale rule for stock fluctuations.

Chapter 3 presents the statistical characteristics of financial phenomena. The object is to highlight discontinuities in asset price changes on all scales – in small changes as well as in large ones. Recently discovered statistical tests allow us to measure these discontinuities in a number of real assets. These results both validate the use of discontinuities on all scales and allow us to maintain a diffusion process. They also show that an infinite activity (average arrival rate of jumps) is a reasonable representation of the discontinuities in asset price changes.

Having established the presence of discontinuities and of an infinite activity, we tackle in Chapter 4 the technical tools allowing us to set up a probabilistic model where jumps can appear at any time – Lévy processes. The object is to present them as simply and as intuitively as possible. We construct and comment upon the Lévy–Khintchine formula, which is the characteristic function of every Lévy process. We begin by studying the simple Poisson process, then extend the discussion to more elaborate processes. This is followed by a detailed analysis of the components of this formula, in particular of its most difficult component, the Lévy measure. The chapter ends with a calculation of the moments of Lévy processes with finite variation.

The marginal probability distributions of Lévy processes are infinitely divisible. There are two families of distributions: those that are stable by addition, and those that are not. "Stability by addition" is defined as follows: given two or more random variables each of which is stable by addition, their sum will have the same distribution as the original variables, but with appropriately adjusted parameters. This is the fractal property of stable distributions that defines the fractal structure of markets. This implies the uniform nature of risk on all scales of observation and for all investment horizons. However, this fractal hypothesis is too strong to apply to real-world financial markets. That is why financial models that used stable distributions needed to be truncated with exponential distributions – the exception being Mandelbrot's first model. In Chapter 5 we introduce stable distributions and the random variables drawn from them. These allow us to build one class of financial models.

Chapter 6 introduces another family of continuous distributions, which are not, however, stable by addition. These are Laplace distributions. The reason for choosing Laplace distributions, as we explain at the start of Chapter 6, is the First Law of Errors discovered by Laplace in 1774, four years before Gauss and Laplace simultaneously discovered the Second Law (the normal distribution) in the context of the method of least squares. Taking note of the close relation of the two Laws but also noting the differences in their probabilistic descriptions, we begin by setting out the criterion for a choice of such a Law best describing the multiple observations of the real world. We then describe the properties of the Laplace distribution. This sets the scene for a family of Laplace distributions generalized by asymmetry and then by convolution. The different stochastic processes generated by various Laplace distributions – simple or general – are then examined. We introduce a new, more general, terminology which allows us to show that several financial models are special instances of Laplace processes. The chapter ends with the connection between Laplace distributions and hyperbolic distributions.

The notion of social time of exchanges was introduced in Chapter 2. One property of a Lévy process is that its characteristics can be changed by altering the notion of time. Any Lévy process in calendar time can be written as Brownian motion in trading time. Chapter 7 deals mainly with the alteration of time. We display the basic model of stochastic time. We start with a simple example to demonstrate the principle of different time measures. Thereafter we generalize and present the calculation for the transformation of moments using different clocks. We examine Brownian motion using a gamma clock and show that it is transformed into a Laplace process in calendar time. If we measure market activity by the volume of shares traded, it may be that in certain cases volumes show persistence, that is, successive trading volumes are not statistically independent. This case is covered at the end of the chapter.

Chapter 8 tackles the tails of distributions. Remember that our method is to suppose the existence of discontinuities at all scales of variation, in contrast to pure "extreme value theory" which separates the market into "calm" and unpredictable "catastrophic" periods. Nevertheless, since large-scale discontinuities show up as large changes, it is useful to present extreme value theory in this chapter. But true to the theme of this book, we do so in the context of Lévy processes. We then show statistical evidence for a well-known (by market professionals) but poorly researched phenomenon: the concentration of portfolio returns due to a small number of stocks and/or a small number of trading days. We end Chapter 8 with a response to one of the problems of extreme value theory: what is the threshold for "extreme"? We present a new method for automatically detecting the optimal threshold.

One of the purposes of this book is to improve the reliability of risk budgets using models that incorporate discontinuities on all scales. We will use Lévy processes to calculate rapid and robust measures of risk. Chapter 9 shows how risk budgets are calculated with the traditional measures (VaR, TCE, CVaR) put forward by European directives (Basel III, Solvency II, UCITS IV). These measures are calculated using Laplace processes, first in calendar time and then in trading time.

Chapter 10 covers decision-making tools for asset allocation. Whether in a monoperiodic or dynamic context, the choice of a portfolio asset allocation requires an appreciation and classification of risks. After a brief review of the notions of utility functions and the certainty equivalents of random variables, this chapter concentrates on the notion of risk premium. We shall use the Arrow–Pratt principle at orders 2, 3, and 4, which will allow us to introduce the notions of risk aversion, prudence, and temperance respectively. We give a simple example to illustrate decision-makers' different attitudes to statistical moments. The chapter concludes with a study of Hyperbolic Absolute Risk Aversion (HARA) utility functions and their different orders of risk aversion.

The next two chapters deal with the asset allocation procedure. Chapter 11 presents a single-period optimizer given a risk-free asset and one or more risky assets, taking into account the first four moments of the return distribution. After a general discussion of optimizers, we show various solutions taking different moments into account. The chapter is structured to move from the simple to the complex. We begin by reviewing the well-known mean-variance model before introducing higher moments. We then solve the model for three moments (mean, variance, skewness). Later we introduce the fourth moment and show the solution. These models are specified initially with one risky asset and then with several risky assets. Finally, we show simplifications that allow us to make optimizers operational. Chapter 11 concludes with an appendix on how to treat uncertainty in a static framework.

Chapter 12 deals with the above in a multi-period framework. We display in a unified manner the solution to Hamilton–Jacobi–Bellman (HJB) equations with an arbitrary number of risky assets, each of which follows a process that has a diffusion component and a jump component. This chapter is a review of portfolio construction given the diffusion process, allied to a jump process in a dynamic framework. We begin by specifying the general problem of dynamic optimization using Hamilton–Jacobi–Bellman equations. We then model the asset returns of the portfolio's constituents, modeling in particular the Dynkin operator of each asset. After inserting the results into the HJB equation we obtain an integro-differential equation whose solution yields the weights of the dynamic portfolio. It turns out that classic portfolio choice theory in continuous time (the Merton models of 1969 and 1973, reviewed in this chapter) is a special case of our general model. A second example, this time with jumps, is the model of Aït-Sahalia, Cacho-Diaz, and Hurd (2009). The appendix treats the problem of parameter uncertainty in a multi-period framework.

The final appendix examines the connections between Lévy processes, risk management and portfolio management. The problem of portfolio diversification versus portfolio concentration is examined from a Value at Risk perspective assuming asset returns follow a Lévy process. The conclusions – which may be surprising – depend on the nature of the underlying risks.

Chapter 2

Market Framework

This chapter describes the analytical framework for financial assets and the system of notation used in this book. It also presents the debate on the measure of the time of markets. We begin by proposing a simple and general formalized convention for assets and portfolios. The notation introduced will be applied throughout this book. In some cases, the traditional notation found in the academic literature or in finance textbooks has been preserved. In other situations, we have used a more intuitive convention. We introduce in a rigorous way connections between empirical professional practices and quantitative finance theories. For example, we link the decomposition of performance to the notion of the stochastic integral. The question of the choice of simple or logarithmic returns is then tackled. We also perform the computation of the corresponding moments.

The second part of this chapter presents two possibilities for measuring the time of markets: calendar time, which corresponds to a physical clock, and the time of exchanges, which is driven by the market clocks of transactions and volumes. The two measures of time define two distinct analytical frameworks where the relevant questions are, respectively, choosing a resolution scale in calendar time and choosing a social clock in the time of exchanges. The transition from a short horizon to a long one requires the rules for the change of scale that will be presented at the end of the chapter.

2.1 Studied Quantities

We start by presenting the main quantities related to individual financial assets and portfolios and then recall the main parameters used by finance practitioners.

2.1.1 *Financial Assets*

We will denote in general the value of any financial asset (stock, bond, or other) at time t by $S(t)$ or S_t. If we need to be more specific regarding the studied asset, as in the case of the management of a portfolio, an indexation by j will be used, either as $S_j(t)$ or as S_t^j. The choice between these conventions will be justified by considerations of simplicity or context.

Investors and risk managers are interested in practice in three quantities that they try to model appropriately and describe statistically: gains (or losses), their simple arithmetic returns, and/or their continuous equivalent returns. We now present these quantities.

The first and most natural quantity is a gain (or loss) in euros incurred between times 0 and t. A gain is defined as a simple difference of asset prices:

$$G(t) = S(t) - S(0) \qquad (2.1)$$

The second quantity is the simple return of an asset expressed as a percentage of euros between times 0 and t. This rate, denoted by $R(t)$, is defined as the simple ratio:

$$R(t) = \frac{S(t) - S(0)}{S(0)} \qquad (2.2)$$

If for instance $S(0) = 100$ euros and $S(t) = 108$ euros for $t = 1$ year, it is clear that $R(t) = 8\%$ so that the gain is 8 euros for 100 euros invested over 1 year: rigorously 8%/year including the time unit in the same way as a speed can be expressed in km/h. This simple return represents the natural space for financial computations because it corresponds to real money gains or losses, so to the money that the investor gains or loses at the end of his investment. In the case of an insurance company, 100 euros can represent the premiums received by the insurer and the question becomes that of a portfolio choice comprising payments of benefits in euros: the natural space of computations is again that where euros (or other cash units) are used.

The third quantity is the continuous return equivalent to the simple return of an asset. This rate, denoted by $X(t)$, is defined by the difference of the logarithms of asset prices:

$$X(t) = \ln S(t) - \ln S(0) \tag{2.3}$$

This continuous equivalent rate represents the limit at infinity of a periodic compounding over periods that tend to zero. For example, if $S(0) = 100$ euros and $S(t) = 108$ euros for $t = 1$ year, then $X(t) = 7.70\%$. This computation convention differs from the case of simple returns, representing instead a toolbox for financial calculus. It is well known that one of the advantages of this convention is that it makes rises and falls symmetrical for an identical amount of euros. Indeed, if the price decreases from 108 euros to 100 euros, then $R(t) = -7.41\%$ and $X(t) = -7.70\%$: using logarithms symmetrizes appreciations and depreciations. This symmetrization is necessary in the case of forex rates where there is no reason to favor one direction over the other. It is useful in other situations. When modeling stock dynamics, we will use the quantity $X(t)$.

The link between $R(t)$ and $X(t)$ is direct:

$$X(t) = \ln \frac{S(t)}{S(0)} = \ln\left(1 + \frac{S(t) - S(0)}{S(0)}\right) = \ln(1 + R(t)) \tag{2.4}$$

Recalling that $\ln(1+x) \simeq x$ when x is small, it is clear that if the ratio $(S_t - S_0)/S_0$ is small, then $X(t) \simeq R(t)$. If the period $[0, t]$ is long, however, this rate can be quite different from the preceding rate: the result depends on the direction of the market's evolution. Conversely:

$$R(t) = e^{X(t)} - 1 \tag{2.5}$$

Financial computations are usually performed using $X(t)$. In particular, it is the stochastic process of $X(t)$ that will be modeled. However, in practice, portfolio choices are made based on previsions in euros (therefore using $R(t)$) and require the moments of $R(t)$. The relationship (2.5) will be used to move from the moments of $X(t)$ to those of $R(t)$.

We can move directly from the asset price at time 0 to the asset price at time t by using Eqs (2.2) and (2.3). In the natural space:

$$S(t) = S(0)\left(1 + R(t)\right) \tag{2.6}$$

and in the modeling space:

$$S(t) = S(0)\, e^{X(t)} \tag{2.7}$$

The gain at time t is as follows:

$$G(t) = S(0)\, R(t) \tag{2.8}$$

Table 2.1 Simple interest rates and their continuous equivalents

Interest rate	Daily	Monthly	Annual	Decennial
Simple (R)	0.020%	0.400%	5.000%	50.000%
Continuous equivalent (X)	0.020%	0.399%	4.879%	40.546%

The collections of values $(G(t), t \geq 0)$, $(S(t), t \geq 0)$, $(R(t), t \geq 0)$, and $(X(t), t \geq 0)$ represent processes for the gains, simple returns, and continuous returns of asset S for t positive. Graphically, $(S(t), t \geq 0)$ represents the trajectory of the asset price for t positive. In the analysis of stock fluctuations, we see that it is possible to examine either the logarithms of prices or the cumulative continuous returns, which are identical up to a constant $S(0)$.

In the next stage of our analysis, we think only in euros. As already seen, the natural space of computations is that of the simple return $R(t)$, because it is a simple transposition of euros into percentages. However, the modeling of stock dynamics is made based on the continuous equivalent $X(t) = \ln(1 + R(t))$ of the simple return. In Table 2.1, we display illustrations of this distinction conducted with respect to the investment horizon t.

2.1.2 Portfolios

We now concentrate on the description of portfolios.

2.1.2.1 Investment Strategy

The market is described by a set of assets whose prices are denoted by $S_j(t)$ or S_t^j at time t, for $j = \{1, \cdots, d\}$. The integer quantity d represents the size or dimension of the market. As previously observed, the return of the asset j over $[0, t]$ is denoted by $X_j(t)$ and defined as:

$$X_j(t) = \ln S_j(t) - \ln S_j(0) \qquad (2.9)$$

We examine the distributions of $X(t)$ with the parameters $\lambda(t)$. Each investment horizon t is at the origin of a set of parameters $\lambda(t)$ relative to the distribution of $X(t)$. When determining the composition of a portfolio, the decision process can be illustrated in the following way:

$$\boxed{\text{horizon } t \to X(t) \to \{\lambda(t)\} \stackrel{?}{\to} \text{portfolio at time } 0}$$

The first choice to be made is that of the investment horizon t. This horizon determines the return $X(t)$, whose analysis requires a study of the parameters $\lambda(t)$. The value of these latter parameters depends on the investment horizon. Then, following a given optimization procedure, we obtain the portfolio composition at time $t = 0$. For practical computations, this is the representation in euros selecting the relationship $R(t) = e^{X(t)} - 1$. We now set out these ideas in more detail.

An investment strategy is a collection of quantities of assets bought at time 0 for the horizon t. Several scales exist for the constitution of a portfolio, as presented below.

A first possible scale is a global one where strategic decisions are made. These decisions define an investment strategy or strategic asset allocation that is characterized by a distribution of assets by geographical zones, by industrial sectors, or by types of asset. These strategic decisions are often made when investing the wealth of an individual or the assets of an insurance company. The investment strategy is defined by large blocks, in percentage, with a given time horizon in mind.

Then, in some situations, the fund manager has the ability to vary the global allocation. These second-degree investment decisions are called tactical asset allocations. By analogy with military terminology, and in comparison with the strategy determined by the investment committee, the tactic chosen by the manager represents his ability to adapt to constraints or to changes in the environment. The manager operates on intermediate scales.

Finally, there exist local scales where the actual composition (stock picking) of the global masses defined above is decided. Note that management can choose to operate from the global to the local scale and *vice versa*. The corresponding investment strategies are called top down and bottom up. If most practitioners have preferred the top down approach over the past 30 years, an increasing number of managers today choose bottom up approaches. This amounts to recognizing the added value that lies in the choice of assets with respect to global investment allocations.

Two parameterizations exist for an investment strategy: by weights (percentages of the total portfolio) and by quantities (number of purchased assets). These two parameterizations are presented below.

Investment strategies parameterized by quantities: An investment strategy between times 0 and t is a vector $\theta = (\theta_1, \theta_2, \cdots, \theta_d)$ where θ_j is the number of shares of asset j within the portfolio. We first assume that the numbers of shares are unchanged between times 0 and t.

The value of the portfolio θ at time t is denoted by $V(t)$ and defined as:

$$V(t) = \sum_{j=1}^{d} \theta_j S_t^j \qquad (2.10)$$

We suppose that $V(0) > 0$: an initial input of funds exists and the manager is never in a short position.

In some situations, it is useful to display the bank account part of the portfolio separately, for instance, when dealing with mutual funds. This account aggregates the outcome of operations, such as brokerage fees, entry or lapse costs, subscriptions, and surrenders. The value of the bank account at time t is denoted by x_t.

Using this notation, the liquidation value of a portfolio $V(t)$ is such that at time t:

$$V(t) = x_t + \sum_{j=1}^{d} \theta_j S_t^j \qquad (2.11)$$

The right-hand side of Eq. (2.11) displays the decomposition of $V(t)$ into two components: cash and investments.

As before, we define the portfolio gain (or loss) in euros incurred at time t:

$$G(t) = V(t) - V(0) \qquad (2.12)$$

and the continuous return in percentage of euros at time t:

$$X(t) = \ln V(t) - \ln V(0) \qquad (2.13)$$

The relationships (2.9), (2.10), and (2.13) yield:

$$V_0 \, e^{X_t} = \sum_{j=1}^{d} \theta_j S_0^j \, e^{X_t^j}$$

so:

$$e^{X_t} = \sum_{j=1}^{d} \left(\frac{\theta_j S_0^j}{V_0} \right) e^{X_t^j}$$

The quantity $\theta_j S_0^j / V_0$ represents the proportion of the portfolio invested at time 0 in asset j. Let us denote it by w_0^j:

$$w_0^j = \frac{\theta_j S_0^j}{V_0} \qquad (2.14)$$

We obtain the relationship between the return of the portfolio $X(t)$ and the returns of the constituent assets $X_j(t)$:

$$e^{X_t} = \sum_{j=1}^{d} w_0^j e^{X_t^j} \qquad (2.15)$$

The value of the portfolio at time t can be expressed as a function of the expected returns of the individual assets:

$$V_t = V_0 \sum_{j=1}^{d} w_0^j e^{X_t^j}$$

To move back to simple returns, it is sufficient to recall that $X(t)$ is the continuous return equivalent to the simple return $R(t)$ such that $e^{X(t)} = 1 + R(t)$. The relationship (2.15) becomes:

$$1 + R(t) = \sum_{j=1}^{d} w_0^j \left(1 + R^j(t)\right)$$

so:

$$R(t) = \sum_{j=1}^{d} w_0^j R^j(t) \qquad (2.16)$$

because $\sum_{j=1}^{d} w_0^j = 1$.

This leads naturally to the second parameterization.

Investment strategies parameterized by weights: An investment strategy between the two times 0 and t is a vector $w = (w_1, \cdots, w_d)$ where w_j is the weight of asset j pertaining to the portfolio at time 0.

2.1.2.2 *Sources of Performance*

We now consider attribution analysis, which is the decomposition of performance into fundamental components related to investment decisions, either by asset class or at the asset scale. The goal is to attribute performance to each of the decisions made by the investment committee or the portfolio manager. This computation is in general performed over a given investment period. This is a periodic decomposition or attribution of performance.

We now present the decomposition principle in which we parameterize investment strategies by quantities. In particular, we assume from now on that these quantities are modified by investors over a discrete set of dates (indexed on $\mathbb{N}$ for the sake of simplicity) while asset prices are given at all times (so are indexed on $\mathbb{R}^+$).

Between any two dates $t-1$ and t, the process modeling the value of the portfolio takes three distinct values: V_{t-1} at time $t-1$, then, due to the natural evolution of the market, V_{t-} at time t^- before any decision of purchase or sale of assets, and finally V_t at time t, after the execution of the purchase and sales decisions. This evolution in value is illustrated below:

$$\underbrace{\cdots \longrightarrow V_{t-1}}_{\text{time } t-1} \longrightarrow \underbrace{V_{t-} \longrightarrow V_t}_{\text{time } t}$$

This evolution produces two types of variation. First:

$$V_{t-} - V_{t-1}$$

which corresponds to the natural evolution of the market. Second:

$$V_t - V_{t-}$$

which corresponds to surrender or subscription operations. The sum of these two terms gives the total variation in value:

$$\Delta V_t = V_t - V_{t-1} \tag{2.17}$$

In general, the two values V_{t-} and V_t are different. This is because mutual fund subscriptions and surrenders increase or decrease the size of the total assets independently from the market. In the absence of such effects, the manager must sell some assets if he wants to buy other assets. An investment policy without inflows or outflows of cash is called self-financed. In this situation, $V_{t-} = V_t$.

In the case of a self-financed strategy, the only variations in portfolio value are due to the gains or losses that affect portfolio lines or to various moves affecting the bank account. These effects can sometimes be quite large and represent an important part of the variation in portfolio value when the dates $t-1$ and t are close. We now present the decomposition of the variation in value.

At time $t-1$, after all transactions have been performed, the value of the portfolio is:

$$V_{t-1} = x_{t-1} + \sum_{j=1}^{d} \theta_{t-1}^j S_{t-1}^j \tag{2.18}$$

where the θ_{t-1}^j are set at time $t-1$ for the period $[t-1, t)$. The value of the portfolio can then be expressed at time t^- as follows:

$$V_{t-} = x_t + \sum_{j=1}^{d} \theta_{t-1}^j S_t^j \tag{2.19}$$

where we assume for the sake of simplicity that $x_{t^-} = x_t$ and $S^j_{t^-} = S^j_t$: the bank account and asset prices do not vary at the lapse/subscription times. The value of the portfolio can then be expressed at time t as:

$$V_t = x_t + \sum_{j=1}^{d} \theta^j_t S^j_t \tag{2.20}$$

To sum up, we can write the value process as:

$$\cdots \longrightarrow \underbrace{x_{t-1} + \sum_{j=1}^{d} \theta^j_{t-1} S^j_{t-1}}_{\text{time } t-1} \longrightarrow \underbrace{x_t + \sum_{j=1}^{d} \theta^j_{t-1} S^j_t \longrightarrow x_t + \sum_{j=1}^{d} \theta^j_t S^j_t}_{\text{time } t}$$

The variation in value of the portfolio over a period can then be decomposed as follows:

$$\Delta V_t = (x_t - x_{t_1}) + \sum_{j=1}^{d} (\theta^j_t - \theta^j_{t-1}) S^j_t + \sum_{j=1}^{d} \theta^j_{t-1} (S^j_t - S^j_{t-1}) \tag{2.21}$$

so, using the notation Δ:

$$\Delta V_t = \Delta x_t + \underbrace{\sum_{j=1}^{d} \Delta \theta^j_t S^j_t}_{\text{sum 1}} + \underbrace{\sum_{j=1}^{d} \theta^j_{t-1} \Delta S^j_t}_{\text{sum 2}} \tag{2.22}$$

Three terms appear: the variation in cash and two sums that we call respectively "sum 1" and "sum 2". Let us now analyze these three terms.

The first variation is that of cash (or bank account of the mutual fund) between the two dates $t-1$ and t. It is defined by the formula:

$$\Delta x_t = x_t - x_{t-1} \tag{2.23}$$

where Δx_t represents the cash variation related to all the management operations on the portfolio: subscriptions and lapses, revaluations of assets in the case of swaps, taxes, brokerage, management, and financial fees. Note that in some situations, when we want to compare the performance of a managed portfolio with that of a benchmark, this cash variation can be negative and constitute a non-negligible difference with respect to the benchmark.

The second variation ("sum 1") is produced by changes in the quantities of assets – keeping prices constant. The variation of the quantity of assets j in the portfolio is given by the formula:

$$\Delta \theta^j_t = \theta^j_t - \theta^j_{t-1} \tag{2.24}$$

This variation is the direct consequence of management decisions (or asset allocation changes). Therefore, the sum 1:

$$\sum_{j=1}^{d} \Delta\theta_t^j S_t^j$$

represents the effect of all the purchases and sales for all of the assets in the portfolio. This is a "quantity effect" when asset prices are given.

Equipped with this definition, we can return to the notion of a self-financed strategy. This is a management strategy where no inflows or outflows of funds occur. In this case, the only variations in value of the portfolio result from market fluctuations and there is no quantity effect. This allows us to relate the definition of any self-financed strategy to the condition "sum 1 = 0", so:

$$\sum_{j=1}^{d} \Delta\theta_t^j S_t^j = 0 \qquad (2.25)$$

which is another way of writing the condition $V_t = V_{t^-}$.

The other possible variation in the value of the portfolio stems from the "market effect", described by sum 2. This sum is composed of the variations of the market prices of all assets between times $t - 1$ and t. Written differently, it comprises the gains or losses on all assets during the period studied. The price variation is defined for asset j by the formula:

$$\Delta S_t^j = S_t^j - S_{t-1}^j \qquad (2.26)$$

and the sum 2:

$$\sum_{j=1}^{d} \theta_{t-1}^j \Delta S_t^j$$

represents the total gain or loss of the portfolio on its investments.

We can now compute the cumulative gain that results from the dynamic management of line j between times 0 and T. This gain is equal to:

$$\sum_{t=1}^{T} \theta_{t-1}^j \Delta S_t^j \qquad (2.27)$$

This latter expression is very interesting because if the price of asset j can be considered a random variable, then (2.27) represents a stochastic integral in discrete time. We see with this example the advantage of the convention Δ that was chosen at the beginning of this presentation.

Equipped with these results, we can express the total cumulative gain of the portfolio over the period of study, for a self-financed strategy, as:

$$G_t = \sum_{t=1}^{T} \Delta x_t + \sum_{j=1}^{d} \sum_{t=1}^{T} \theta_{t-1}^{j} \Delta S_t^{j} \qquad (2.28)$$

2.1.3 Distribution Parameters

In classic finance, the risk associated with the possession of a given asset is quantified by the probability distribution of gains (or losses) in euros (monetary units), so of the gains or losses of $S(t)$, equivalently of $R(t)$. We stress the interpretation of the quantity $R(t)$ and its parameters: it is the cumulative return of an asset over the period $[0, t]$. Seen from time 0, the date t is an investment horizon for the assets whose return evolution is modeled by the stochastic process $(R(t), t \geq 0)$. As noted before, when performing financial computations, the space of $X(t)$ is used for the various workings. In order to quantify a risk between times 0 and t, we will concentrate on the distribution of $X(t)$ and on its parameters.

2.1.3.1 Moments and Pearson–Fisher Coefficients

We assume from now on that, for any asset associated with the return $X(t)$ between times 0 and t, the following quantities exist when k is a non-negative integer.

(1) The moments of order k denoted by $\mu_k(t)$:

$$\mu_k(t) = \mathbb{E}[(X(t))^k] \qquad (2.29)$$

implying the existence of the expectation:

$$\mu(t) = \mathbb{E}[X(t)] \qquad (2.30)$$

and of the following moments.

(2) The centered moments of order k denoted by $m_k(t)$:

$$m_k(t) = \mathbb{E}\left[(X(t) - \mu(t))^k\right] \qquad (2.31)$$

(a) in particular the variance:

$$\text{var}(X(t)) = m_2(t) = \mathbb{E}\left[(X(t) - \mu(t))^2\right] \qquad (2.32)$$

(b) therefore the standard deviation (or volatility):

$$\sigma(t) = \sqrt{\operatorname{var}(X(t))} \qquad (2.33)$$

(3) The Pearson–Fisher asymmetry (or skewness S) coefficient:

$$S(t) = \frac{m_3(t)}{\sigma(t)^3} \qquad (2.34)$$

This coefficient expresses the renormalization of the centered moment of order 3 by the standard deviation of the distribution elevated to the power of 3. This normalization allows us to compare the asymmetry of any two distributions with differing standard deviations.
If the distribution is symmetrical, the odd centered moments are null, and therefore $S = 0$.

(4) The Pearson–Fisher tail thickness (or excess kurtosis K) coefficient:

$$K(t) = \frac{m_4(t)}{\sigma(t)^4} - 3 \qquad (2.35)$$

This coefficient expresses the renormalization of the centered moment of order 4 by the standard deviation of the distribution elevated to the power of 4. This normalization allows us compare the tail thickness of any two distributions with different standard deviations.

If the distribution is Gaussian, the even centered moments can be written as $m_{2p} = 2^p \Gamma(1/2 + p)/\Gamma(1/2)$, where Γ is the Euler gamma function. Thus, in the Gaussian setting, $m_4/m_2^2 = 4\Gamma(5/2)/\Gamma(1/2) = 3$, and $K = 0$.

Recall that the gamma function introduced by Euler in 1729 is defined, for $\alpha > 0$, by the equation:

$$\Gamma(\alpha) = \int_0^\infty x^{\alpha-1} e^{-x} dx \qquad (2.36)$$

In particular $\Gamma(1) = 1$ and, for all $\alpha > 1$, $\Gamma(\alpha + 1) = \alpha \Gamma(\alpha)$. If n is an integer, then $\Gamma(n+1) = n!$. Because the factorial number $n!$ does not exist for n non-integer, an intuitive interpretation of the gamma function is that it naturally extends the notion of factorial numbers to real (non-integer) numbers.

2.1.3.2 The Natural Computation Space

Let us come back to the problem of portfolio choice. As already noted, the goal is to examine $R(t)$ but not $X(t)$. Indeed:

$$m_k\left(R(t)\right) = \mathbb{E}\left[\left(R(t) - \mathbb{E}\left[R(t)\right]\right)^k\right]$$
$$= \mathbb{E}\left[\left(e^{X(t)} - 1 - \mathbb{E}\left[e^{X(t)} - 1\right]\right)^k\right]$$
$$= \mathbb{E}\left[\left(e^{X(t)} - \mathbb{E}\left[e^{X(t)}\right]\right)^k\right]$$
$$= m_k\left(e^{X(t)}\right)$$

Two situations arise for the computation of $m_k\left(R(t)\right)$, depending on whether $X(t)$ follows a marginal Gaussian law. In the first simple case where $X(t)$ follows a Gaussian distribution, the formulas allowing us to move from the moments of $X(t)$ to those of $R(t)$ are as follows:

$$\mathbb{E}\left[e^{X(t)}\right] = \exp\left(\mathbb{E}[X(t)] - \frac{1}{2}\text{var}(X(t))\right) \tag{2.37}$$

$$\text{var}\left(e^{X(t)}\right) = \left(e^{\text{var}(X(t))} - 1\right) e^{2\,\mathbb{E}[X(t)] + \text{var}(X(t))} \tag{2.38}$$

$$S\left(e^{X(t)}\right) = \left(e^{\text{var}(X(t))} + 2\right)\left(e^{\text{var}(X(t))} - 1\right)^{\frac{1}{2}} \tag{2.39}$$

$$K\left(e^{X(t)}\right) = \left(e^{4\text{var}(X(t))} + 2e^{3\text{var}(X(t))} + 3e^{2\text{var}(X(t))} - 3\right) \tag{2.40}$$

In any other situation, the following formulas can be used to compute the moments numerically:

$$\mathbb{E}\left[e^{X(t)}\right] = \int_{-\infty}^{+\infty} e^u\, f_{X(t)}(u) du \tag{2.41}$$

$$m_k\left(e^{X(t)}\right) = \int_{-\infty}^{+\infty} \left(e^u - \mathbb{E}\left[e^{X(t)}\right]\right)^k f_{X(t)}(u) du \tag{2.42}$$

where f_X is the density function of the random variable X.

We will come back in the following to the notion of asymmetry. We want only to stress here that the debate on the presence or the absence of asymmetry ($m_3 \neq 0$) concerns the marginal distribution of $X(t)$ but not that of $R(t)$. Knowing the transformation rule $R(t) = e^{X(t)} - 1$, the use of $R(t)$ creates an asymmetry on the shape of the distribution, but it stems only from the passage from a continuous rate to a simple rate. In the rest of this book, except when specified, we will only examine the asymmetry affecting $X(t)$.

Table 2.2 Symmetry or asymmetry?

Ω	$P(\omega)$	$X(t)$	$R(t) = e^{X(t)} - 1$	$S(t)$	$R(t) - \hat{\mu}$
ω_1	0.33	-25%	-22.12%	77.88	-24.21%
ω_2	0.33	0	0	100	-2.09%
ω_3	0.33	$+25\%$	$+28.40\%$	128.40	$+26.31\%$
$\hat{\mu}$		0%	$+2.09\%$	102.09	0%
S		0	0.15	0.15	0.15
K		1.50	1.50	1.50	1.50

Table 2.2 displays a trivial example showing how an asymmetry on $R(t)$ can appear when the fluctuations of $X(t)$ are nevertheless symmetrical. The distribution of $X(t)$ is assumed symmetrical about 0, at time t: $X(t,\omega_1) = -25\%$, $X(t,\omega_2) = 0\%$, and $X(t,\omega_3) = +25\%$. Recall that $S(t) = S(0) e^{X(t)} = S(0)(1 + R(t))$. Moving to the natural computation space yields here a positive asymmetry coefficient equal to 0.15.

2.2 The Question of Time

Let us now consider the description of the measurement of time. This question is essential for the characterization of the exchange phenomenon and of financial practices because it puts forward distribution probabilities (and therefore risk measures) with different shapes depending on the chosen measure.

In fact, two possibilities exists for measuring time: either by using a calendar clock (a quote index at every n time units) or by using a stochastic clock (a quote index at every n quotes for instance), which corresponds to a "social" clock, meaning that the intrinsic time of the market is driven by the pace of exchanges. We now present these two approaches. Then, for the case of the physical time, we discuss the problem of the change of time scale.

2.2.1 Choosing the Measure of Time

To tackle the question of the choice of a measure of time within financial models, it is necessary to begin with the precise description of a stock exchange without neglecting any of the variables that constitute the exchange. This leads us to consider three types of data: quotes, volumes, and numbers of transactions between any two given dates.

The value of any asset S at time t is denoted by $S(t)$ or S_t. The fundamental data of the exchange is the quotes. The asset value corresponding to the i-th quote is denoted by S_i and the number of securities exchanged at the time of this quote (or volume) is denoted by v_i. The total number of transactions occurring between time $t = 0$ and any given date t is denoted by $N(t)$. Put differently, between two dates 0 and t, we have three sets of data: times of quotes $t_0, t_1, t_2, \cdots, t_{N(t)}$, quotes $S_1, S_2, \ldots, S_{t_{N(t)}}$, and the number of securities exchanged for each quote i (volumes) $v_1, v_2, \ldots, v_{t_{N(t)}}$. The analysis of the financial phenomenon must therefore examine three different processes that are described below.

The first process, $(N(t), t \in \{t_0, t_1, t_2, \cdots, t_{N(t)}\})$, describes the number of quotes between times 0 and T. This is a counting process of trades. At any time t, $N(t)$ quotes were performed since time $t = 0$.

The second process, $(S(t), t \in \{t_0, t_1, t_2, \cdots, t_{N(t)}\})$, expresses the value of the asset between times 0 and T. At a given date t, the asset price is obtained by summing the past variations that we denote by $\Delta S_i = S_i - S_{i-1}$:

$$S(t) = S(0) + \sum_{i=1}^{N_t} \Delta S_i \qquad (2.43)$$

As previously seen, the cumulative return process is defined as $X(t) = \ln S(t) - \ln S(0)$.

The third process, $(\mathcal{V}(t), t \in \{t_0, t_1, t_2, \cdots, t_{N(t)}\})$, describes the number of assets traded between times 0 and T. This is a counting process of securities. At at given date t, $\mathcal{V}(t)$ assets have been exchanged since time $t = 0$. The total volume of assets traded between times 0 and t is simply the sum of the volumes attached to all trades:

$$\mathcal{V}(t) = \sum_{i=1}^{N_t} v_i \qquad (2.44)$$

Table 2.3 displays an example of quotes for Bouygues stock from the NYSE Euronext exchange. The transactions presented took place on November 24, 2008 between 10 h 51 min 06 s and 10 h 52 min 24 s.

Let us summarize these discussions. The price quoted at any time t results from three processes, namely trades, price variations, and volumes. Specifically, the price at time t is a consequence of the joint effect of these three factors.

The left panel of Fig. 2.1 displays the values of the fourth column of Table 2.3: the inter-quote durations (from bottom to top). We denote this duration by $\Delta t_k = t_k - t_{k-1}$. The first seven values of Δt_k appear from

Table 2.3 Bouygues stock transactions on November 24, 2008

Quote time	Quote (in euros)	Number of exchanged securities	Duration (in seconds)
10 : 51 : 06	26.060	660	
10 : 51 : 27	26.010	350	21
10 : 51 : 52	26.005	90	25
10 : 52 : 10	26.010	194	18
10 : 52 : 15	26.010	233	05
10 : 52 : 16	25.995	510	06
10 : 52 : 23	25.995	362	07
10 : 52 : 24	25.990	291	01

bottom to top, in seconds: $\{1, 7, 6, 5, 18, 25, 21\}$. The right panel of Fig. 2.1 represents the number of quotes $N(t_k)$ between 10 h 51 min 06 s and 10 h 52 min 24 s in social time, i.e. separated by the durations displayed in the left panel of the figure.

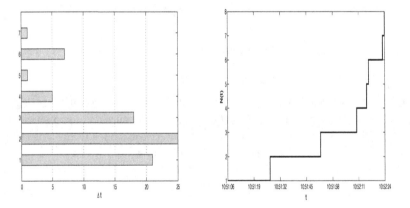

Figure 2.1 Evolution of Bouygues stock. Left panel: time Δt between two quotes, for the first 7 quotes. Right panel: number of quotes $N(t)$

At each quote time t_k, for $k = 1, \cdots, 7$, the number increases by one, explaining the stepped trajectory of the process N. Between two quote dates, there are no trades and the curve of the process N remains flat. For the value and cumulative return trajectories (not shown here) we would observe jumps with the respective amplitudes $\Delta S_k = S_k - S_{k-1}$ and $\Delta X_k = X_k - X_{k-1}$. Between two quote dates, the stock price $S(t)$ and the return $X(t)$ remain constant.

The trajectory that corresponds to the Table 2.3 data is displayed in Fig. 2.2. The inter-quote duration is not defined by the deterministic clock of physical time, but by the pace of trades that empties order books. This is the time of events, as opposed to the calendar time of the physical clock, which would be a time without events. This time of exchanges is defined in the social space of the market; this is why we use the term "social time". Because this social time is intrinsic to the market, everything happens as if markets were creating their own time. The clock that counts this time is modulated by the unpredictable arrival of buy and sell orders. This is therefore a stochastic clock. In order to model this clock, we choose a stochastic process (see Chapter 7).

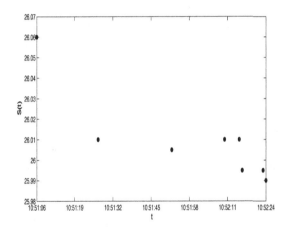

Figure 2.2 Trajectory of Bouygues stock in physical time

2.2.2 *Choosing the Scale of Time*

We now turn to the question of the determination of the scale of time that allows the risk of a financial investment to be measured adequately. We reformulate this question as follows. Does an appropriate time scale – a specific characteristic time – exist for the analysis of the financial phenomenon? Also, when several scales coexist, how can we move from one scale (for instance high frequency) to another (for instance low frequency)? These questions are developed below.

2.2.2.1 The Change of Scale Rule

This question can be formulated precisely in the following manner. Let $X(t)$ be the return of an asset (investment, portfolio, etc.) at time t. We recall that, seen from time 0, time t represents an investment horizon for an asset whose return dynamics are modeled by the stochastic process $(X(t),\ t \geq 0)$. In most practical situations, it is only possible to obtain the parameters of $X(t)$ (for example by statistical methods) for some values of t.

In many situations, it is necessary to characterize the parameters of a probability distribution for a very long term investment (a decade or more) from what can be estimated over a shorter period (one week or one month). In other situations, we may want to invest over the short term (say for one week) in a mutual fund for which we only know the yearly net asset values. Whatever the practical situation encountered, the question remains the same: we need to be able to move from one horizon t to another horizon at where a can be any multiplier or divider of t. In summary, the question is that of the passage from the distribution of $X(t)$ to that of $X(at)$:

$$\boxed{X(t) \xleftrightarrow{\ ?\ } X(at)}$$

To solve this problem in practice, we need to find the passage law that allows us to move from the horizon (scale) t to the horizon (scale) at. This passage law defines the change of scale relationships for the parameters of $X(t)$.

Let us consider the example of the computation of annualized volatilities. Assuming that is is possible to compute a weekly volatility (so, over a horizon equal to $t = 1$ week) and that the aim is to obtain a one year volatility (over 52 weeks), we need to know the passage rule (for the change of scale) for the volatility with $a = 52$. This question also arises for the asymmetry and kurtosis coefficients of $X(t)$: the passage rule defines a deformation of asymmetry and kurtosis. Is asymmetry dampened or reinforced (and how?) when changing scale? Is the distribution more or less leptokurtic when changing scale? This question arises most often for all of the moments and parameters of financial distributions.

Let us formulate this question mathematically. Let $\lambda(t)$ be a parameter of the distribution of $X(t)$. The goal is to know how to move from $\lambda(t)$ to $\lambda(at)$. This can be formulated as:

$$\boxed{\lambda(t) \xleftrightarrow{\ ?\ } \lambda(at)}$$

Specifically, the problem of the change of scale is difficult to solve for the moments because the passage rule from $m_k(t)$ to $m_k(at)$ for all k is not direct and depends on hypotheses regarding the aggregation of marginal distributions. These hypotheses correspond to different approaches to the uncertainty that affects the future returns of financial assets.

2.2.2.2 *The Hypotheses for the Change of Scale Rule*

To obtain the moments at a scale at knowing the moments at scale t, it is necessary to describe the relationships between the consecutive probability distributions of the asset considered. Let us study the return for the horizon n (for n integer). With $X_0 = 0$ (classical hypothesis), it is clear that:

$$X_n = (X_1 - X_0) + (X_2 - X_1) + \cdots + (X_n - X_{n-1}) = \sum_{k=1}^{n}(X_k - X_{k-1})$$

so, with the notation introduced before:

$$X_n = \sum_{k=1}^{n} \Delta X_k$$

We now introduce the probability distributions of consecutive returns. The probability distribution of the return ΔX_k is denoted by $P_k = P(\Delta X_k)$ whereas the probability distribution of the return X_n on the horizon n is denoted by $P = P(X_n)$. It thereby follows:

$$P(X_n) = P\left(\sum_{k=1}^{n} \Delta X_k\right) \tag{2.45}$$

We need to characterize this probability distribution. Let us try to reach an intuitive idea of the problem. Assume we want to compute the probability that the return at horizon n is equal to 20%, so that $\Pr(X_n = 20\%)$. Using formula (2.45), we can present the problem in the following form:

$$\Pr\left(\sum_{k=1}^{n} \Delta X_k = 20\%\right) \tag{2.46}$$

It is clear that the 20% target can be obtained in different ways, based on the values of the successive periodic returns, as long as their sum is indeed equal to 20%. However, these successive values depend on the assumptions that are made about the evolution of the cumulative return and therefore the relationship between the probability distributions of successive returns:

are increments independent? Are they stationary? We see that to solve this problem, we need to specify the characteristics of the between successive periodic returns. Several possibilities exist, as examined below.

In a first potential situation, successive periodic returns are neither independent nor stationary. This is the most general case in which it is not possible to obtain a simple solution for the computation of P: the value of $\Pr(X_n = 20\%)$ depends on all of the relationships between the periodic return distributions. It is therefore necessary to know the joint law of the n-uplet $(\Delta X_1, \cdots, \Delta X_n)$.

To simplify the problem, we can assume that periodic returns are independent. In this case, the computation of the probability distribution of the return over the horizon n is easier, because if two events are independent, the probability of their joint occurrence is equal to the product of their marginal probabilities of occurrence. In this context, the mathematical operator describing the behavior of a sum is the convolution product. Thus, if the successive periodic returns ΔX_k are independent, the value of $\Pr(X_n = 20\%)$ is obtained as the convolution product of the distributions of the ΔX_k. The convolution operator is denoted by $\otimes$. The probability distribution at scale n is given by the formula:

$$P = P_1 \otimes P_2 \otimes \cdots \otimes P_n \qquad (2.47)$$

where the independence assumption consequently allows us to simplify the computation of P.

We can simplify the problem further by assuming that periodic returns are not only independent but also stationary. In this situation, the distribution remains constant over time and $P_1 = P_2 = \cdots = P_n$. The formula (2.47) is then transformed into:

$$P = P_1 \otimes \cdots \otimes P_1 = P_1^{\otimes n} \qquad (2.48)$$

where the convolution operation is performed n times on the same probability distribution. When the independence and stationarity assumptions are simultaneously satisfied, and when the considered dynamics are defined for all positive t, then $(X(t), t \geq 0)$ is a Lévy process (see Chapter 4).

If the Lévy process is alpha-stable, then (see Chapter 5, p. 79) the distribution of returns at time at (so of $X(at)$) can be deduced from that at time t (so of $X(t)$) by the scale relationship:

$$X(at) \stackrel{d}{=} a^{1/\alpha} X(t) \qquad (2.49)$$

where the symbol $\stackrel{d}{=}$ denotes the equality in distribution and where $\alpha \leq 2$. The increments of $X(t)$ have non-Gaussian alpha-stable distributions defined later on in formula (5.2).

If periodic returns are independent, stationary, and alpha-stable, then it is possible to simplify the convolution as follows:

$$P = P_1^{\otimes n} = n^{1/\alpha} P_1 \qquad (2.50)$$

In the particular case where $\alpha = 2$, $(X(t), t \geq 0)$ is a Brownian motion and the scale relationship (2.49) becomes:

$$X(at) \stackrel{d}{=} \sqrt{a}\, X(t) \qquad (2.51)$$

This scale relationship justifies the empirical rule for the transformation of market volatilities in time, known as the "square root of time law", from Jules Regnault (1863):

$$\sigma(at) = \sqrt{a} \times \sigma(t) \qquad (2.52)$$

In particular, this is used in the formula for computing the annual volatility from the weekly one:

$$\text{annual volatility} = \text{weekly volatility} \times \sqrt{52}$$

We see with this example that the annualization of volatility, as performed on financial markets, is related to the fractal nature of Brownian motion. We summarize in Table 2.4 the change of scale rules that prevail in all situations.

Table 2.4 Aggregation of financial randomness.

Nature of increments	Financial models	Change of scale	Mechanism of time aggregation
non i.i.d.	Joint distribution	$P = \int dP_1 dP_2 \cdots dP_n$	
i. non i.d. **Additive process**	Convolution product 2000s	$P = P_1 \otimes \cdots \otimes P_n$	
i.i.d. non stable **Lévy process**	Convolution product 1990s	$P = P_1^{\otimes n}$	
i.i.d.-α-stable **α-stable motion**	Fractal Invariance B. Mandelbrot (1963)	$P = a^{1/\alpha} P_1$	
i.i.d.-2-stable **Brownian motion**	Fractal Invariance L. Bachelier (1900)	$P = \sqrt{a} P_1$	"**Square root of time law**" Jules Regnault (1863)

Chapter 3

Statistical Description of Markets

The statistical description of markets is a prerequisite to any probabilist construction and to any financial computation of risk budgets or portfolio allocations. After an overview of the representations of financial data, this chapter recalls the basic notions that are applied in the construction of tests, using the example of normality tests.

We then tackle three important questions. First, we confirm that a representation taking jumps into account is appropriate for the modeling of market data. Further, we examine if the introduction of jumps is accompanied by the suppression of the continuous component within probabilistic representations, or if we need to keep this continuous component. Finally, it is important to test whether the activity (the total arrival rate of jumps) should be modeled as finite or infinite, as this has a strong influence on the nature of financial models.

3.1 Construction of a Representation

Before presenting the statistical tests examined in this chapter, we set out the traditional debate on the construction of probabilistic representations of stock returns.

3.1.1 Role of the Statistical Description

The statistical description of stock dynamics has played an important role in the design of new theoretical concepts, such as the informational efficiency concept elaborated by Eugene Fama between 1965 and 1976, and the design of options models by Fisher Black, Myron Scholes, and Robert Merton in 1973. The mathematical construction of these financial models was conducted based on a continuous representation of stock price fluctuations. This representation was introduced indirectly: it was the Gaussian hypothesis that was postulated for modeling the marginal distributions of stock returns.

The chronology of the construction of these models is important: theoretical concepts could emerge only after statistical evidence of what seemed to validate, at that time, the Gaussian hypothesis and, implicitly, the continuity hypothesis. It is only after it had been admitted from a statistical viewpoint that stock fluctuations could be represented as a first approximation by Gaussian random walks, that the idea of the informational efficiency of markets could be forged between the years 1950 and 1965. To Fama, Black, Scholes, and Merton, the representation of market fluctuations was necessarily Brownian. In the absence of such a hypothesis, their models would not have been solvable. This example illustrates the importance of an appropriate statistical description of stock fluctuations for the elaboration of models dealing with market behavior.

3.1.2 Continuous or Discontinuous Representations

The debate on the continuity or discontinuity of trajectories can be conducted in statistical terms.

3.1.2.1 The Debate on Continuity

In the classic case of the Brownian representation of fluctuations, trajectories are continuous. However, a stock price trajectory is by construction discontinuous, because it comprises jumps at all quote times. The classic

Brownian representation views quotes as points sampled in a continuous trajectory, whereas in this book we postulate a representation in terms of (non-Brownian) Lévy processes – we model stock dynamics by discontinuous trajectories.

One could object that, in the case of a Brownian representation, quote jumps are proportional to the volatility of Brownian motion and that it is not necessary to change representation. This objection correctly addresses the question of the existence of jumps. In a given representation, are the points sampled on a trajectory separated by distances that are consistent with the postulated model for paths? In the case of a Brownian representation, are the observed quote jumps consistent with the diffusive nature of Brownian motion, or are they too important, and do they therefore indicate that it is preferable to select a jump model for trajectories?

Our issue here is therefore that of the implementation of a test allowing the choice between the continuous and discontinuous representations.

3.1.2.2 Related Tests

As strange as it might seem, this question had not been tackled in finance literature until very recently. While normality tests have been well known for many years, it was not until the contributions of Yacine Aït-Sahalia, Ole Barndorff-Nielsen, and Jean Jacod that appropriate tests for the detection of jumps were constructed. This design does not come without difficulties when the chosen probabilistic representation is that of a jump process. We will come back to certain concepts related to the jumps of Lévy processes in Chapter 4.

In a representation with jumps, we need to determine whether it is necessary to add a jump component to a classic diffusive component, or whether considering only a jump component is sufficient to model the discrete trajectories of stock prices. In the first models with jumps, such as the mixed processes of Merton, the diffusive component was the origin of normal return fluctuations, whereas the Poissonian component produced deviations from normality, explained for example by the occurrence of non-expected news on the markets.

However, it is possible with these models, which consist in adding a Poisson process to a Brownian motion, that the number of jumps is not sufficient to render a very rapid succession of quotes, because the activity of these processes is finite (on this notion, see Chapter 4, p. 63). For example, the study conducted in the previous chapter on Bouygues stock has shown that quotes are distant by about ten seconds and that jumps are very small (see Table 2.3 and Fig. 2.2, p. 25). A representation based on a finite activity process, such as a mixed process for instance, can appear too limitative in this case. The observation of these limitations is at the origin of the choice in the early 2000s by Hélyette Geman, Peter Carr, Dilip Madan, Ernst Eberlein, and Marc Yor of infinite activity processes that allowed them to avoid the use of a Brownian component. For these processes, each trajectory is by definition discontinuous everywhere.

The literature on the pricing of derivatives by Lévy processes did not tackle the following reciprocal question. If, after performing *ad hoc* tests, it becomes certain that one should incorporate discontinuities (with finite or infinite activity, and with finite or infinite variation) into the modeling of stocks, does this mean that we should not incorporate a continuous component in these models? Put differently, it is also necessary to design a test for the presence of a continuous component in market dynamics. Such a test has been constructed by Jacod and Shiryaev. We end our study of the recent contributions of these authors by examining their test characterizing the activity of market dynamics.

3.2 Normality Tests

We start by briefly recalling the definitions of the classic normality tests. We construct a sample of values between the two dates $t = 0$ and $t = T$ (terminal date). This construction takes place in five steps.

(1) Choice of analysis period, i.e. choice of the two dates 0 and T.
(2) Choice of the frequency of quote sampling. This choice determines the characteristic time of the analysis that we denote by convention τ (this is the resolution scale of the market). The size of the sample is therefore $n = T/\tau$.
(3) Introduction of the n intermediate dates $t_k = k\tau$, $k = 1, \cdots, n$, such that $0 = t_0 < t_1 < \cdots < t_n = T$.
(4) Storing the quotes at each date t_k. We obtain a sample of size $n+1$ comprising the values $\{S_{t_0}, S_{t_1}, \cdots, S_{t_n}\}$ or a sample of size n of the returns $\{X_{t_1}, X_{t_2}, \cdots, X_{t_n}\}$.
(5) Computation of periodic return rates. These latter quantities constitute the sample of size n that we are aiming to find:

$$\{\Delta X_1, \Delta X_2, \cdots, \Delta X_n\} \tag{3.1}$$

Statistical tests will be performed on these values.

In order to simplify the notation when it is not necessary to recall the characteristics (τ, T, Δ) of the difference process, we denote by $\Delta X_k = x_k$ the k-th periodic return obtained on a chosen scale τ. We work with a sample of size n:

$$\{x_1, x_2, \cdots, x_n\}$$

whose distribution we want to characterize.

3.2.1 The Pearson–Fisher Coefficients

To implement normality tests, we first need to compute the following quantities.

(1) The mean of periodic returns over the period analyzed. It is denoted by $\hat{\mu}$ and is computed using the estimator:

$$\hat{\mu} = \frac{1}{n} \sum_{k=1}^{n} x_k \tag{3.2}$$

(2) The empirical centered moments.

   (a) The empirical centered moment of order 2 taken to the power of 1/2:

   $$\hat{m}_2^{1/2} = \left(\sum_{k=1}^{n} (x_k - \hat{\mu})^2 \right)^{1/2} \tag{3.3}$$

   This is the periodic volatility over the period analyzed. The unbiased, efficient, and convergent estimator of the standard deviation $\hat{\sigma}$ is:

   $$\hat{\sigma} = \hat{m}_2^{1/2} \times \frac{1}{\sqrt{n-1}} \tag{3.4}$$

   (b) The empirical centered moment of order 3 taken to the power of 1/3:

   $$\hat{m}_3^{1/3} = \left(\sum_{k=1}^{n} (x_k - \hat{\mu})^3 \right)^{1/3} \tag{3.5}$$

   (c) The empirical centered moment of order 3 taken to the power of 1/4:

   $$\hat{m}_4^{1/4} = \left(\sum_{k=1}^{n} (x_k - \hat{\mu})^4 \right)^{1/4} \tag{3.6}$$

(3) The two Pearson–Fisher coefficients:

(a) The coefficient of asymmetry defined in (2.34). Its estimator is denoted by $\hat{S}$ and can be computed as:

$$\hat{S} = \frac{n}{(n-1)(n-2)} \times \sum_{k=1}^{n} \left(\frac{x_k - \hat{\mu}}{\hat{\sigma}} \right)^3 \qquad (3.7)$$

(b) The kurtosis coefficient defined in (2.35). Its estimator is denoted by $\hat{K}$ and can be computed as:

$$\hat{K} = \frac{n(n+1)}{(n-1)(n-2)(n-3)} \times \sum_{k=1}^{n} \left(\frac{x_k - \hat{\mu}}{\hat{\sigma}} \right)^4 - 3 \frac{(n-1)^2}{(n-2)(n-3)} \qquad (3.8)$$

If a tested distribution is Gaussian then $\hat{S} = \hat{K} = 0$.

To assess how far the obtained value is from the theoretical value 0, we need a distribution of estimations. For this, we use the following results, when $n \to +\infty$:

$$\frac{\hat{S}}{\sqrt{\text{var}(\hat{S})}} \sim \mathcal{N}(0,1) \quad \text{and} \quad \frac{\hat{K}}{\sqrt{\text{var}(\hat{K})}} \sim \mathcal{N}(0,1) \qquad (3.9)$$

The variances of the estimators $\hat{S}$ and $\hat{K}$ of the Pearson–Fisher coefficients of a sample of size n are respectively:

$$\text{var}(\hat{S}) = \frac{6n(n-1)}{(n-2)(n-1)(n+3)} \simeq \frac{6}{n} \qquad (3.10)$$

$$\text{var}(\hat{K}) = \frac{24n(n-1)^2}{(n-3)(n-2)(n+3)(n+5)} \simeq \frac{24}{n} \qquad (3.11)$$

where the symbol $\simeq$ means here "is equivalent, for n large, to".

3.2.1.1 Asymmetry and Kurtosis Tests

Equipped with these results, we can now define the prevailing decision rules for normality tests. Once the fundamental hypothesis $\mathbf{H_0}$ is chosen, the critical region of a test of significance α of a parameter λ is defined by the bound c such that $\Pr(\lambda > c | \mathbf{H_0}) = \alpha$, where c is the quantile of order $1 - \alpha/2$ of $\mathcal{N}(0,1)$. We have from (3.9) that, asymptotically:

$$\Pr\left(\frac{\hat{S}}{\sqrt{\text{var}(\hat{S})}} > c \,|\, \mathbf{H_0} \right) = \alpha \quad \text{and} \quad \Pr\left(\frac{\hat{K}}{\sqrt{\text{var}(\hat{K})}} > c \,|\, \mathbf{H_0} \right) = \alpha \qquad (3.12)$$

Using (3.11), and again asymptotically:

$$\Pr\left(\frac{\hat{S}}{\sqrt{\frac{6}{n}}} > c \mid \mathbf{H_0}\right) = \alpha \quad \text{and} \quad \Pr\left(\frac{\hat{K}}{\sqrt{\frac{24}{n}}} > c \mid \mathbf{H_0}\right) = \alpha \quad (3.13)$$

When $\alpha = 5\%$, we have $1 - 0.05/2 = 0.975$ so $c = 1.96$. The two normality tests are as follows.

Asymmetry test: under the hypothesis $\mathbf{H_0}$ (normality), $S \sim \mathcal{N}(0,1)$. When $\alpha = 5\%$, the critical region (of rejection of $\mathbf{H_0}$) is given by:

$$\mathcal{R} = \left\{ \frac{\hat{S}}{\sqrt{\frac{6}{n}}}, \frac{\hat{S}}{\sqrt{\frac{6}{n}}} > 1.96 \right\}$$

This test is reliable for $8 \leq n \leq 5\,000$.

Kurtosis test: under the hypothesis $\mathbf{H_0}$ (normality), $K \sim \mathcal{N}(0,1)$. When $\alpha = 5\%$, the critical region (of rejection of $\mathbf{H_0}$) is given by:

$$\mathcal{R} = \left\{ \frac{\hat{K}}{\sqrt{\frac{24}{n}}}, \frac{\hat{K}}{\sqrt{\frac{24}{n}}} > 1.96 \right\}$$

This test is reliable for $7 \leq n \leq 1\,000$.

3.2.1.2 Jarque–Bera Test

Take the convention $S = \hat{S}/\sqrt{\text{var}(\hat{S})}$ and $K = \hat{K}/\sqrt{\text{var}(\hat{K})}$. The Jarque–Bera test unites the two previous tests by means of the normality property of S and K. If S and K follow an $\mathcal{N}(0,1)$ Gaussian distribution, then $S^2 + K^2$ follows a χ^2 distribution with two degrees of freedom. At the significance level of 5%, the bound of the critical region is $c = 5.99$.

Several illustrations of these tests using actual market data are given below.

3.2.2 Kolmogorov Test

The Kolmogorov test relies conventionally on the computation of a distance that allows us to distinguish two regions in the space of answers, on both sides of a critical value or threshold. This distance is defined from empirical and theoretical cumulative distribution functions of the random variable studied.

Let X be a real random variable of unknown distribution L^X and c.d.f. F_X, presumed continuous. From an i.i.d. sample of X, we wish to adjust the unknown law L to a given law L^0, of known c.d.f.

Recall that the empirical cumulative distribution function (or cumulated frequency) denoted by $F_n(x)$ of a random variable X for a sample of size n is defined for all $x \in \mathbb{R}$ by:

$$F_n(x) = \frac{1}{n} \operatorname{card}\{i : 1 \leq i \leq n,\ X_i \leq x\} = \frac{1}{n} \sum_{i=1}^{n} \mathbf{1}_{\{X_i \leq x\}}(x_i) \quad (3.14)$$

where it is known that $F_n(x)$ converges almost surely toward $F(x)$.

Thus, we consider the following problem:

$$\begin{cases} \text{Hypothesis } \mathbf{H_0} & F = F_0 \quad F_0 \text{ given} \\ \text{Hypothesis } \mathbf{H_1} & F \neq F_0 \end{cases} \quad (3.15)$$

To construct this test, we define the following statistics:

$$\begin{cases} K_n^+ = \sup_x (F_n(x) - F_0(x)) \\ K_n^- = \sup_x (F_0(x) - F_n(x)) \\ K_n = \sup_x |F_n(x) - F_0(x)| = \max(K_n^+, K_n^-) \end{cases} \quad (3.16)$$

In practice, denoting by $x_{(i)}$ the i-th element of the sample ranked in increasing order, the test can be performed with:

$$\begin{cases} K_n^+ = \max_i \left(\frac{i-1}{n} - F_0(x_{(i)})\right) \\ K_n^- = \max_i \left(F_0(x_{(i)}) - \frac{i}{n}\right) \end{cases} \quad (3.17)$$

Under the hypothesis $\mathbf{H_0}$ ($F = F_0$), the statistic $K_n\sqrt{n}$ converges in law, when n tends to infinity, toward a random variable of c.d.f. Q:

$$Q(x) = 1 - \sum_{i=1}^{+\infty} (-1)^{k-1} e^{-2k^2 x^2}$$

The critical region of the test is:

$$\mathcal{R} = \{K_n\,;\, K_n > c\} \qquad \Pr(\mathcal{R}|\mathbf{H_0}) = \alpha$$

where $c\sqrt{n}$ is the quantile of order $1 - \alpha$ of the asymptotic distribution Q of $K_n\sqrt{n}$. The exact and asymptotic laws of K_n are tabulated. A candidate law will be accepted if the statistic K_n takes "small" values according to a chosen level α.

3.3 Discontinuity Test

In this section, we test the presence of a jump component in the representation of asset dynamics. This test aims to answer the following question: should we introduce jumps to the models that represent the dynamics of securities? We consider here a test constructed by Aït-Sahalia and Jacod (2009) and describe its inner functioning. We introduce the estimators of the test and give the corresponding confidence intervals. An illustration of the test concludes this section.

3.3.1 *Definition of Estimators*

We start by setting a time step for the analysis of the market. This time step denoted by τ allows us to construct the following quote dates $t_k = k\tau$. Define:

$$\Delta X(t_k, \tau) = X(t_k) - X(t_k - \tau) = X(k\tau) - X((k-1)\tau) = \Delta X_k(\tau)$$

to be the increment of X over the period of length τ. The interpretation of $\Delta X_k(\tau)$ is as follows: these are the observed increments of the process X. These increments are different from the jumps at any time s that are by definition instantaneous and non-observable. Jumps are denoted by ΔX_s and defined as follows:

$$\Delta X_s = X_s - X_{s^-}$$

To conduct a discontinuity test, an asymptotic statistic is necessary, for τ tending to zero. The ability of the test to choose between two representations – a unique continuous component or a prospective continuous component plus a jump component – is directly related to the quality of the asymptotic statistic.

The starting idea is to use integer powers of the increments of processes. Indeed, it has been shown (see for instance Barndorff-Nielsen and Shepard (2006)) that the elevation to integer powers allows the continuous part of increments to be distinguished from their jump part. The larger this power p, the more large jumps are highlighted. According to this idea, the variability of the continuous component is related to its volatility σ and defined for any integer number p as follows:

$$A_t(p) = \int_0^t |\sigma_s|^p ds \qquad (3.18)$$

and the variability of the discontinuous component as:

$$B_t(p) = \sum_{s \leq t} |\Delta X_s|^p \tag{3.19}$$

This latter indicator can be approximated in some situations by the following estimator, computed from the increments of the dynamics:

$$\hat{B}_t(p, \tau_n) = \sum_{k=1}^{\lfloor t/\tau_n \rfloor} |\Delta X_k(\tau_n)|^p \tag{3.20}$$

where τ_n is an element of a sequence of characteristic times converging to zero. The asymptotic behavior of $\hat{B}$ depends on the value of p and of the prospective continuity of the process X. For dynamics displaying jumps and for p strictly superior to 2, we have:

$$\lim_{\tau_n \to 0} \hat{B}_t(p, \tau_n) = B_t(p) \tag{3.21}$$

whereas for p strictly inferior to 2 when X displays jumps, or for any p when X is continuous, we have:

$$\lim_{\tau_n \to 0} \frac{\tau_n^{1-p/2}}{\sqrt{\frac{2^p}{\pi}} \Gamma\left(\frac{p+1}{2}\right)} \hat{B}_t(p, \tau_n) = A_t(p) \tag{3.22}$$

Let us comment briefly on the renormalization term. The absolute moment of order p of a centered reduced Gaussian random variable denoted by U can be presented as:

$$\mu_p = \mathbb{E}\left[|U|^p\right] = \sqrt{\frac{2^p}{\pi}} \, \Gamma\left(\frac{p+1}{2}\right) \tag{3.23}$$

where we can recognize the denominator of the above renormalization term.

Equipped with these results, we can now give the intuition behind the test of the presence of jumps. This test relies on the comparison of the asymptotic values of $\hat{B}_t$ at various resolution scales of the market (for various time steps τ_n). The goal is to compare $\hat{B}_t(p, \tau_n)$ and $\hat{B}_t(p, k\tau_n)$, where this latter quantity corresponds to a sampling k times thinner than that of $\hat{B}_t(p, \tau_n)$. We denote by $\hat{S}_t(p, k, \tau_n)$ the statistics of the test, defined as:

$$\hat{S}_t(p, k, \tau_n) = \frac{\hat{B}_t(p, k\tau_n)}{\hat{B}_t(p, \tau_n)} \tag{3.24}$$

Equations (3.21) and (3.22) lead to:

$$\lim_{\tau_n \to 0} \hat{S}_t(p, k, \tau_n) = \lim_{\tau_n \to 0} \frac{\hat{B}_t(p, k\tau_n)}{\hat{B}_t(p, \tau_n)} = \frac{B_t(p)}{B_t(p)} = 1 \tag{3.25}$$

for dynamics displaying jumps (and potentially a continuous component) when $p > 2$, and to:

$$\lim_{\tau_n \to 0} k^{1-p/2} \hat{S}_t(p, k, \tau_n) = \lim_{\tau_n \to 0} \frac{\frac{(k\tau_n)^{1-p/2}}{\sqrt{\frac{2^p}{\pi}} \Gamma(\frac{p+1}{2})} \hat{B}_t(p, k\tau_n)}{\frac{\tau_n^{1-p/2}}{\sqrt{\frac{2^p}{\pi}} \Gamma(\frac{p+1}{2})} \hat{B}_t(p, \tau_n)} = \frac{A_t(p)}{A_t(p)} = 1 \quad (3.26)$$

for continuous dynamics when $p > 2$.

We deduce the decision rule of Aït-Sahalia and Jacod (2009) that is linked to the statistic $\hat{S}_t$ in the two situations examined below:

$$\begin{cases} \lim_{\tau_n \to 0} \hat{S}_t(p, k, \tau_n) = 1 & \text{presence of jumps} \\ \lim_{\tau_n \to 0} \hat{S}_t(p, k, \tau_n) = k^{p/2-1} & \text{absence of jumps} \end{cases} \quad (3.27)$$

Choosing the values $p = 4$ and $k = 2$, the decision rule becomes:

$$\begin{cases} \lim_{\tau_n \to 0} \hat{S}_t(p, k, \tau_n) = 1 & \text{presence of jumps} \\ \lim_{\tau_n \to 0} \hat{S}_t(p, k, \tau_n) = 2 & \text{absence of jumps} \end{cases} \quad (3.28)$$

However, these results are not sufficient to define the test exhaustively. It is also necessary to have the convergence rules of the estimator and the asymptotic variances in the two situations of continuity and discontinuity. We now concentrate on these latter elements.

3.3.2 Confidence Intervals

We start by introducing the central limit theorem that was obtained by Aït-Sahalia and Jacod (2009) for the estimator of the jump test and we then concentrate on the use of this theorem in order to obtain confidence intervals (on the presence or absence of jumps). We assume here that $p > 3$ and that the volatility of the process does not display jumps that are correlated with the process itself.

The goal of the central limit theorem that is constructed hereafter is to describe the asymptotic behavior of the estimator $\hat{S}_t(p, k, \tau)$ introduced in Eq. (3.24). We start by giving a few preliminary definitions.

For U and V, two independent centered reduced Gaussian random variables, we can define:

$$\mu_{k,p} = \mathbb{E}\left[|U|^p \cdot |U + \sqrt{k-1}V|^p\right] \quad (3.29)$$

which can be computed as follows:

$$\mu_{k,p} = \frac{2p}{\pi}(k-1)^{p/2} \Gamma\left(\frac{1+p}{2}\right)^2 F_{2,1}\left(-\frac{p}{2}, \frac{p+1}{2}, \frac{1}{2}, \frac{-1}{k-1}\right) \quad (3.30)$$

where $F_{2,1}$ is the Gaussian hypergeometric function (see for example Abramowitz and Stegun (1965)). In practice, the Monte Carlo method also allows the quantities $\mu_{k,p}$ to be computed very quickly. Note that μ_p defined in (3.23) satisfies:

$$\mu_p = \mathbb{E}\left[|U|^p\right] = \mu_{1,\frac{p}{2}}$$

Based on these expressions, we further define:

$$M(p,k) = \frac{k^{p-2}(1+k)\,\mu_{2p} + k^{p-2}(k-1)\,\mu_p^2 - 2\,k^{\frac{p}{2}-1}\mu_{k,p}}{\mu_p^2}$$

Then, we introduce two new discretized quantities. Let α be a strictly positive arbitrary number and ω any number pertaining to the interval $]0,1[$. We construct first $\hat{A}_t(p,\tau_n)$ as follows:

$$\hat{A}_t(p,\tau_n) = \frac{\tau_n^{1-p/2}}{\mu_p} \sum_{k=1}^{\lfloor t/\tau_n \rfloor} |\Delta X_k(\tau_n)|^p \mathbf{1}_{\{|\Delta X_k(\tau_n)| \leq \alpha \tau_n^\omega\}}$$

Take now a sequence of integers k_n such that $\lim_{n \to +\infty} k_n = +\infty$ and such that $\lim_{n \to +\infty} k_n \tau_n = 0$ (we can take for instance $k_n = \lfloor K/\sqrt{\tau_n} \rfloor$ or $k_n = \lfloor K/\tau_n^{1/4} \rfloor$ where K is a constant, because $\lim_{n \to +\infty} \tau_n = 0$). This allows to define $I_{n,t}(k) = \{j \in \mathbf{N} \mid j \neq k, 1 \leq j \leq [t/\tau_n], |k-j| \leq k_n\}$. We then construct $\hat{D}_t(p,\tau_n)$ according to:

$$\hat{D}_t(p,\tau_n) = \frac{1}{k_n \tau_n} \sum_{k=1}^{\lfloor t/\tau_n \rfloor} |\Delta X_k(\tau_n)|^p \sum_{j \in I_{n,t}(k)} (\Delta X_j(\tau_n))^2 \mathbf{1}_{\{|\Delta X_j(\tau_n)| \leq \alpha \tau_n^\omega\}}$$

We now have all the necessary elements to construct a confidence interval for the jump test.

3.3.2.1 Central Limit Theorem of Discontinuities

Introducing the variance term $V_{n,t}^{\text{dis}}$ (where "dis" stands for a "discontinuous" model):

$$V_{n,t}^{\text{dis}} = \frac{\tau_n(k-1)\,p^2\,\hat{D}_t(2p-2,\tau_n)}{2\,\hat{B}_t(p,\tau_n)^2}$$

we obtain the asymptotic result for the estimator $\hat{S}_t(p,k,\tau_n)$:

$$\frac{\hat{S}_t(p,k,\tau_n) - 1}{\sqrt{V_{n,t}^{\text{dis}}}} \to \mathcal{N}(0,1) \qquad (3.31)$$

If the stochastic process considered is continuous, then the variance term $V_{n,t}^{\text{con}}$ (where "con" stands for the "continuous" hypothesis):

$$V_{n,t}^{\text{con}} = \frac{\tau_n\, M(p,k)\, \hat{A}_t(2p,\tau_n)}{\hat{A}_t(p,\tau_n)^2}$$

can be used to construct the asymptotic result for the estimator $\hat{S}_t(p,k,\tau_n)$:

$$\frac{\hat{S}_t(p,k,\tau_n) - k^{p/2-1}}{\sqrt{V_{n,t}^{\text{con}}}} \to \mathcal{N}(0,1) \qquad (3.32)$$

We conclude that the jump estimator displayed above, which tends to 1 or $k^{p/2-1}$ if the process is a jump-diffusion or a diffusion, is asymptotically Gaussian and presents an asymptotic variance given respectively by $V_{n,t}^{\text{dis}}$ and $V_{n,t}^{\text{con}}$, at the order n.

The following rules for the construction of confidence intervals stem from the central limit theorems presented above.

Consider the null hypothesis $\mathbf{H}_0$: *no jumps in the dynamics.*

In this case, $\hat{S}_t(p,k,\tau_n)$ tends to $k^{p/2-1}$, which is strictly larger than 1 for usual values of p and k. The bound of the critical region is:

$$c_{n,t}^{\text{con}} = k^{p/2-1} - z_\alpha \sqrt{V_{n,t}^{\text{con}}}$$

where z_α is the quantile of order α of a centered reduced Gaussian random variable (the number such that for $U \sim \mathcal{N}(0,1)$, we have $P(U > z_\alpha) = \alpha$).

The critical region is given by:

$$\mathcal{R}_1 = \{\hat{S}_t(p,k,\tau_n) < c_{n,t}^{\text{con}}\} \qquad (3.33)$$

Assuming the absence of jumps, this means that the probability of the estimator $\hat{S}_t(p,k,\tau_n)$ being inferior to $c_{n,t}^{\text{con}}$ is α. This implies in practice that when $\hat{S}_t(p,k,\tau_n)$ is inferior to $c_{n,t}^{\text{con}}$, we cannot conclude that the dynamics are continuous.

Consider the null hypothesis $\mathbf{H}_0$: *jumps in the dynamics.*

In this case, $\hat{S}_t(p,k,\tau_n)$ tends to 1. The bound of the critical region is:

$$c_{n,t}^{\text{dis}} = 1 + z_\alpha \sqrt{V_{n,t}^{\text{dis}}}$$

where z_α is the quantile of order α of the centered reduced Gaussian distribution.

Table 3.1 Discontinuous component test (BNP)

Day	$\hat{S}$	c^{con}	c^{dis}
01/21/09	0.55	1.79	1.05
01/20/09	0.86	1.92	1.04
01/19/09	0.98	1.91	1.05
01/16/09	0.80	1.90	1.05
01/15/09	0.97	1.90	1.05
01/14/09	1.44*	1.90	1.05
01/13/09	1.01	1.89	1.03
01/12/09	1.12*	1.86	1.06
01/09/09	0.93	1.89	1.06
01/08/09	0.95	1.88	1.09
01/07/09	0.96	1.90	1.06
01/06/09	0.94	1.91	1.07
01/05/09	0.79	1.89	1.04
01/02/09	1.05	1.85	1.07
12/31/08	0.91	1.82	1.09
12/30/08	1.03	1.88	1.06
12/29/08	1.01	1.85	1.04
12/24/08	1.08	1.81	1.10
12/23/08	0.99	1.89	1.04
12/22/08	0.95	1.91	1.05

The critical region is given by:

$$\mathcal{R}_2 = \{\hat{S}_t(p, k, \tau_n) > c_{n,t}^{\text{dis}}\} \qquad (3.34)$$

Assuming the presence of jumps, this means that the probability of the estimator $\hat{S}_t(p, k, \tau_n)$ being superior to $c_{n,t}^{\text{dis}}$ is α. This implies in practice that when $\hat{S}_t(p, k, \tau_n)$ is superior to $c_{n,t}^{\text{dis}}$, we cannot conclude that the dynamics display jumps.

3.3.2.2 Illustration

We conduct the discontinuity test on the dynamics of BNP stock. To do so, we use high frequency quotes (sampled every five seconds) between December 22, 2008 and January 21, 2009. We set $p = 4$ and $k = 2$ and perform, for each day between these dates, the test described in (3.28). We observe in Table 3.1 that the indicator for the presence of jumps is close to one almost every day, which is an indication of the importance of incorporating jumps into models.

If we examine the critical region $\mathcal{R}_1$ described in (3.33), we observe that each point in the table belongs to it. Indeed, we always have $\hat{S} < c^{\text{con}}$, meaning that we can exclude in all instances purely continuous dynamics. Concerning now the critical region $\mathcal{R}_2$ described in (3.34), we observe that there are almost no estimations belonging to this region (except two, distinguished by an asterix). It is not possible to exclude the presence of jumps in the dynamics of BNP stock.

Our study on the high frequency quotes of the BNP stock shows that the best continuous-time model to represent the empirical dynamics of this stock is a model that incorporates jumps. Similar tests led to identical results for the other CAC40 stocks.

3.4 Continuity Test

In this section, we test for the presence of a continuous component in the representation of market dynamics. The test presented here was constructed by Aït-Sahalia and Jacod (2010). We introduce the estimators of the test and then give the corresponding confidence intervals. An illustration concludes the section.

3.4.1 *Definition of the Estimators*

The Aït-Sahalia and Jacod (2010) test investigates the presence of a continuous component in the representation of stock dynamics. We introduce first the variability of small jumps:

$$b_t(p, u) = \sum_{s \leq t} |\Delta X_s|^p \, \mathbf{1}_{|\Delta X_s| \leq u} \qquad (3.35)$$

where u is an arbitrary cutoff number and we can construct an approximation of this variability in terms of the increments as follows:

$$\hat{b}_t(p, u_n, \tau_n) = \sum_{k=1}^{\lfloor t/\tau_n \rfloor} |\Delta X_k(\tau_n)|^p \, \mathbf{1}_{|\Delta X_k(\tau_n)| \leq u_n} \qquad (3.36)$$

where u_n and τ_n are elements of sequences tending to zero.

Denote by $\hat{s}_t(p, k, u_n, \tau_n)$ the statistic of the test, defined as:

$$\hat{s}_t(p, k, u_n, \tau_n) = \frac{\hat{b}_t(p, u_n, \tau_n)}{\hat{b}_t(p, u_n, k\tau_n)} \qquad (3.37)$$

The decision rule linked to the statistic $\hat{s}_t$ can then be written, when $k \geq 2$ and $p \in]1, 2[$, as:

$$\begin{cases} \lim_{u_n,\tau_n \to 0} \hat{s}_t(p,k,u_n,\tau_n) = k^{1-p/2} & \text{presence of Brownian motion} \\ \lim_{u_n,\tau_n \to 0} \hat{s}_t(p,k,u_n,\tau_n) = 1 & \text{absence of Brownian motion} \end{cases} \quad (3.38)$$

Note first that if the test for the presence of jumps relies on the use of an exponent $p > 2$ that allows the importance of jumps to be increased in the asymptotic statistic, the test for the presence of a Brownian component relies on the use of an exponent $p < 2$ yielding an increase of the importance of the diffusive component in the asymptotic statistic.

Note also that the test that is mentioned here is more restrictive than the jump test. Indeed, we test the hypothesis "presence of Brownian motion" against the hypothesis "absence of Brownian motion *and* presence of jumps of a stable-like process". Technical constraints also exist regarding the speed of the limits to zero of u and τ, which we will not discuss further here.

3.4.2 Confidence Interval

We now come to the construction of the confidence interval for the test of the presence of a continuous component in the representation of market dynamics.

3.4.2.1 A New Indicator

Introducing the indicator:

$$N(p,k) = \frac{k^{2-p}(1+k)\mu_{2p} + k^{2-p}(k-1)\mu_p^2 - 2k^{3-3p/2}\mu_{k,p}}{\mu_{2p}}$$

allows us to define the variance term $V_{n,t}^{\text{cbis}}$ ("cbis" standing for "continuous bis") as:

$$V_{n,t}^{\text{cbis}} = \frac{N(p,k)\,\hat{b}_t(2p,u_n,\tau_n)}{\hat{b}_t(p,u_n,\tau_n)^2}$$

which can be used to obtain the following asymptotic result for the estimator $\hat{s}_t(p,k,u_n,\tau_n)$:

$$\frac{\hat{s}_t(p,k,u_n,\tau_n) - k^{1-p/2}}{\sqrt{V_{n,t}^{\text{cbis}}}} \to \mathcal{N}(0,1) \quad (3.39)$$

From these elements, we deduce the frontier of the confidence interval (or bound of the critical region):

$$c_{n,t}^{\text{cbis}} = k^{1-p/2} - z_\alpha \sqrt{V_{n,t}^{\text{cbis}}}$$

where z_α is the quantile of order α of a centered reduced Gaussian random variable.

Under the null hypothesis $\mathbf{H}_0$: "a continuous component exists in the dynamics", the critical region is given by:

$$\mathcal{R}_3 = \{\hat{s}_t(p, k, u_n, \tau_n) < c_{n,t}^{\text{cbis}}\} \tag{3.40}$$

In other words, the probability of the estimator $\hat{s}_t(p, k, u_n, \tau_n)$ being inferior to $c_{n,t}^{\text{cbis}}$ is α. In practice, this means that when $\hat{s}_t(p, k, u_n, \tau_n)$ is inferior to $c_{n,t}^{\text{cbis}}$, we can reject, with only a small probability of error, the hypothesis of the presence of a continuous component in the representation of the dynamics considered.

3.4.2.2 Illustration

We now wish to determine if a continuous component is needed to represent the dynamics of BNP stock, or if a model having only jumps is sufficient. We use the same sample as in the previous illustration. We set $p = 1.6$ and $k = 2$ and perform the test described in (3.38), where $k^{1-p/2} = 1.1487$, for each day between December 22, 2008 and January 21, 2009. We observe in Table 3.2 that $\hat{s}$, the indicator of the presence of jumps, is in general approximately equal to 1.2, so is closer to $k^{1-p/2}$ than to 1. This is a clear indication that a continuous component should be included in the representation of the dynamics of BNP stock.

Let us now examine the critical region $\mathcal{R}_3$ described in (3.40). We observe that no point in Table 3.2 belongs to this critical region: we always have $\hat{s} > c^{\text{cbis}}$, meaning that we can never exclude the presence of a continuous component in the representation of the dynamics.

The conclusion of this second study on the high frequency data of BNP stock is that the best continuous-time model for this stock incorporates a continuous component. This latter component should be added to the jump component, which is also necessary, as shown previously. Similar tests led to analogous results for other CAC40 stocks.

3.5 Testing the Finiteness of the Activity

We end this chapter by studying the finiteness of the activity of the jumps affecting stock dynamics. If the activity is finite, then a model based on jump-diffusions will be suitable. On the other hand, if the activity is infinite, a model with (at least) Lévy processes will be required.

Table 3.2 Continuous component test (BNP)

Day	$\hat{s}$	c^{cbis}
01/21/09	1.34	1.08
01/20/09	1.30	1.12
01/19/09	1.23	1.12
01/16/09	1.21	1.12
01/15/09	1.20	1.12
01/14/09	1.21	1.12
01/13/09	1.20	1.11
01/12/09	1.23	1.11
01/09/09	1.25	1.11
01/08/09	1.22	1.11
01/07/09	1.25	1.12
01/06/09	1.21	1.12
01/05/09	1.24	1.11
01/02/09	1.30	1.10
12/31/08	1.23	1.09
12/30/08	1.24	1.11
12/29/08	1.19	1.10
12/24/08	1.30	1.09
12/23/08	1.26	1.12
12/22/08	1.29	1.12

3.5.1 *Construction of the Tests*

The estimator of the test of the finiteness of activity is the inverse of the estimator (3.37) of the test of the presence of a continuous component. It is denoted by:

$$\tilde{s}_t = \frac{1}{\hat{s}_t} = \frac{\hat{b}_t(p, u_n, k\tau_n)}{\hat{b}_t(p, u_n, \tau_n)}$$

and is defined from the quantity (3.36).

The decision rule linked to the statistic $\tilde{s}_t$ is applied over a range of parameters that differs from that used for the test of the presence of a continuous component. In particular, we assume here that $k \geq 2$ and $p > 2$. In such a setting, we have:

$$\begin{cases} \lim_{u_n, \tau_n \to 0} \tilde{s}_t(p, k, u_n, \tau_n) = k^{p/2-1} & \text{finite activity} \\ \lim_{u_n, \tau_n \to 0} \tilde{s}_t(p, k, u_n, \tau_n) = 1 & \text{infinite activity} \end{cases} \quad (3.41)$$

Using the quantity:

$$\tilde{N}(p, k) = \frac{k^{p-2}(1+k)\,\mu_{2p} + k^{p-2}(k-1)\,\mu_p^2 - 2\,k^{p/2-1}\mu_{k,p}}{\mu_{2p}}$$

we define the variance term $V_{n,t}^{\text{actf}}$ ("actf" stands for "finite activity") by:

$$V_{n,t}^{\text{actf}} = \frac{\tilde{N}(p,k)\,\hat{b}_t(2p,u_n,\tau_n)}{\hat{b}_t(p,u_n,\tau_n)^2}$$

yielding the frontier of the confidence interval:

$$c_{n,t}^{\text{actf}} = k^{p/2-1} - z_\alpha \sqrt{V_{n,t}^{\text{actf}}}$$

Under the null hypothesis $\mathbf{H_0}$: "the activity is finite", the critical region is then given by:

$$\mathcal{R}_4 = \{\tilde{s}_t(p,k,u_n,\tau_n) < c_{n,t}^{\text{actf}}\} \tag{3.42}$$

Let us now come to the reverse test, where the null hypothesis is "the activity is infinite". Set $\gamma > 1$ and $p' > p > 2$. The estimator of this test can be written as:

$$\tilde{\tilde{s}} = \frac{\hat{b}_t(p',\gamma u_n,\tau_n)\,\hat{b}_t(p,u_n,\tau_n)}{\hat{b}_t(p',u_n,\tau_n)\,\hat{b}_t(p,\gamma u_n,\tau_n)}$$

The decision rule is as follows:

$$\begin{cases} \lim_{u_n,\tau_n \to 0} \tilde{\tilde{s}}_t(p,k,u_n,\tau_n) = \gamma^{p'-p} & \text{infinite activity} \\ \lim_{u_n,\tau_n \to 0} \tilde{\tilde{s}}_t(p,k,u_n,\tau_n) = 1 & \text{finite activity} \end{cases} \tag{3.43}$$

We define now the variance term $V_{n,t}^{\text{acti}}$ ("acti" stands for "infinite activity") by:

$$\frac{V_{n,t}^{\text{acti}}}{\gamma^{2p'-2p}} = \frac{\hat{b}_t(2p,u_n,\tau_n)}{\hat{b}_t(p,u_n,\tau_n)^2} + (1-2\gamma^{-p})\frac{\hat{b}_t(2p,u_n,\tau_n)}{\hat{b}_t(p,\gamma u_n,\tau_n)^2} + \frac{\hat{b}_t(2p',u_n,\tau_n)}{\hat{b}_t(p',u_n,\tau_n)^2}$$

$$+ (1-2\gamma^{-p'})\frac{\hat{b}_t(2p',\gamma u_n,\tau_n)}{\hat{b}_t(p',\gamma u_n,\tau_n)^2} - 2\frac{\hat{b}_t(p+p',u_n,\tau_n)}{\hat{b}_t(p,u_n,\tau_n)\hat{b}_t(p',u_n,\tau_n)}$$

$$- 2(1-\gamma^{-p}-\gamma^{-p'})\frac{\hat{b}_t(p+p',\gamma u_n,\tau_n)}{\hat{b}_t(p,\gamma u_n,\tau_n)\hat{b}_t(p',\gamma u_n,\tau_n)} \tag{3.44}$$

Table 3.3 Activity test (BNP)

Day	$\tilde{s}$	c^{actf}	$\tilde{\tilde{s}}$	c^{acti}
01/21/09	0.64	1.60	3.16	2.92
01/20/09	0.81	1.84	3.23	2.97
01/19/09	0.87	1.84	2.71	2.97
01/16/09	0.93	1.81	3.35	2.97
01/15/09	0.95	1.80	3.06	2.96
01/14/09	0.95	1.81	2.82	2.96
01/13/09	0.89	1.76	2.84	2.94
01/12/09	0.94	1.73	2.95	2.95
01/09/09	0.84	1.78	2.88	2.96
01/08/09	0.88	1.76	2.95	2.96
01/07/09	0.87	1.82	2.75	2.96
01/06/09	0.92	1.82	2.79	2.97
01/05/09	0.89	1.77	2.89	2.94
01/02/09	0.81	1.72	3.25	2.94
12/31/08	0.82	1.70	2.99	2.95
12/30/08	0.87	1.75	2.96	2.96
12/29/08	0.90	1.73	2.94	2.95
12/24/08	0.73	1.58	2.92	2.91
12/23/08	0.87	1.80	3.22	2.96
12/22/08	0.86	1.82	3.35	2.97

which defines the frontier of the confidence interval:

$$c_{n,t}^{\text{acti}} = \gamma^{p'-p} - z_\alpha \sqrt{V_{n,t}^{\text{acti}}}$$

Under the null hypothesis $\mathbf{H_0}$: "the activity is infinite", the critical region is then given by:

$$\mathcal{R}_5 = \{\tilde{\tilde{s}}_t(p, k, u_n, \tau_n) < c_{n,t}^{\text{acti}}\} \quad (3.45)$$

3.5.2 Illustration

We now wish to determine whether the activity of the dynamics of BNP stock is finite or infinite. We use the same sample as for the preceding illustrations. We set $p = 4$ and $k = 2$ and perform the test described in (3.41), where $k^{p/2-1} = 2$, for each day between December 22, 2008 and January 21, 2009. We observe in Table 3.3 that $\tilde{s}$, the first indicator of the finiteness of the activity, is in general approximately equal to 0.85. It is therefore closer to 1 than to $k^{p/2-1}$, indicating that activity is infinite in the case of BNP stock.

If we look at the critical region $\mathcal{R}_4$ described in (3.42), we observe that all the points of the table belong to it. We always have $\tilde{s} < c^{\text{actf}}$, meaning that we can always exclude the finiteness of the activity for the dynamics considered.

We now set $\gamma = 3$, $p = 4$ and $p' = 5$ and perform the test described in (3.43), where $\gamma^{p'-p} = 3$, for each day between December 22, 2008 and January 21, 2009. We observe in Table 3.3 that $\tilde{\tilde{s}}$, the second indicator of finiteness of the activity, is in general approximately equal to 3, i.e. closer to $\gamma^{p'-p}$ than to 1. This confirms that the activity should be considered infinite in the case of BNP stock.

If we consider the critical region $\mathcal{R}_5$ described in (3.45), we observe that half of the points of the table do not belong to it, while the other half of the points do belong to it, but only barely. In such a context, it seems difficult to exclude the possibility that the activity is infinite, especially taking into consideration all of the results presented in this illustration.

We conclude from this third illustration on the high frequency data of BNP stock that the best continuous-time model for the representation of the empirical dynamics of this stock incorporates jumps whose activity is infinite. Similar results where obtained for the dynamics of other CAC40 stocks.

Recalling the results of the preceding illustrations, a realistic model for the representation of stock dynamics incorporates both a continuous and a jump component, with this latter component being characterized by an infinite activity. The following chapters will set out several classes of Lévy processes that are pure jump processes with infinite activity. It is always possible and straightforward to add a continuous Brownian component when implementing such processes.

Chapter 4

Lévy Processes

Lévy processes constitute the class of stochastic processes with independent and stationary increments. With the exception of Brownian motion with drift, they consist entirely of jumps. These processes are used throughout this book to represent the evolution of the returns of financial instruments. The books by Bertoin (1996), Satō (1999), and Applebaum (2009) present many results about Lévy processes and describe many of their properties. Schoutens (2003) and Cont and Tankov (2004) also give a mathematical exposition of these processes, but then focus on the pricing of financial derivatives.

In this chapter we aim to present Lévy processes in the simplest and most intuitive way possible. After providing certain definitions, we develop the intricate Lévy–Khintchine formula. This formula gives the form of the characteristic function of any Lévy process. The development is done step by step: it begins with an analysis of the simple Poisson process, and then extends to the study of increasingly elaborate processes. Next, we give a detailed analysis of the Lévy–Khintchine formula on the one hand, and the Lévy measure on the other hand. Finally, the chapter concludes with the calculation of the moments of a Lévy process.

4.1 Definitions and Construction

We give here several fundamental definitions regarding Lévy processes. Then, we introduce the building blocks that allow an intuitive construction of these processes.

4.1.1 The Characteristic Exponent

Recall that the characteristic function of a random variable Y is the Fourier transform of its density function, i.e.:

$$\Phi_Y(u) = \mathbb{E}\left[e^{iuY}\right] = \underbrace{\sum_{k=0}^{\infty} e^{iuk} \Pr(Y = k)}_{\text{discrete r.v.}} = \underbrace{\int_{-\infty}^{+\infty} e^{iux} f_Y(x) dx}_{\text{continuous r.v.}}$$

where $f_Y(x)$ is the density function. The characteristic function can be expressed in the following way:

$$\Phi_Y(u) = \exp\left(\Psi_Y(u)\right) \tag{4.1}$$

where $\Psi_Y(u)$ is called the characteristic exponent of Y.

In all subsequent chapters of this book, we will use Ψ and Φ interchangeably to characterize the marginal distributions of Lévy processes. In the case of Lévy processes, the exponent Ψ is proportional to time. This follows from a very important property of these processes: their marginal distributions are infinitely divisible. This property will be explained in more detail later on.

4.1.2 Infinitely Divisible Distributions

Roughly speaking, the property of infinite divisibility states that it is possible to write a random variable as a sum of n identical independent random variables, for any given n. More precisely, for any Lévy process X and any positive integer n, it is clear that:

$$X_n = X_1 + (X_2 - X_1) + \cdots + (X_n - X_{n-1})$$

and therefore:

$$\begin{aligned}
\Phi_{X_n}(u) &= \mathbb{E}\left[e^{iuX_n}\right] \\
&= \mathbb{E}\left[e^{iu(X_1 + (X_2 - X_1) + \cdots + (X_n - X_{n-1}))}\right] \\
&= \mathbb{E}\left[e^{iuX_1} \times e^{iu(X_2 - X_1)} \times \cdots \times e^{iu(X_n - X_{n-1})}\right]
\end{aligned}$$

From the independence of the increments of X, it follows that:

$$\Phi_{X_n}(u) = \mathbb{E}\left[e^{iuX_1}\right] \times \mathbb{E}\left[e^{iu(X_2-X_1)}\right] \times \cdots \times \mathbb{E}\left[e^{iu(X_n-X_{n-1})}\right]$$

The stationarity of the increments of X then implies that:

$$\Phi_{X_n}(u) = \underbrace{\mathbb{E}\left[e^{iuX_1}\right] \times \mathbb{E}\left[e^{iuX_1}\right] \times \cdots \times \mathbb{E}\left[e^{iuX_1}\right]}_{n \text{ times}} = \left(\mathbb{E}\left[e^{iuX_1}\right]\right)^n$$

Finally, we obtain:

$$\Phi_{X_n}(u) = \left(\Phi_{X_1}(u)\right)^n \tag{4.2}$$

This result can be generalized in the case of a continuous time t (real, positive) to:

$$\Phi_{X_t}(u) = \left(\Phi_{X_1}(u)\right)^t \tag{4.3}$$

If we now consider characteristic exponents, it follows immediately from (4.3) that:

$$\exp\left(\Psi_{X_t}(u)\right) = \left(\exp\left(\Psi_{X_1}(u)\right)\right)^t = \exp\left(t\Psi_{X_1}(u)\right)$$

hence the relationship between the distribution of the process X at time 1 and at time t:

$$\Psi_{X_t}(u) = t\,\Psi_{X_1}(u) \tag{4.4}$$

This is the reason why the characteristic exponent of a Lévy process is equal to t times that of its underlying infinitely divisible distribution, which is in fact the distribution of X_1.

4.1.3 *A Construction with Poisson Processes*

We give some examples of how to obtain the characteristic exponent in some simple special cases that are based on Poisson distributions and Poisson processes. We show the calculations in detail so that the construction of the characteristic exponent appears clearly.

4.1.3.1 *The Simple Poisson Distribution*

Let us begin with the very simple example of a Poisson distribution. Recall that this distribution is defined by:

$$\Pr(X = k) = \frac{\lambda^k}{k!}e^{-\lambda}$$

Calculating the characteristic function gives:

$$\Phi_X(u) = \mathbb{E}\left[e^{iuX}\right] = \sum_{k=0}^{\infty} e^{iuk} \frac{\lambda^k}{k!} e^{-\lambda} = e^{-\lambda} \sum_{k=0}^{\infty} \frac{(\lambda e^{iu})^k}{k!}$$

From the Taylor series of the exponential function we now obtain:

$$\sum_{k=0}^{\infty} \frac{(\lambda e^{iu})^k}{k!} = e^{\lambda e^{iu}}$$

and therefore:

$$\Phi_X(u) = e^{-\lambda} e^{\lambda e^{iu}} = e^{\lambda(e^{iu}-1)}$$

which allows us to recognize the characteristic exponent of the Poisson distribution as:

$$\Psi_X(u) = \lambda\left(e^{iu} - 1\right) \qquad (4.5)$$

4.1.3.2 The Centered Poisson Distribution

We can similarly try to find the characteristic function of a centered Poisson distribution, which we denote $\tilde{X}$. Since the expectation of a Poisson distribution with parameter λ is λ, we have:

$$\mathbb{E}\left[e^{iu\tilde{X}}\right] = \mathbb{E}\left[e^{iu(X-\lambda)}\right] = \sum_{k=0}^{\infty} e^{iu(k-\lambda)} \frac{\lambda^k}{k!} e^{-\lambda} = e^{\lambda(e^{iu}-1-iu)}$$

which gives, in the case of a centered Poisson distribution:

$$\Psi_{\tilde{X}}(u) = \lambda\left(e^{iu} - 1 - iu\right) \qquad (4.6)$$

4.1.3.3 The Poisson Process

Let us now consider a Poisson process N where, for all t, $N(t)$ represents the number of events that have happened between times 0 and t. Previously, we introduced the Poisson process in an intuitive fashion. We shall now present it rigorously. It is a process such that:

- $N_0 = 0$,
- for all $0 = t_0 < t_1 < t_2 < \cdots < t_n$, the variables:

$$N_{t_1} - N_{t_0}, N_{t_2} - N_{t_1}, \cdots, N_{t_n} - N_{t_{n-1}}$$

are independent,

- for all $0 < s < t$, the variable $N_t - N_s$ representing the number of jumps that occur between the dates s and t is Poisson distributed with parameter λ, i.e.:

$$\Pr(N_t - N_s = k) = \frac{(\lambda(t-s))^k}{k!} e^{-\lambda(t-s)}$$

In particular, we have:

$$\Pr(N_t = k) = \frac{(\lambda t)^k}{k!} e^{-\lambda t}$$

and it immediately follows that:

$$\Psi_{N_t}(u) = \lambda t \left(e^{iu} - 1\right) \tag{4.7}$$

4.1.3.4 The Compensated Poisson Process

If we try to center a Poisson process by its mean λt, we obtain a new process:

$$\tilde{N}(t) = N(t) - \lambda t$$

called a compensated Poisson process (compensated by the mean). In the same way as for the centered Poisson distribution, we obtain:

$$\Psi_{\tilde{N}_t}(u) = \lambda t \left(e^{iu} - 1 - iu\right) \tag{4.8}$$

Let us remark at this point that a compensated Poisson process is a martingale, just as Brownian motion is, in contrast to a simple Poisson process. In particular:

$$\mathbb{E}\left[\tilde{N}(t)\right] = \mathbb{E}[N(t) - \lambda t] = \mathbb{E}[N(t)] - \lambda t = \lambda t - \lambda t = 0 = \tilde{N}(0)$$

whereas for a simple Poisson process we have:

$$\mathbb{E}\left[N(t)\right] = \lambda t \neq N(0) = 0$$

4.1.3.5 The Compound Poisson Processes

Let us now consider a compound Poisson process, which is defined as a process whose jumps, of size Y_k, occur at times t_k, themselves generated by a simple Poisson process.

The jump sizes Y_k are independent of each other and also of the simple Poisson process (the sizes of the jumps and the number of jumps are independent), and they have the same distribution, which we denote f_Y.

The value a compound Poisson process takes at time t is therefore given by:

$$X_t = \sum_{k=1}^{N(t)} Y_k \qquad (4.9)$$

We calculate its characteristic function:

$$\Phi_{X_t}(u) = \mathbb{E}\left[e^{iuX_t}\right] = \mathbb{E}\left[e^{iu \sum_{k=1}^{N(t)} Y_k}\right] = \mathbb{E}\left[\mathbb{E}\left[e^{iu \sum_{k=1}^{N_t} Y_k} | N_t\right]\right]$$

Inside the conditional expectation, N_t is no longer random. The Y_k are independent and stationary. These are N_t i.i.d. copies of any given Y. It follows:

$$\mathbb{E}\left[e^{iu(Y_1+Y_2+\cdots+Y_{N_t})} | N_t\right] = \mathbb{E}\left[e^{iuY_1} e^{iuY_2} \cdots e^{iuY_{N_t}} | N_t\right] = \left(\mathbb{E}\left[e^{iuY}\right]\right)^{N_t}$$

and:

$$\Phi_{X_t}(u) = \mathbb{E}\left[\left(\mathbb{E}\left[e^{iuY}\right]\right)^{N_t}\right]$$

By definition:

$$\mathbb{E}\left[e^{iuY}\right] = \Phi_Y(u)$$

so we can write:

$$\Phi_{X_t}(u) = \mathbb{E}\left[\Phi_Y(u)^{N_t}\right]$$

Calculating the expectation gives:

$$\Phi_{X_t}(u) = \sum_{k=0}^{\infty} \Phi_Y(u)^k \Pr(N_t = k) = \sum_{k=0}^{\infty} \Phi_Y(u)^k \frac{(\lambda t)^k}{k!} e^{-\lambda t}$$

hence:

$$\Phi_{X_t}(u) = \sum_{k=0}^{\infty} \frac{(\Phi_Y(u)\lambda t)^k}{k!} e^{-\lambda t} = e^{\Phi_Y(u)\lambda t} e^{-\lambda t} = e^{\lambda t(\Phi_Y(u)-1)}$$

The characteristic exponent of a compound Poisson process can therefore be written as:

$$\Psi_{X_t}(u) = \lambda t \left(\Phi_Y(u) - 1\right)$$

We can rewrite this exponent by using the definition of the characteristic function of the random variable Y in terms of its density f_Y:

$$\Phi_Y(u) - 1 = \int_{-\infty}^{+\infty} e^{iux} f_Y(x) dx - 1 = \int_{-\infty}^{+\infty} e^{iux} f_Y(x) dx - \int_{-\infty}^{+\infty} f_Y(x) dx$$

where the second equality follows from the fact that the total probability is equal to one. We therefore obtain:

$$\Psi_{X_t}(u) = \lambda t \int_{-\infty}^{+\infty} (e^{iux} - 1) f_Y(x) dx \qquad (4.10)$$

Another way to write this is by using the notation $f_Y(x)dx = f_Y(dx)$. This reveals a new quantity called the Lévy measure, which we will examine in more detail in the following paragraph. In this particular case it is defined as:

$$\nu(dx) \stackrel{\text{déf}}{=} \lambda f(dx)$$

Using this notation, we finally obtain the characteristic exponent of a compound Poisson process:

$$\Psi_{X_t}(u) = t \int_{-\infty}^{+\infty} (e^{iux} - 1) \nu(dx) \qquad (4.11)$$

4.1.3.6 The Compensated Compound Poisson Process

To conclude, we can try to compensate the compound Poisson process in the same way we compensated the Poisson process. The expectation of a compound Poisson process is $\lambda t \mathbb{E}[Y_1]$, where $\mathbb{E}[Y_1]$ is the average size of a representative jump Y_1. If X is a compound Poisson process, then we have:

$$\tilde{X}_t = X_t - \lambda t \, \mathbb{E}[Y_1] = \sum_{k=1}^{N(t)} Y_k - \lambda t \, \mathbb{E}[Y_1]$$

From the preceding arguments it follows that the characteristic exponent of a compensated compound Poisson process is:

$$\Psi_{\tilde{X}_t}(u) = \lambda t \left(\Phi_Y(u) - 1 - iu\mathbb{E}[Y_1] \right) \qquad (4.12)$$

Recall that by definition $E[Y_1] = \int x f_Y(x) dx$, so that we obtain the integral form of the characteristic exponent:

$$\Psi_{\tilde{X}_t}(u) = \lambda t \int_{-\infty}^{+\infty} (e^{iux} - 1 - iux) f(x) dx \qquad (4.13)$$

which, using the Lévy measure, we can write as:

$$\Psi_{\tilde{X}_t}(u) = t \int_{-\infty}^{+\infty} (e^{iux} - 1 - iux) \nu(dx) \qquad (4.14)$$

Table 4.1 summarizes the characteristic exponents of various Poisson distributions and processes.

Having obtained the characteristic exponent of several types of Poisson process, we can now proceed to finding the characteristic exponent of a Lévy process. In other words, we will present the Lévy–Khintchine formula.

Table 4.1 Poissonian characteristic exponents

Distribution/process	Characteristic exponent
Poisson distribution	$\lambda \left(e^{iu} - 1\right)$
Centered Poisson distribution	$\lambda \left(e^{iu} - 1 - iu\right)$
Poisson process	$\lambda t \left(e^{iu} - 1\right)$
Compensated Poisson process	$\lambda t \left(e^{iu} - 1 - iu\right)$
Compound Poisson process	$\lambda t \int_{-\infty}^{+\infty} (e^{iux} - 1) f(x) dx$
Compensated compound Poisson process	$\lambda t \int_{-\infty}^{+\infty} (e^{iux} - 1 - iux) f(x) dx$

4.2 The Lévy–Khintchine Formula

We now introduce the representation of Lévy–Khintchine, which is the general form of the characteristic exponent of a Lévy process, by giving an intuitive interpretation and a precise mathematical description of its components. Once this is achieved, we concentrate on several quantities derived from the Lévy measure, which characterizes the jumps of a process bearing the same name.

4.2.1 *Form of the Characteristic Exponent*

In general, the characteristic exponent of any Lévy process X can be decomposed as follows:

$$\Psi_{X_t}(u) = t \underbrace{\left(i\mu u - \frac{1}{2}\sigma^2 u^2\right)}_{\substack{\text{Gaussian law} \\ \text{DIFFUSION}}} +$$

$$\underbrace{t \underbrace{\int_{|x|<1} \left(e^{iux} - 1 - iux\right) \nu(dx)}_{\text{small jumps}} + t \underbrace{\int_{|x|\geq 1} \left(e^{iux} - 1\right) \nu(dx)}_{\text{large jumps}}}_{\text{JUMP COMPONENT}} \quad (4.15)$$

The three terms of this formula represent the characteristic exponents of the normal distribution (diffusive component), a generalized (in the sense that the arrival rate of jumps may be infinite) compensated compound Poisson process, and that of a compound Poisson process, respectively. The second term corresponds to modeling small jumps, i.e. (by arbitrary construction) those jumps of size less than 1 in absolute value, and the third one to large jumps.

The last two terms can be written together in the same integral using an indicator function:

$$\Psi_{X_t}(u) = t\left(i\mu u - \frac{1}{2}\sigma^2 u^2\right) + t\int_{\mathbb{R}^*}\left(e^{iux} - 1 - iux\,\mathbf{1}_{|x|<1}(x)\right)\nu(dx)$$
(4.16)

This is the Lévy–Khintchine formula.

The characteristic exponent is indeed proportional to time:

$$\begin{cases} \Psi_{X_1}(u) = i\mu u - \frac{1}{2}\sigma^2 u^2 + \int_{\mathbb{R}^*}\left(e^{iux} - 1 - iux\,\mathbf{1}_{|x|<1}(x)\right)\nu(dx) \\ \Psi_{X_t}(u) = t\,\Psi_{X_1}(u) \end{cases}$$
(4.17)

A Lévy process is therefore completely determined by the following three quantities:

(1) The expectation μ of the diffusion component. This is the drift of the process.
(2) The diffusion coefficient σ. This coefficient functions as a scale of the diffusive fluctuations.
(3) The measure $\nu(dx)$ which determines the fluctuations of the jumps, as well as the skewness and the kurtosis of the process. Note that for the integral in Eq. (4.16) to be defined, we impose the following restriction:

$$\int_{\mathbb{R}}(x^2 \wedge 1)\nu(dx) < +\infty$$

The first two terms of the Lévy–Khintchine formula correspond to (arithmetic) Brownian motion. Indeed, if f is the density of a normal distribution, then:

$$f(x) = \frac{1}{\sigma\sqrt{2\pi}}\exp\left(-\frac{1}{2}\left(\frac{x-\mu}{\sigma}\right)^2\right)$$

The corresponding characteristic function follows:

$$\Phi(u) = \int_{-\infty}^{+\infty} e^{iux}\,\frac{e^{-\frac{1}{2}\left(\frac{x-\mu}{\sigma}\right)^2}}{\sigma\sqrt{2\pi}}\,dx$$

which can be written as:

$$\Phi(u) = \int_{-\infty}^{+\infty}\frac{e^{-\frac{1}{2}\left(\frac{x-\mu}{\sigma} - iu\sigma\right)^2}}{\sigma\sqrt{2\pi}}\,e^{\frac{i^2 u^2 \sigma^2}{2}}\,e^{\frac{\mu iu\sigma}{\sigma}}\,dx$$

making the change of variable $y = \frac{x-\mu}{\sigma} - iu\sigma$, we obtain:

$$\Phi(u) = e^{-\frac{u^2\sigma^2}{2}}\,e^{\mu iu}\int_{-\infty}^{+\infty}\frac{e^{-\frac{y^2}{2}}}{\sqrt{2\pi}}\,dy$$

Table 4.2 Comparison of Characteristic Exponents

Process / characteristic exponent	$\int_{-\infty}^{+\infty} \nu(dx)$
Compound Poisson:	finite
$t \int_{-\infty}^{+\infty} (e^{iux} - 1)\nu(dx)$	
Compensated compound Poisson:	finite
$t \int_{-\infty}^{+\infty} (e^{iux} - 1 - iux)\nu(dx)$	
Lévy pure jump:	infinite
$t \int_{|x|\geq 1} (e^{iux} - 1)\nu(dx) + t \int_{|x|<1} (e^{iux} - 1 - iux)\nu(dx)$	
$t \int_{\mathbf{R}^*} \left(e^{iux} - 1 - iux\,\mathbf{1}_{|x|<1}(x) \right) \nu(dx)$	
Lévy:	infinite
$t \left(i\mu u - \frac{1}{2}\sigma^2 u^2 \right) + t \int_{\mathbf{R}^*} \left(e^{iux} - 1 - iux\,\mathbf{1}_{|x|<1}(x) \right) \nu(dx)$	

where the integral on the right-hand side is the total probability of a standard normal distribution, and therefore equal to 1. To conclude, this gives:

$$\Phi(u) = \exp\left(i\mu u - \frac{1}{2}\sigma^2 u^2\right)$$

which is equivalent to:

$$\Psi(u) = i\mu u - \frac{1}{2}\sigma^2 u^2 \qquad (4.18)$$

Table 4.2 gives an overview of the jump component in the Lévy–Khintchine formula of the Poisson processes we have seen so far, which are written in terms of their Lévy measure.

4.2.2 The Lévy Measure

The important term of the characteristic exponent is the Lévy measure. To explain its meaning intuitively, we choose the example, given in the preceding paragraph, of a compound Poisson process with intensity λ (the Poisson parameter) as our starting point. We assume this process has the density function $f_Y(x)$ for the jump size given by Y. The Lévy measure can then be written as:

$$\nu(x) = \lambda f_Y(x) \qquad (4.19)$$

Intuitively, the Lévy measure gives the average number of jumps per unit of time in terms of their size. We can therefore view the Lévy measure as the mathematical object that quantifies the arrival of jumps and their size. It can be used to create discontinuities in the paths of the stochastic processes that represent fluctuations on the stock market.

For the Poisson processes introduced previously, the average number λ of jumps per unit of time is finite. However, in a very general way, we can arbitrarily adjust both the frequency with which jumps occur and their size by using the Lévy measure. We call the average number of jumps per unit of time the intensity of the process (also called activity of the process in analogy to turbulence). Then, if we also know the density function that determines the size of the jumps, we can join them together:

$$\text{intensity} \times \text{density}$$

which is exactly the Lévy measure in the case when it is finite.

Let us now define this measure and explain its role in financial modeling.

4.2.2.1 The Activity of a Process

Consider the integral $\int_{-\infty}^{+\infty} \nu(dx)$. There are two possible cases. Either we have, as for a compound Poisson process:

$$\lambda = \int_{-\infty}^{+\infty} \nu(dx) < +\infty \tag{4.20}$$

We see that a compound Poisson process with intensity λ is defined by a finite activity. In this situation we can completely separate the intensity λ from the density f, as in Eq. (4.19). Or we accept the possibility that the average number of small jumps per unit of time is infinite. In other words, we lift the previous restriction and have:

$$\int_{-\infty}^{+\infty} \nu(dx) = +\infty$$

The Lévy process then has infinite activity. In this case, it is no longer possible to separate the intensity λ from the density f: both are bound together in the Lévy measure, which therefore gives the number of jumps per unit of time as a function of their size.

We find that there are two possibilities for a Lévy process:

(1) Finite activity: this indicates that the average number of jumps per unit of time is finite. Usually a Brownian component is added to allow for fluctuations between jumps.

(2) Infinite activity: this indicates that the average number of jumps per unit of time is infinite. In this case it seems less important to add a Brownian component.

Let us now consider the way compound and compensated compound Poisson processes are given in terms of their measure. We can extend their interpretation to the case of infinite activity, which will generalize the notion of compound and compensated compound Poisson processes. For example:

$$t \int_{|x|\geq 1} \left(e^{iux} - 1\right) \nu(dx) \tag{4.21}$$

resembles a compound Poisson process, with the difference that the Lévy measure can give rise to infinite activity. In the same manner:

$$t \int_{|x|<1} \left(e^{iux} - 1 - iux\right) \nu(dx) \tag{4.22}$$

resembles, in its form, a compensated compound Poisson process, but now with infinite activity. This allows us to give an intuitive interpretation to the terms that appear in formula (4.16).

4.2.2.2 The Variation of a Process

An average distance between two points of these paths may or may not exist. The following quantity is called the variation of a function f on an interval $[a, b]$:

$$\sup \sum_{i=1}^{n} |f(t_i) - f(t_{i-1})|$$

where the supremum is taken over all possible partitions $a = t_0 < t_1 < \cdots < t_{n-1} < t_n = b$ of the interval $[a, b]$. Intuitively, this represents the greatest possible distance covered along the path of the process. The variation of a Lévy process depends on the Lévy measure through the quantity:

$$\int_{|x|\leq 1} |x|\nu(dx)$$

where two cases are possible: the variation is finite or infinite.

The classification of Satō (1999) defines three classes of processes according to whether their variation is finite or infinite. Following this classification, it is possible to characterize a Lévy process by its pair (activity, variation). Distinguishing between Lévy processes according to the double criterion of their activity (finite or infinite) and their variation (finite or

Table 4.3 Lévy processes in financial modeling

Activity	Variation (of jump part)	Examples of models
Finite	Finite	Press (1967), Merton (1976)
		Cox and Ross (1976)
Infinite	Finite	Barndorff-Nielsen (1997), Prause (1999)
		Madan and Milne (1991), Carr et al. (2002)
		Eberlein et al. (1998), Carr and Wu (2003)
Infinite	Infinite	Mandelbrot (1963), Schoutens (2003)

infinite), we obtain Table 4.3, which allows us to appreciate the use made of Lévy processes in financial modeling.

Let us summarize what we have presented so far. The role played by the Lévy measure is decisive. It determines both the values of the moments of the distributions (if they exist) and the behavior of their tails.

This means that the Lévy measure contains all information that is necessary to characterize the paths of a Lévy process, with the exception of the drift μ and the scale σ of the diffusive fluctuations. It is the quantity that shapes the form of the randomness, the structure of the distribution tails, and the model for the non-diffusive fluctuations.

Figure 4.1 displays four different Lévy measure densities. The first panel (reading the graphs from left to right and top to bottom) shows the Lévy density of a simple Poisson process. It is a Dirac delta function, i.e. an infinite mass given at just one abscissa, in this case $x = 1$: only jumps of size one can occur, which is in keeping with the nature of a counting process such as the simple Poisson process. Next the Lévy density of a compound Poisson process is shown. By definition of such a process, its measure is of finite activity. We have chosen to display $\nu(x) = \exp(x)\mathbf{1}_{x \leq 0} + \exp(-3x)\mathbf{1}_{x \geq 0}$, which is the Lévy density of a Kou process. For a process of this kind, the arrival rate of the jumps is given by a decreasing double exponential function.

The third Lévy density shown in Fig. 4.1 is that of a general Lévy process. We have chosen to display the Lévy density of a CGMY process, which is given by $\nu(x) = \frac{\exp(x)}{(-x)^{1,5}}\mathbf{1}_{x \leq 0} + \frac{\exp(-3x)}{x^{1,5}}\mathbf{1}_{x \geq 0}$. The fourth and last density is that of a stable process. As an example, we have chosen the Lévy density given by $\nu(x) = \frac{1}{(-x)^{1,5}}\mathbf{1}_{x \leq 0} + \frac{1}{x^{1,5}}\mathbf{1}_{x \geq 0}$. The models presented here, and in particular those with Lévy densities that can be written as the product of a power function and an exponential function, will be studied in detail in

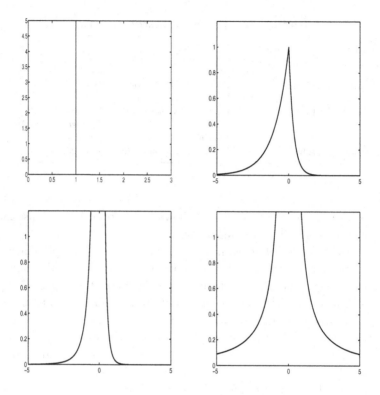

Figure 4.1 Four examples of Lévy densities

what follows. For now we point out to the reader that Lévy measures are not probability measures, since they do not necessarily integrate to one.

4.2.2.3 Financial Models

The classic models of random movements on stock markets are based on a single Brownian component. In other words, they only use the diffusive part of a Lévy process. If we let X denote the total return of a given asset, the classic models are given by the equation:

$$X(t) = \mu t + \sigma W(t) \tag{4.23}$$

where W is a standard Wiener process. In the space of characteristic functions, this leads to:

$$\Psi_{X_t}(u) = t\left(i\mu u - \frac{1}{2}\sigma^2 u^2\right) \tag{4.24}$$

Table 4.4 Adding jumps to a Brownian model

Samuelson (1965)	$X_t = \mu t + \sigma W_t$
Press (1967)	$X_t = \mu t + \sigma W_t + \sum_{k=1}^{N_t} Y_k$

Press in 1967 and then Merton in 1976 (and more recently Kou (2002)) had the idea to add a compound Poisson process to the diffusive Brownian component. In the space of characteristic functions, this leads to:

$$\Psi_{X_t}(u) = t\left(i\mu u - \frac{1}{2}\sigma^2 u^2\right) + \lambda t \int_{-\infty}^{+\infty} (e^{iux} - 1) f_Y(x) dx \quad (4.25)$$

Coming back to the notation for the return on the stock market, it is given by the equation:

$$X(t) = \mu t + \sigma W(t) + \sum_{k=1}^{N(t)} Y_k \quad (4.26)$$

where the jump sizes Y_k are modeled by Gaussian random variables. In these early types of jump models (also called jump-diffusion processes), the total return X is generated by a large number of small movements (Brownian part) on the one hand, and a small number of large movements (Poisson part) on the other hand. The activity and the variation of the process remain finite in these simple examples of Lévy processes. In certain models, such as that of Carr, Geman, Madan, and Yor (2002), the activity is infinite. This last model, as well as several other models with infinite activity, will be presented in more detail later on.

4.3 The Moments of Lévy Processes of Finite Variation

Compared to Brownian motion, Lévy processes of infinite activity and finite variation allow for an interesting generalization, without however giving up the concept – and hence the use – of moments to classic financial theory. In other words, in this class of Lévy processes, the choice of a particular measure will lead to the appearance of jumps and skewness, and therefore to non-trivial higher moments.

We now need to understand how to get from the Lévy density, assuming it exists, to the values of these moments. Thus, our goal is to obtain a general expression for the moments of a Lévy process in terms of the Lévy density. The only assumption we make is that the jump component has finite variation. The idea of the calculation is to show the moments by means of the characteristic function.

4.3.1 Existence of the Moments

Let us briefly recall how the moments of a distribution can be obtained from its characteristic function. Let $\Phi_X(u)$ be the characteristic function of a random variable X with density function f_X:

$$\Phi_X(u) = \mathbb{E}\left[e^{iuX}\right] = \int_{-\infty}^{+\infty} e^{iux} f_X(x) dx$$

Using a Lebesgue derivation theorem, the k-th derivative w.r.t. u of the characteristic function can be written as follows:

$$\Phi_X^{(k)}(u) = \int_{-\infty}^{+\infty} (ix)^k e^{iux} f(x) dx = i^k \int_{-\infty}^{+\infty} x^k e^{iux} f(x) dx$$

For $u = 0$, we obtain:

$$\Phi_X^{(k)}(0) = i^k \int_{-\infty}^{+\infty} x^k f(x) dx = i^k \mathbb{E}\left[X_t^k\right] = i^k \mu_k \qquad (4.27)$$

which makes the non-centered moments μ_k of order k appear. In other words, we have for any random variable X:

$$\mu_k = i^{-k} \Phi_X^{(k)}(0) \qquad (4.28)$$

Like the moment generating function, the characteristic function completely determines a probability distribution. Therefore, given the characteristic function of a Lévy process by the Lévy–Khintchine formula, we must calculate its k-th derivatives at $u = 0$. This means that in order to obtain the first four moments, we need to calculate:

$$\Phi_{X_t}^{(1)}(u), \Phi_{X_t}^{(2)}(u), \Phi_{X_t}^{(3)}(u), \Phi_{X_t}^{(4)}(u)$$

Recall that:

$$\Phi_{X_t}(u) = \exp\left(\Psi_{X_t}(u)\right)$$

where $\Psi_{X_t}(u)$ is the characteristic exponent. In the case of a process of finite variation, the compensation term corresponding to small jumps is removed from the Lévy–Khintchine formula (4.16), and the characteristic exponent can be simplified to:

$$\Psi_{X_t}(u) = t\left(i\mu u - \frac{1}{2}\sigma^2 u^2\right) + t \int_{-\infty}^{+\infty} \left(e^{iux} - 1\right) \nu(dx) \qquad (4.29)$$

hence:

$$\Phi_{X_t}(u) = \exp\left(t\left(i\mu u - \frac{1}{2}\sigma^2 u^2\right) + t \int_{-\infty}^{+\infty} \left(e^{iux} - 1\right) \nu(dx)\right) \qquad (4.30)$$

Table 4.5 Moments of a Lévy process of finite variation

Trigonometric circle	Eq. (4.27) of the characteristic function	Moment	Centered moment
i	$\Phi_{X_t}^{(1)}(0) = i\,\mathbb{E}[X_t]$	μ_1	
$i^2 = -1$	$\Phi_{X_t}^{(2)}(0) = -\mathbb{E}[X_t^2]$	μ_2	m_2
$i^3 = -i$	$\Phi_{X_t}^{(3)}(0) = -i\,\mathbb{E}[X_t^3]$	μ_3	m_3
$i^4 = +1$	$\Phi_{X_t}^{(4)}(0) = \mathbb{E}[X_t^4]$	μ_4	m_4

Recall the well-known formula to calculate the derivative of the exponential of a function:

$$\left(e^{\Psi(u)}\right)' = \Psi'(u)\,e^{\Psi(u)}$$

which gives here, since by definition $e^{\Psi(u)} = \Phi(u)$:

$$\left(e^{\Psi(u)}\right)' = \Psi'(u)\,\Phi(u)$$

Table 4.5 summarizes how we can obtain the moments of a Lévy process of finite variation.

4.3.2 Calculating the Moments

We perform these calculations in detail, assuming that the Lévy density exists and writing therefore $\nu(dx) = \nu(x)dx$.

4.3.2.1 Calculation of the Mean (k = 1)

Equation (4.27) becomes:

$$\Phi_{X_t}^{(1)}(0) = i\,\mathbb{E}[X_t] \tag{4.31}$$

Differentiating the expression (4.30) once w.r.t. u gives:

$$\Phi_{X_t}^{(1)}(u) = \left(it\mu - tu\sigma^2 + t\int_{-\infty}^{+\infty} ixe^{iux}\nu(x)dx\right)\Phi_{X_t}(u)$$

For $u = 0$ (using $\Phi_{X_t}(0) = 1$) we have:

$$\Phi_{X_t}^{(1)}(0) = it\mu + it\int_{-\infty}^{+\infty} x\nu(x)dx = it\left(\mu + \int_{-\infty}^{+\infty} x\nu(x)dx\right)$$

From Eq. (4.31) it follows immediately that:

$$\mathbb{E}[X_t] = t\left(\mu + \int_{-\infty}^{+\infty} x\nu(x)dx\right) \tag{4.32}$$

which gives the mean in terms of the Lévy density, assuming the integral in the right part of the equation exists.

The first term corresponds to the drift of the process and the second one to the jumps. Therefore, the jumps have an immediate effect on the value of the mean.

4.3.2.2 Calculation of the Variance (k = 2)

Equation (4.27) becomes

$$\Phi_{X_t}^{(2)}(0) = -\mathbb{E}\left[X_t^2\right] \tag{4.33}$$

Differentiating the expression (4.30) twice w.r.t. u gives:

$$\Phi_{X_t}^{(2)}(u) = \Phi_{X_t}(u)\left(-t\sigma^2 + t\int_{-\infty}^{+\infty}(ix)^2 e^{iux}\nu(x)dx\right)$$
$$+ \Phi_{X_t}(u)\left(it\mu - tu\sigma^2 + t\int_{-\infty}^{+\infty} ixe^{iux}\nu(x)dx\right)^2$$

For $u = 0$ (using $\Phi_{X_t}(0) = 1$ and $i^2 = -1$) we have:

$$\Phi_{X_t}^{(2)}(0) = -\left(t\sigma^2 + t\int_{-\infty}^{+\infty} x^2\nu(x)dx\right) - \left(t\mu + t\int_{-\infty}^{+\infty} x\nu(x)dx\right)^2$$

From Eq. (4.33), and observing that the square term on the right corresponds to $(\mathbb{E}[X_t])^2$, it follows that:

$$\mathbb{E}\left[X_t^2\right] = \left(t\sigma^2 + t\int_{-\infty}^{+\infty} x^2\nu(x)dx\right) + (\mathbb{E}[X_t])^2$$

Since $\text{var}(X_t) = \mathbb{E}\left[X_t^2\right] - (\mathbb{E}[X_t])^2$, we have:

$$\text{var}(X_t) = t\left(\sigma^2 + \int_{-\infty}^{+\infty} x^2\nu(x)dx\right) \tag{4.34}$$

which gives the variance in terms of the Lévy density. Just as before, the first term corresponds to the classic diffusion and the second one to the jumps, whose effect is immediately apparent.

4.3.2.3 Calculation of the Skewness (k = 3)

Equation (4.27) becomes:

$$\Phi_{X_t}^{(3)}(0) = -i\,\mathbb{E}\left[X_t^3\right] \tag{4.35}$$

Differentiating the expression (4.30) three times w.r.t. u gives:

$$\Phi^{(3)}_{X_t}(u) = \Phi_{X_t}(u)\left(t\int_{-\infty}^{+\infty}(ix)^3 e^{iux}\nu(x)dx\right)$$

$$+ \Phi_{X_t}(u)\left(-t\sigma^2 + t\int_{-\infty}^{+\infty}(ix)^2 e^{iux}\nu(x)dx\right) \times$$

$$\left(it\mu - tu\sigma^2 + t\int_{-\infty}^{+\infty}ixe^{iux}\nu(x)dx\right)$$

$$+ 2\,\Phi_{X_t}(u)\left(-t\sigma^2 + t\int_{-\infty}^{+\infty}(ix)^2 e^{iux}\nu(x)dx\right) \times$$

$$\left(it\mu - tu\sigma^2 + t\int_{-\infty}^{+\infty}ixe^{iux}\nu(x)dx\right)$$

$$+ \Phi_{X_t}(u)\left(it\mu - tu\sigma^2 + t\int_{-\infty}^{+\infty}ixe^{iux}\nu(x)dx\right)^3$$

For $u = 0$ (using $\Phi_{X_t}(0) = 1$ and $i^3 = -i$) we have:

$$\Phi^{(3)}_{X_t}(0) = -i\left(t\int_{-\infty}^{+\infty}x^3\nu(x)dx\right)$$

$$- 3i\left(t\sigma^2 + t\int_{-\infty}^{+\infty}x^2\nu(x)dx\right)\left(t\mu + t\int_{-\infty}^{+\infty}x\nu(x)dx\right)$$

$$- i\left(t\mu + t\int_{-\infty}^{+\infty}x\nu(x)dx\right)^3$$

and, recognizing the terms for the mean and the variance:

$$\Phi^{(3)}_{X_t}(0) = -i\left(t\int_{-\infty}^{+\infty}x^3\nu(x)dx + 3\,\text{var}(X_t)\,\mathbb{E}[X_t] + (\mathbb{E}[X_t])^3\right)$$

From Eq. (4.27) it follows that:

$$\mathbb{E}\left[X_t^3\right] = \left(t\int_{-\infty}^{+\infty}x^3\nu(x)dx\right) + 3\,\text{var}(X_t)\,\mathbb{E}[X_t] + (\mathbb{E}[X_t])^3$$

Since:

$$\mathbb{E}\left[(X_t - \mathbb{E}(X_t))^3\right] = \mathbb{E}\left[X_t^3\right] - 3\,\text{var}(X_t)\,\mathbb{E}[X_t] - (\mathbb{E}[X_t])^3$$

we obtain immediately:

$$\mathbb{E}\left[(X_t - \mathbb{E}[X_t])^3\right] = t\int_{-\infty}^{+\infty}x^3\nu(x)dx \qquad (4.36)$$

which gives the third centered moment in terms of the Lévy density. This moment depends directly and entirely on the presence of jumps. It is possible to reduce it and to calculate the *skewness* S of Pearson–Fisher. This gives:

$$S(X_t) = \frac{1}{\sqrt{t}} \frac{\int_{-\infty}^{+\infty} x^3 \nu(x) dx}{\left(\sigma^2 + \int_{-\infty}^{+\infty} x^2 \nu(x) dx\right)^{3/2}} \quad (4.37)$$

We note that on the one hand, the value of this coefficient depends on the jump component, and that on the other hand, a factor given by the square root of time leads from $S(X_1)$ to $S(X_t)$:

$$S(X_t) = \frac{S(X_1)}{\sqrt{t}}$$

which is consistent with the relationship we obtained for the skewness in the case of i.i.d. random variables: any asymmetry disappears over time (temporal diversification is, in fact, an aggregation). The passing of time plays the same role here as that of diversification in the case of i.i.d. random variables (spatial diversification) for a static approach.

4.3.2.4 Calculation of the Kurtosis ($k = 4$)

Equation (4.27) becomes:

$$\Phi_{X_t}^{(4)}(0) = \mathbb{E}\left[X_t^4\right] \quad (4.38)$$

Differentiating the expression (4.30) four times w.r.t. u gives:

$$\Phi_{X_t}^{(4)}(u) = A(u) + B(u) + C(u)$$

where:

$$A(u) = \Phi_{X_t}(u) \left(t \int_{-\infty}^{+\infty} (ix)^4 e^{iux} \nu(x) dx \right)$$
$$+ \Phi_{X_t}(u) \left(t \int_{-\infty}^{+\infty} (ix)^3 e^{iux} \nu(x) dx \right) \times$$
$$\left(it\mu - tu\sigma^2 + t \int_{-\infty}^{+\infty} ixe^{iux} \nu(x) dx \right)$$

and:
$$B(u) = 3\,\Phi_{X_t}(u)\left(t\int_{-\infty}^{+\infty}(ix)^3 e^{iux}\nu(x)dx\right)\times$$
$$\left(it\mu - tu\sigma^2 + t\int_{-\infty}^{+\infty}ixe^{iux}\nu(x)dx\right)$$
$$+ 3\,\Phi_{X_t}(u)\left(-t\sigma^2 + t\int_{-\infty}^{+\infty}(ix)^2 e^{iux}\nu(x)dx\right)^2$$
$$+ 3\,\Phi_{X_t}(u)\left(-t\sigma^2 + t\int_{-\infty}^{+\infty}(ix)^2 e^{iux}\nu(x)dx\right)\times$$
$$\left(it\mu - tu\sigma^2 + t\int_{-\infty}^{+\infty}ixe^{iux}\nu(x)dx\right)^2$$

as well as:
$$C(u) = 3\,\Phi_{X_t}(u)\left(-t\sigma^2 + t\int_{-\infty}^{+\infty}(ix)^2 e^{iux}\nu(x)dx\right)\times$$
$$\left(it\mu - tu\sigma^2 + t\int_{-\infty}^{+\infty}ixe^{iux}\nu(x)dx\right)^2$$
$$+ \Phi_{X_t}(u)\left(it\mu - tu\sigma^2 + t\int_{-\infty}^{+\infty}ixe^{iux}\nu(x)dx\right)^4$$

Let us calculate the values for $u = 0$ (where $\Phi_{X_t}(0) = 1$ and $i^4 = 1$):
$$A(0) = \left(t\int_{-\infty}^{+\infty}x^4\nu(x)dx\right)$$
$$+ \left(t\int_{-\infty}^{+\infty}x^3\nu(x)dx\right)\left(t\mu + t\int_{-\infty}^{+\infty}x\nu(x)dx\right)$$
$$= \left(t\int_{-\infty}^{+\infty}x^4\nu(x)dx\right) + S(X_t)\,\mathrm{var}(X_t)^{3/2}\,\mathbb{E}[X_t]$$

and:
$$B(0) = 3\left(t\int_{-\infty}^{+\infty}x^3\nu(x)dx\right)\left(t\mu + t\int_{-\infty}^{+\infty}x\nu(x)dx\right)$$
$$+ 3\left(t\sigma^2 + t\int_{-\infty}^{+\infty}x^2\nu(x)dx\right)^2$$
$$+ 3\left(t\sigma^2 + t\int_{-\infty}^{+\infty}x^2\nu(x)dx\right)\left(t\mu + t\int_{-\infty}^{+\infty}x\nu(x)dx\right)^2$$
$$= 3\,S(X_t)\,\mathrm{var}(X_t)^{3/2}\,\mathbb{E}[X_t] + 3\,\mathrm{var}(X_t)^2 + 3\,\mathrm{var}(X_t)\,(\mathbb{E}[X_t])^2$$

Table 4.6 Moments of a finite variation Lévy process ($t = 1$)

Moment	Value for X_1	Remark
μ_1	$\mu + \int_{-\infty}^{+\infty} x\,\nu(x)dx$	drift + jump component
m_2	$\sigma^2 + \int_{-\infty}^{+\infty} x^2\nu(x)dx$	diffusion + jump component
m_3	$\int_{-\infty}^{+\infty} x^3\nu(x)dx$	jump component
m_4	$3\,m_2^2 + \int_{-\infty}^{+\infty} x^4\nu(x)dx$	mixed base + jump component

as well as:

$$C(0) = 3\left(t\sigma^2 + t\int_{-\infty}^{+\infty} x^2\nu(x)dx\right)\left(t\mu + t\int_{-\infty}^{+\infty} x\nu(x)dx\right)^2$$
$$+ \left(t\mu + t\int_{-\infty}^{+\infty} x\nu(x)dx\right)^4$$
$$= 3\operatorname{var}(X_t)\,(\mathbb{E}[X_t])^2 + (\mathbb{E}[X_t])^4$$

Since:

$$\Phi_{X_t}^{(4)}(0) = A(0) + B(0) + C(0)$$

it follows from Eq. (4.27) that:

$$\mathbb{E}[X_t^4] = t\int_{-\infty}^{+\infty} x^4\nu(x)dx + 4\,S(X_t)\operatorname{var}(X_t)^{3/2}\mathbb{E}[X_t]$$
$$+ 3\operatorname{var}(X_t)^2 + 6\operatorname{var}(X_t)\,(\mathbb{E}[X_t])^2 + (\mathbb{E}[X_t])^4$$

Hence we obtain:

$$\mathbb{E}\left[(X_t - \mathbb{E}[X_t])^4\right] = t\int_{-\infty}^{+\infty} x^4\nu(x)dx + 3\operatorname{var}(X_t)^2 \qquad (4.39)$$

which gives the fourth centered moment in terms of the Lévy density.

Again, this moment depends directly and entirely on the presence of jumps. It is possible to reduce it in order to calculate the kurtosis K of Pearson–Fisher. This gives:

$$K(X_t) = 3 + \frac{1}{t}\frac{\int_{-\infty}^{+\infty} x^4\nu(x)dx}{\left(\sigma^2 + \int_{-\infty}^{+\infty} x^2\nu(x)dx\right)^2} \qquad (4.40)$$

We note that on the one hand, the value of this coefficient depends on the jump component, and that on the other hand, a factor given by the inverse of time leads from $K(X_1) - 3$ to $K(X_t) - 3$:

$$K(X_t) - 3 = \frac{K(X_1) - 3}{t}$$

Table 4.7 Moments & coefficients: finite variation Lévy process

Moment / coefficient	Value for X_t	Formula
$\mathbb{E}[X_t]$	$t\left(\mu + \int_{-\infty}^{+\infty} x\nu(x)dx\right)$	(4.32)
$\text{var}(X_t)$	$t\left(\sigma^2 + \int_{-\infty}^{+\infty} x^2\nu(x)dx\right)$	(4.34)
$S(X_t)$	$\dfrac{1}{\sqrt{t}} \dfrac{\int_{-\infty}^{+\infty} x^3\nu(x)dx}{\left(\sigma^2 + \int_{-\infty}^{+\infty} x^2\nu(x)dx\right)^{3/2}}$	(4.37)
$K(X_t)$	$3 + \dfrac{1}{t} \dfrac{\int_{-\infty}^{+\infty} x^4\nu(x)dx}{\left(\sigma^2 + \int_{-\infty}^{+\infty} x^2\nu(x)dx\right)^{2}}$	(4.40)

Here again, any non-normality disappears over time, which is consistent with the relationship we obtained for the kurtosis in the case of diversification with i.i.d. random variables.

Tables 4.6 and 4.7 summarize the values taken by the moments and coefficients for a Lévy process of finite activity. These results also hold for Lévy processes of infinite activity, with the exception of the first moment, which needs to be modified by a term related to the compensation of the small jumps of a process of this type.

Chapter 5

Stable Distributions and Processes

The marginal distributions of a Lévy process are infinitely divisible. There are two families of distributions of this type: those that are stable under addition, and those that are not. The property of stability under addition of random variables means that when two or more random variables having this property are added, the resulting distribution remains the same, up to the scale of the parameters.

The aim of this chapter is to present stable distributions, as well as the stochastic processes that can be built from them. For a more detailed exposition, the reader should consult the works of Janicki and Weron (1993), Samorodnitsky and Taqqu (1994) and, for an introduction in French, Lévy-Véhel and Walter (2002). The first part of this chapter describes the main properties of stable distributions, how to obtain the associated Lévy-Khintchine formula and the simulation of stable processes, as well as the conditions under which the moments of stable distributions exist. The second part of the chapter describes several financial applications of stable distributions and presents a model which is derived from these distributions and used in the physical sciences as well as in finance: the model of truncated stable distributions.

5.1 Definitions and Properties

Most stable distributions do not have an analytically given density function and are defined via their characteristic function, which can take one of two forms: the first using the Lévy measure (or non-integrated form), the second being the integrated form.

We will present these two forms in what follows by developing the calculations that lead from one to the other in detail. This allows for a clear interpretation of the two parameterizations, in the Lévy measure and in integrated form. The additional benefit of a presentation using the two-fold parameterization is that the Lévy measure directly gives the behavior of the tails of the density. It is therefore very simple to go from the characteristic function to the domains of attraction.

5.1.1 *Definitions*

The main interest of stable distributions lies precisely in the fact that they are stable. Their property of being stable under addition is presented in the following.

5.1.1.1 *Stability Under Addition and Self-Similarity*

Let X, Y, and Z be three independent identically distributed random variables, and A, B, C, and D four real numbers (the first three being strictly positive). Then if the following relationship holds:

$$AX + BY \stackrel{d}{=} CZ + D$$

the random variable X (as well as its independent copies) is said to be stable.

This generalizes for all $n \geq 2$ to:

$$X_1 + X_2 + \cdots + X_n \stackrel{d}{=} C_n X + D_n$$

where X_1, X_2, ..., X_n, and X are independent identically distributed random variables, and C_n and D_n are real numbers (the first of which is strictly positive). If this last relationship holds, then X (as well as its independent copies) is said to be stable.

Stable distributions are described by four parameters $(\alpha, \beta, \gamma, \mu)$, which correspond to the fatness of the tails, the asymmetry, the width, and the mean of these distributions, respectively (when this last quantity does not exist, μ is interpreted as a shift parameter). We will denote stable distributions by $\mathcal{S}(\alpha,\beta,\gamma,\mu)$ in the following. The interpretation of their parameters will be refined in this chapter.

More generally, a stable process is a process whose underlying distributions, marginal or joint, are stable. In this chapter, we will consider several sub-classes of these processes. For simplicity, we call a Lévy process (with ensuing independent and stationary increments) whose increments have stable distributions, a stable process. Furthermore, we call a stable process whose marginal distributions are symmetric and normalized, an alpha-stable process such as:

$$X_t - X_s \sim \mathcal{S}(\alpha, 0, |t-s|^{\frac{1}{\alpha}}, 0)$$

These symmetric and normalized alpha-stable distributions are often denoted by $S\alpha S$.

If X is an alpha-stable process, then it has the following property:

$$X(at) - X(as) \stackrel{d}{=} a^{1/\alpha}(X(t) - X(s)) \qquad (5.1)$$

for all $\alpha \in]0,2]$, where we use $\stackrel{d}{=}$ to denote equality in distribution. In this case, the stochastic process X is said to be self-similar with exponent $1/\alpha$, or $1/\alpha$–self-similar with stationary increments ($1/\alpha$–sssi), or also fractal. This means that the distribution of the return at date at (hence $X(at)$), where a determines some multiple or fraction of t (for example a year-fraction, or a non-integer multiple of a year, if a year is the unit of time), is related to the distribution at date t by simple invariance of scale. Graphically, the dilation of the time-axis is equivalent to a dilation of the axis of cumulative returns, up to normalization (and stressing the fact that this is a dilation of the distributions, and not of the paths).

5.1.1.2 *Morphology of Alpha-Stable Distributions*

The tails of alpha-stable distributions become fatter and fatter as the characteristic exponent α decreases. The increase in value of the quantiles for a given value x is quite rapid. Table 5.1 shows the values of the higher quantiles q_c^α of the $S\alpha S$ as a function of α.

For example:

- for $c = 95\%$, the standard Gaussian quantile is $q_{0.95} = 1.6448$. The equivalent 2-stable quantile is $q_{0.95}^2 = q_{0.95}\sqrt{2} = 2.3262$, whereas the 1.4-stable quantile is $q_{0.95}^{1.4} = 3.6999$.

Table 5.1 Higher quantiles q_c^α as a function of α

c	1	1.2	1.4	1.6	1.8	2	$\mathcal{N}(0,1)$
99%		16.160	9.6576	6.2768	4.2766	3.2899	2.3263
98%	15.895	9.1635	6.0613	4.3601	3.3915	2.9044	2.0537
97%	10.578	6.5924	4.6571	3.5780	2.9785	2.6598	1.8808
96%	7.9155	5.2277	3.8778	3.1247	2.7079	2.4758	1.7507
95%	6.3119	4.3687	3.6999	2.8143	2.5049	2.3262	1.6448

- for $c = 99\%$, we observe a strong increase of $q_{0.99}^\alpha$ as α decreases towards the value 1 (Cauchy distribution). In particular, for $\alpha = 1.6$, which is a value often found in tests carried out on financial markets, $q_{0.99}^{1.6} = 6.2768$, i.e. 1.9 times more than $q_{0.99}^2$.

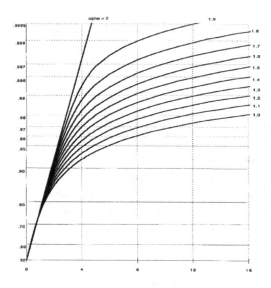

Figure 5.1 Deviation of alpha-stable distributions from the standard normal distribution in terms of α

We can directly compare the cumulative probability given by each distribution and observe how a decrease in the characteristic exponent α affects the shape of the distribution tail, and therefore the probabilities. For $x = 4$, $F_2(4) = 0.9977$ (Gaussian), whereas $F_{1.5}(4) = 0.9694$, and $F_1(4) = 0.9220$ for a Cauchy distribution ($\alpha = 1$). We also note that for large values of x, for example $x = 8$, while the value of the cumulative probability is close to 1 for a Gaussian distribution, this value is still far from 1 for a non-Gaussian stable distribution (for example 0.9764 for $\alpha = 1.2$).

Figure 5.1 gives an illustration of the fatness of these tails in terms of the Henry line. The curve of the Gaussian distribution function is transformed into a line if it is drawn in the appropriate coordinate system with an arithmetic abscissa and a Gaussian ordinate. The fatness of the tails of alpha-stable distributions can then be seen in terms of the deviation of the obtained points from the line. For each value $F_\alpha(x)$ taken by the distribution function F_α of an alpha-stable distribution for a given value x, we calculate $F^{-1}(F_\alpha(x))$, using the inverse of a normal distribution function F with mean 0, to obtain the corresponding quantile of this normal distribution. For example, let us consider an alpha-stable distribution with exponent $\alpha = 1.4$. For $x = 3.9$, we have $F_{1.4}(x) = 0.9604$. Therefore, we search the quantile x' of the normal distribution such that $F_2(x') = 0.9604$. We find $x' = 2.4824$. The point (3.9, 2.4824) is then marked in the graph. For a normal distribution, we naturally would have found 3.9, and the point would have lain on the Henry line.

5.1.2 Characteristic Function and Lévy Measure

We are interested in the form of the characteristic function of a stable distribution. To obtain the characteristic function of the marginal distribution of a stable process at a given time t, we take the t-th power of the characteristic function of a reference stable distribution taken at time $t = 1$ (see formula (4.3)).

5.1.2.1 The Lévy–Khintchine Representation

In the general case, the density function is not known in closed form, and the characteristic function is used. As seen before, we know that for any random variable the characteristic function is given in terms of the characteristic exponent by:

$$\Phi_X(u) = \exp\left(\Psi_X(u)\right)$$

and for a stable random variable X we have by definition:

$$\Psi_X(u) = i\mu u + \int_{-\infty}^{0} \psi(u,x) \frac{C_-}{|x|^{1+\alpha}} dx + \int_{0}^{+\infty} \psi(u,x) \frac{C_+}{x^{1+\alpha}} dx \quad (5.2)$$

with:

$$\psi(u,x) = e^{iux} - 1 - iux\,\mathbf{1}_{|x|<1} \quad (5.3)$$

The parameter μ gives the drift, while the term α describes the form of the distribution and the force of the jumps affecting the paths. The tails of the distribution become fatter and fatter as α decreases. C_- and C_+ are two constants, called the tail amplitudes, that represent scale parameters for the distribution tails. In practice, C_- and C_+ give the order of magnitude of large fluctuations (negative or positive). Comparing Eq. (5.2) to the Lévy–Khintchine formula (4.16), we see that the Lévy measure of a stable distribution is:

$$\nu(dx) = \frac{C_-}{|x|^{1+\alpha}} \mathbf{1}_{(-\infty,0)}(x)dx + \frac{C_+}{x^{1+\alpha}} \mathbf{1}_{(0,+\infty)}(x)dx \qquad (5.4)$$

or, more intuitively:

$$\nu(x) = \begin{cases} \frac{C_-}{|x|^{1+\alpha}} & \text{if } x < 0 \\ \frac{C_+}{x^{1+\alpha}} & \text{if } x > 0 \end{cases} \qquad (5.5)$$

which is the equation also found in Samorodnitsky and Taqqu (1994).

An important characteristic evident in this form is a decrease given by a power law, which is slow for large values of x compared to the fast decrease observed for exponential laws. This slow decrease of the distribution tails is a key characteristic of Pareto distributions. Here, the Lévy measure is defined by a Pareto Type I distribution, or simple Pareto distribution (see Chapter 8).

5.1.2.2 Integrated Form of the Characteristic Function

The integrals appearing in (5.2) will be calculated in two steps. We will first calculate this function (and with it the characteristic exponent) in the case where the random variable is of finite variation, i.e. for $0 < \alpha < 1$. Then we will treat the case $1 < \alpha < 2$.

The case $0 < \alpha < 1$ Starting from (5.2), which we recall here:

$$\Psi_X(u) = i\mu u + \int_{-\infty}^{0} \psi(u,x) \frac{C_-}{|x|^{1+\alpha}} dx + \int_{0}^{+\infty} \psi(u,x) \frac{C_+}{x^{1+\alpha}} dx$$

we must calculate:

$$\int_{-\infty}^{0} \left(e^{iux} - 1\right) \frac{C_-}{(-x)^{1+\alpha}} dx + \int_{0}^{+\infty} \left(e^{iux} - 1\right) \frac{C_+}{x^{1+\alpha}} dx \stackrel{\text{def}}{=} I^- + I^+ \qquad (5.6)$$

where the function (5.3) is reduced to its first two terms, since the random variable we are considering is of finite variation. Let us begin by calculating

I^+. Integration by parts gives:

$$I^+ = C_+ \left[\left(e^{iux} - 1\right) \frac{x^{-\alpha}}{-\alpha} \right]_0^{+\infty} - C_+ \int_0^{+\infty} iu e^{iux} \frac{x^{-\alpha}}{-\alpha} dx$$

or:

$$I^+ = \frac{iuC_+}{\alpha} \int_0^{+\infty} e^{iux} x^{-\alpha} dx$$

This follows because, for the boundary at $+\infty$, with u being a real number, $\left(e^{iux} - 1\right)$ is bounded while $x^{-\alpha}/-\alpha$ goes to 0. And for the boundary at 0, we have $\left(e^{iux} - 1\right) \approx iux$, which divided by x^α goes to 0 as x goes to 0, since $\alpha < 1$. We make a change of variable and define $y = -iux$. We obtain for $u > 0$:

$$I^+ = \frac{iuC_+}{\alpha} \int_{-iu.0}^{-iu.(+\infty)} e^{-y} \left(\frac{y}{-iu}\right)^{-\alpha} \frac{dy}{-iu}$$

or:

$$I^+ = -\frac{C_+(-iu)^\alpha}{\alpha} \int_{-i.0}^{-i.(+\infty)} e^{-y} y^{-\alpha} dy \qquad (5.7)$$

We define:

$$J^+ = \int_{-i.0}^{-i.(+\infty)} e^{-y} y^{-\alpha} dy$$

This term resembles a Γ-function, but given on the complex half-line $[-i.0, -i.(+\infty))$. We now show how this can be related to the Γ-function using well-known results from complex analysis.

Consider the function $e^{-y} y^{-\alpha}$ in the lower right quadrant of the complex plane. In Fig. 5.2, we integrate this function along the quarter circle with radius r (small and arbitrary), then along the line segment $[-i.r, -i.R]$, then along the quarter circle with radius R (large and arbitrary), then along the line segment $[R, r]$. We denote these four integrals by I_1, I_2, I_3 and I_4, respectively.

Since there is no pole in the interior of the contour just constructed (there is a pole at 0, hence the quarter circle of radius r), it follows from the residue theorem (or Cauchy's integral theorem in this case) that for all strictly positive r and R we have:

$$I_1 + I_2 + I_3 + I_4 = 0 \qquad (5.8)$$

Let us see what happens when $r \to 0$ and $R \to +\infty$. To begin, we have directly:

$$\lim_{r \to 0, R \to +\infty} I_2 = \int_{-i.0}^{-i.+\infty} e^{-y} y^{-\alpha} dy = J^+$$

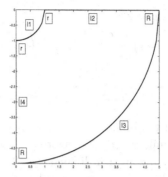

Figure 5.2 Quadrant for the calculation of the stable characteristic exponent

and:
$$\lim_{r \to 0, R \to +\infty} I_4 = \int_{+\infty}^{0} e^{-y} y^{-\alpha} dy = -\int_{0}^{+\infty} e^{-y} y^{-\alpha} dy$$

where we immediately recognize the Γ-function. It follows:
$$\lim_{r \to 0, R \to +\infty} I_4 = -\Gamma(1 - \alpha)$$

Noting that $\lim_{r \to 0} y \left(e^{-y} y^{-\alpha} \right) = 0$, we can use Jordan's lemma for small circles from complex analysis to obtain:
$$\lim_{r \to 0} I_1 = 0$$

Noting that $\lim_{R \to +\infty} y \left(e^{-y} y^{-\alpha} \right) = 0$, we can use Jordan's lemma for large circles from complex analysis to obtain:
$$\lim_{R \to +\infty} I_3 = 0$$

Considering the limit for $r \to 0$ and $R \to +\infty$ of Eq. (5.8), we therefore obtain:
$$J^+ = \Gamma(1 - \alpha)$$

which directly gives:
$$I^+ = -\frac{C_+(-iu)^\alpha}{\alpha} \Gamma(1 - \alpha)$$

Using the notation $-i = e^{-\frac{i\pi}{2}}$, we may also write this as:
$$I^+ = -\frac{C_+ u^\alpha}{\alpha} e^{-\frac{i\pi\alpha}{2}} \Gamma(1 - \alpha) \tag{5.9}$$

which holds for $u > 0$. Let us now calculate I^- under the same hypothesis. The change of variable $z = -x$ allows us to write:

$$I^- = \int_{-\infty}^{0} \left(e^{iux} - 1\right) \frac{C_-}{(-x)^{1+\alpha}} dx = \int_{+\infty}^{0} \left(e^{-iuz} - 1\right) \frac{C_-}{z^{1+\alpha}} d(-z)$$

or also:

$$I^- = \int_{0}^{+\infty} \left(e^{-iuz} - 1\right) \frac{C_-}{z^{1+\alpha}} dz$$

I^- takes the same form as I^+ if we replace C_+ by C_- and u by $-u$ in the expression for I^+. It follows:

$$I^- = -\frac{C_-(-u)^\alpha}{\alpha} e^{-\frac{i\pi\alpha}{2}} \Gamma(1-\alpha)$$

Noting that $(-1)^\alpha = e^{i\pi\alpha}$, this last equation can be rewritten as:

$$I^- = -\frac{C_- u^\alpha}{\alpha} e^{\frac{i\pi\alpha}{2}} \Gamma(1-\alpha) \tag{5.10}$$

Finally, we obtain:

$$I^+ + I^- = -\frac{C_+ u^\alpha}{\alpha} e^{-\frac{i\pi\alpha}{2}} \Gamma(1-\alpha) - \frac{C_- u^\alpha}{\alpha} e^{\frac{i\pi\alpha}{2}} \Gamma(1-\alpha) \tag{5.11}$$

and, by using Euler's formula and collecting terms:

$$I^+ + I^- = -\frac{u^\alpha}{\alpha} \Gamma(1-\alpha) \left[(C_+ + C_-) \cos \frac{\pi\alpha}{2} - i(C_+ - C_-) \sin \frac{\pi\alpha}{2} \right]$$

factoring gives:

$$I^+ + I^- = -\frac{u^\alpha}{\alpha} \Gamma(1-\alpha) (C_+ + C_-) \cos \frac{\pi\alpha}{2} \left[1 - i \frac{C_+ - C_-}{C_+ + C_-} \mathrm{tg} \frac{\pi\alpha}{2} \right] \tag{5.12}$$

Setting:

$$\beta = \frac{C_+ - C_-}{C_+ + C_-}$$

and:

$$\gamma^\alpha = \frac{1}{\alpha} \Gamma(1-\alpha) (C_+ + C_-) \cos \frac{\pi\alpha}{2}$$

we obtain for $u > 0$:

$$I^+ + I^- = -u^\alpha \gamma^\alpha \left[1 - i\beta \mathrm{tg} \frac{\pi\alpha}{2} \right]$$

Similar calculations show that for $u < 0$ we have:

$$I^+ + I^- = -(-u)^\alpha \gamma^\alpha \left[1 + i\beta \mathrm{tg} \frac{\pi\alpha}{2} \right]$$

from which we obtain, setting:
$$u^\alpha = |u|^\alpha \frac{u}{|u|}$$
the following equation for any real, non-zero u:
$$I^+ + I^- = -\gamma^\alpha |u|^\alpha \left(1 - i\beta \frac{u}{|u|} \operatorname{tg} \frac{\pi\alpha}{2}\right) \tag{5.13}$$
and with it the generic value of the characteristic exponent of a stable distribution:
$$\Psi_X(u) = i\mu u - \gamma^\alpha |u|^\alpha \left(1 - i\beta \frac{u}{|u|} \operatorname{tg} \frac{\pi\alpha}{2}\right) \tag{5.14}$$

Integrating Eq. (5.2) leads to Eq. (5.14), which is better known than the non-integrated form explicitly using the Lévy measure. In their integrated form (5.14), stable distributions are therefore given by the four parameters $(\alpha, \beta, \gamma, \mu)$ and are denoted by $\mathcal{S}(\alpha, \beta, \gamma, \mu)$. The parameter $\beta \in [-1, +1]$ is a measure of asymmetry. If $\beta < 0$ (resp. $\beta > 0$), the left (resp. right) tail of the distribution becomes fatter, which is reflected in the presence of some large left (resp. right) values, corresponding to a skewness $S < 0$ (resp. $S > 0$). When $\beta = -1$ (resp. $+1$), the distribution is completely skewed to the left (resp. right). Finally, when $\beta = 0$, the distribution is symmetric. The parameter γ is a generalized volatility which corresponds to the standard deviation, up to a factor of $\sqrt{2}$, in case $\alpha = 2$ (Gaussian); when $\beta = 0$ and $\gamma = 1$, the stable random variable is called standard. The parameter μ is the expectation, when it exists, i.e. if $\alpha > 1$. The Laplace–Gauss normal distribution $\mathcal{N}(0, \sqrt{2})$ with mean 0 is the standard symmetric stable distribution $\mathcal{S}(2, 0, 1, 0)$. The Cauchy distribution is the standard symmetric stable distribution $\mathcal{S}(1, 0, 1, 0)$. The characteristic exponent of a symmetric stable random variable ($\beta = 0$, denoted by $S\alpha S$) takes a particularly simple form:
$$\Psi_X(u) = i\mu u - \gamma^\alpha |u|^\alpha \tag{5.15}$$

The case $1 < \alpha < 2$ In this case, the characteristic exponent can be written in the following form:
$$\Psi_X(u) = \int_{-\infty}^{+\infty} \left(e^{iux} - 1 - iux\right) \nu(x) dx \tag{5.16}$$

To calculate the characteristic exponent in case $1 < \alpha < 2$, we will try to come back to the characteristic exponent from formula (5.6) by deriving the integral in (5.16) with respect to u. We write:
$$\psi_u = \frac{\partial \Psi_X(u)}{\partial u}$$

and obtain:
$$\psi_u = \frac{\partial}{\partial u}\left(\int_{-\infty}^{+\infty}\left(e^{iux}-1-iux\right)\nu(x)dx\right)$$
$$= \int_{-\infty}^{+\infty}\frac{\partial}{\partial u}\left[\left(e^{iux}-1-iux\right)\nu(x)\right]dx$$
$$= \int_{-\infty}^{+\infty}\left(ixe^{iux}-ix\right)\nu(x)dx$$
$$= i\int_{-\infty}^{+\infty}\left(e^{iux}-1\right)x\nu(x)dx$$

We write:
$$x\nu(x) = \begin{cases} \frac{x\,C_-}{|x|^{1+\alpha}} & \forall x < 0 \\ \frac{x\,C_+}{|x|^{1+\alpha}} & \forall x > 0 \end{cases} = \begin{cases} \frac{-C_-}{|x|^{1+\alpha'}} & \forall x < 0 \\ \frac{C_+}{|x|^{1+\alpha'}} & \forall x > 0 \end{cases} \quad (5.17)$$

where $\alpha' = \alpha - 1$ lies in the interval $]0,1[$.

We can calculate ψ_u using the results obtained above for stability (or tail) parameters smaller than one. In particular, using Eq. (5.12) for a stable distribution whose stability parameter is now α', and whose Lévy measure is given by (5.17), we can write for $u > 0$:

$$\psi_u = i\left(-\frac{C_+ u^{\alpha'}}{\alpha'}e^{-\frac{i\pi\alpha'}{2}} + \frac{C_- u^{\alpha'}}{\alpha'}e^{\frac{i\pi\alpha'}{2}}\right)\Gamma(1-\alpha') = i\left(I^+ + I^-\right)$$

and we come back to Eq. (5.11). Let us now replace i with $e^{i\frac{\pi}{2}}$ and α' with $\alpha - 1$ to obtain:
$$\psi_u = -\frac{C_+ u^{\alpha-1}}{\alpha-1}e^{i\frac{\pi}{2}}\,e^{-\frac{i\pi(\alpha-1)}{2}}\,\Gamma(2-\alpha) + \frac{C_- u^{\alpha-1}}{\alpha-1}e^{i\frac{\pi}{2}}\,e^{\frac{i\pi(\alpha-1)}{2}}\,\Gamma(2-\alpha)$$

which simplifies to:
$$\psi_u = \frac{C_+\,u^{\alpha-1}}{\alpha-1}\,e^{-\frac{i\pi\alpha}{2}}\,\Gamma(2-\alpha) + \frac{C_-\,u^{\alpha-1}}{\alpha-1}\,e^{\frac{i\pi\alpha}{2}}\,\Gamma(2-\alpha)$$

To obtain the characteristic exponent of formula (5.16), it is useful to calculate the derivative (which vanishes at zero) of ψ_u:

$$\int_0^u \psi_v\,dv = C_+ u^\alpha\,e^{-\frac{i\pi\alpha}{2}}\,\Gamma(2-\alpha) + C_- u^\alpha\,e^{\frac{i\pi\alpha}{2}}\,\Gamma(2-\alpha) \quad (5.18)$$

We then proceed as we did when we deduced Eq. (5.13) from Eq. (5.12), observing that Eq. (5.18) is Eq. (5.12) multiplied by $-\alpha$, and with terms $\Gamma(2-\alpha)$ instead of $\Gamma(1-\alpha)$. Let us begin by putting:

$$\beta = \frac{C_+ - C_-}{C_+ + C_-} \quad (5.19)$$

and then:

$$\gamma = \left(-\Gamma(2-\alpha)(C_+ + C_-)\cos\left(\frac{\pi\alpha}{2}\right)\right)^{\frac{1}{\alpha}} \qquad (5.20)$$

We therefore find again, for real and non-zero u, the classic form of the characteristic exponent of a stable distribution with mean zero:

$$\Psi_X(u) = -\gamma^\alpha |u|^\alpha \left(1 - i\beta \frac{u}{|u|} \operatorname{tg} \frac{\pi\alpha}{2}\right) \qquad (5.21)$$

5.1.2.3 The Relationship Between the Two Parameterizations

We can observe a difference in these parameterizations. Whereas the non-integrated form (5.2) depends on the four parameters (α, C_-, C_+, μ), the parameterization of the integrated form (5.14) uses the four parameters $(\alpha, \beta, \gamma, \mu)$, of which two are different. In fact, going from one form to the other allows us to make explicit the role of, and the relationship between, these parameters. While α and μ retain the same interpretation (shape of the distribution tails and shift – or expectation, if it exists), C_- and C_+ have been transformed into β and γ. Calculating the integrals in practice allows us to obtain the value and the meaning of these parameters.

From the non-integrated form (5.2) it can be seen that the Lévy measure is asymmetric. This asymmetry is expressed by the two distinct values C_- and C_+ that characterize the weight of the distribution tails.

With regard to this form, the parameter β of the integrated form (5.14) summarizes the asymmetry in the same way as a classic skewness parameter via the relationship:

$$\beta = \frac{C_+ - C_-}{C_+ + C_-}$$

which shows clearly how this parameter models the asymmetry: in case $C_- = C_+$ (symmetry), then $\beta = 0$. However, the parameters C_- and C_+ also contain information about the weights of the distribution tails, namely an amplitude of fluctuations as seen before. This amplitude is given by the parameter γ. It must be remarked here that while the definition of β in terms of C_+ and C_- is independent of the value of α, this is not the case for the definition of γ. Observe that γ is always positive (since $\cos\left(\frac{\pi\alpha}{2}\right)$ is negative for $1 < \alpha < 2$). We conclude from this that C_+ and C_- have a different definition in terms of β and γ for $1 < \alpha < 2$. In case $\alpha < 1$, we have:

$$\gamma^\alpha = \frac{1}{\alpha}\Gamma(1-\alpha)(C_+ + C_-)\cos\frac{\pi\alpha}{2}$$

To obtain explicit values for C_- and C_+ in terms of β and γ, it is sufficient to solve a system of two equations with two unknowns. We obtain:

$$C_+ = \frac{\alpha \gamma^\alpha (1+\beta)}{2\,\Gamma(1-\alpha)\cos\frac{\pi\alpha}{2}} \tag{5.22}$$

and:

$$C_- = \frac{\alpha \gamma^\alpha (1-\beta)}{2\,\Gamma(1-\alpha)\cos\frac{\pi\alpha}{2}} \tag{5.23}$$

Putting these coefficients into Eq. (5.5), we obtain a new formula for the Lévy measure which can be expressed in terms of the conventional parameters of stable distributions. In order to make this form of the Lévy measure appear explicitly, we recall Eq. (5.5):

$$\nu(x) = \begin{cases} \frac{C_-}{|x|^{1+\alpha}} & \text{if } x < 0 \\ \frac{C_+}{x^{1+\alpha}} & \text{if } x > 0 \end{cases}$$

Therefore, this gives:

$$\nu(x) = \begin{cases} \frac{\alpha \gamma^\alpha}{\Gamma(1-\alpha)\cos\frac{\pi\alpha}{2}} \frac{1-\beta}{2} \frac{1}{|x|^{1+\alpha}} & \text{if } x < 0 \\ \frac{\alpha \gamma^\alpha}{\Gamma(1-\alpha)\cos\frac{\pi\alpha}{2}} \frac{1+\beta}{2} \frac{1}{x^{1+\alpha}} & \text{if } x > 0 \end{cases} \tag{5.24}$$

In case $1 < \alpha < 2$, we have:

$$C_+ = \frac{\gamma^\alpha (1+\beta)}{2\,\Gamma(2-\alpha)\cos\frac{\pi\alpha}{2}} \tag{5.25}$$

and:

$$C_- = \frac{\gamma^\alpha (1-\beta)}{2\,\Gamma(2-\alpha)\cos\frac{\pi\alpha}{2}} \tag{5.26}$$

from which we derive a new formula for the Lévy measure in case $1 < \alpha < 2$:

$$\nu(x) = \begin{cases} \frac{\gamma^\alpha}{\Gamma(2-\alpha)\cos\frac{\pi\alpha}{2}} \frac{1-\beta}{2} \frac{1}{|x|^{1+\alpha}} & \text{if } x < 0 \\ \frac{\gamma^\alpha}{\Gamma(2-\alpha)\cos\frac{\pi\alpha}{2}} \frac{1+\beta}{2} \frac{1}{x^{1+\alpha}} & \text{if } x > 0 \end{cases} \tag{5.27}$$

5.1.2.4 Existence of Moments

Let $X \sim \mathcal{S}(\alpha, \beta, \gamma, \mu)$ where $0 < \alpha < 2$. Let k be a real number. When k is smaller than α, the absolute moments of order k are finite:

$$\mathbb{E}(|X|^k) < +\infty \qquad \text{for } 0 < k < \alpha$$

whereas they are infinite when k is equal to or larger than α:

$$\mathbb{E}(|X|^k) = +\infty \qquad \text{for } k \geq \alpha$$

It therefore appears that the choice of a Lévy measure as defined in (5.4) means that it is not possible to use higher moments of distributions in financial calculations as soon as $k \geq \alpha$.

The presence of power functions in the Lévy measure thus has these consequences:

(1) good modeling of tails that decrease hyperbolically in $x^{-\alpha}$,
(2) loss of the use of moments of order k in financial applications for $k \geq \alpha$.

Estimates carried out on financial markets in general lead to values of α between 1 and 2. It follows that all moments of order 2 or higher do not exist. In other words, the variance (and *a fortiori* the skewness and the kurtosis) is infinite. Therefore, no moment apart from the expectation (which itself may be infinite as in the case of the Cauchy distribution seen above) is available for financial calculations. This argument is also valid in the case of markets with little liquidity where $\alpha < 1$. This loss of the use of moments was one of the reasons for the rejection of Mandelbrot's model in the 1970s.

5.1.3 Some Special Cases of Stable Distributions

The densities of stable distributions are not known in analytical form, with three exceptions: the normal distribution, the Cauchy distribution, and the inverse Gaussian distribution. The normal distribution, or Laplace–Gauss distribution, corresponds to the value $\alpha = 2$. The parameter values of a random variable of this type are equal to the usual ones for a standard normal distribution, up to a scaling factor, which comes from the equation $\gamma = \sigma/\sqrt{2}$. The equivalence of a 2-stable distribution with mean 0 and variance 1 and a standard normal distribution is therefore simply obtained as follows:

$$\Pr\left(X_{\mathcal{S}_{2,0}(1,0)} \leq x\right) = \Pr\left(X_{N(0,\sqrt{2})} \leq x\right) = \Pr\left(X_{N(0,1)} \leq x/\sqrt{2}\right)$$

We now come to two other distributions that are of interest for financial applications: the Cauchy distribution and the inverse Gaussian distribution.

5.1.3.1 1-stable or Cauchy Distribution

The Cauchy distribution corresponds to the case where $\alpha = 1$. It can be used in all situations in which extremely abrupt movements can occur suddenly after long periods of inactivity, giving the appearance of non-stationarity to the increments of a process, where in fact we have a very particular case of stationarity of the increments. This distinctive feature is due to the fact that a Cauchy distribution has no moments. In fact, the density function of a Cauchy distribution is:

$$f(x) = \frac{1}{\pi(1+x^2)}$$

and its distribution function is therefore:

$$F(x) = \frac{1}{\pi}\operatorname{Arctg} x + \frac{1}{2}$$

We calculate:

$$\mathbb{E}[|X|] = 2\int_0^{+\infty} \frac{x}{\pi(1+x^2)}\,dx$$

and obtain:

$$\mathbb{E}[|X|] = \lim_{b\to+\infty}\int_0^b \frac{2x}{\pi(1+x^2)}\,dx = \frac{1}{\pi}\lim_{b\to+\infty}\left[\ln(1+x^2)\right]_0^b$$

or:

$$\mathbb{E}[|X|] = \frac{1}{\pi}\lim_{b\to+\infty}\ln(1+b^2) = +\infty$$

Since the expectation is infinite, the Cauchy distribution does not have a mean and, *a fortiori*, higher moments.

In practice, by simulating Cauchy paths, we observe that the empirical mean values obtained always tend to 0, but there always comes a moment where a too far-off value perturbs this convergence: the probability of drawing values far from 0 is very high. Hence the possibility of using this distribution when modeling highly unstable environments. We may conjecture that markets for fine art, or also certain insurance contracts, constitute natural domains for applications of Cauchy distributions.

5.1.3.2 1/2-stable or Inverse Gaussian Distribution

The so-called inverse Gaussian distribution gives the probability distribution of the first passage time of a Brownian motion. It is defined on the interval $[0, +\infty[$. The 1/2-stable distribution corresponds to the limit of an inverse Gaussian distribution when the drift of the Brownian motion is zero. We now elaborate this point. We can get an intuitive idea of the inverse Gaussian distribution in the following way: let us imagine that the total return of a financial asset follows a Brownian motion with a mean of 1% per month (choosing a month as the unit of time) and an arbitrary diffusion (volatility) coefficient. Over a period of four months, the average total return is $4 \times 1\% = 4\%$. The dispersion of the potential total returns is given by the monthly volatility $\times \sqrt{4} = 2$. The distribution of the total return over four months is normal with a mean of 4% and a standard deviation of 2 × the monthly volatility. Now, let us reverse our reasoning: let us ask ourselves after how much time we obtain a return of 4%, given that the mean monthly return is 1%. We can immediately give the answer: four months. However, this is the average period of time. What is the distribution of the possible time periods around this mean? The distribution we find is precisely the inverse Gaussian distribution: it is a distribution of the time periods, whereas the Gaussian distribution is a distribution of the values (here the total returns). Put differently, the normal distribution gives the distribution of the possible values taken by the end point of the path of a Brownian motion at a given date t, while the reciprocal distribution, i.e. the distribution of the amount of time needed to obtain a given value, is the inverse Gaussian one, hence its name. We will see that there is an inverse relationship between the cumulant generating function of this distribution and that of the normal distribution.

Let us make these intuitive ideas mathematically rigorous. Consider a Brownian motion B with mean μ and diffusion coefficient σ, i.e. the marginal distribution $B(t) = \mu t + \sigma W(t)$, where W is a standard Brownian motion with $W(0) = 0$. We are looking for the time period T_a necessary for B to attain a strictly positive value a for the first time, or in other words:

$$T_a = \inf\{0 < t < \infty | B_t = a\}$$

Then T_a follows an inverse Gaussian distribution whose density, written in terms of the parameters of the original Gaussian distribution, is given by:

$$f_{\text{GI}}(x) = \frac{a}{\sigma\sqrt{2\pi x^3}} \exp\left(-\frac{(a - \mu x)^2}{2\sigma^2 x}\right)$$

Table 5.2 Moments of the inverse Gaussian distribution in both parameterizations

Moments	(μ, σ)	(ν, γ)
$\mathbb{E}[T_a]$	$\frac{a}{\mu}$	ν
$\text{var}(T_a)$	$\frac{a\sigma^2}{\mu^3}$	$\frac{\nu^3}{\gamma}$
$S(T_a)$	$\frac{3\sigma}{\sqrt{\mu a}}$	$3\sqrt{\frac{\nu}{\gamma}}$
$K(T_a)$	$3 + \frac{15\sigma^2}{\mu a}$	$3 + \frac{15\nu}{\lambda}$

In general, we will use a different parameterization by setting $\nu = a/\mu$ and $\gamma = a^2/\sigma^2$. With this parameterization, the density of the inverse Gaussian distribution can be written as:

$$f_{\text{GI}}(x) = \sqrt{\frac{\gamma}{2\pi x^3}} \exp\left(-\frac{\gamma(x-\nu)^2}{2\nu^2 x}\right) \quad (5.28)$$

The moments of this distribution are given in Table 5.2 in both notations.

The distribution function $F_{\text{GI}}(x) = \Pr(X_{\text{GI}} < x)$ is related to the normal distribution function Φ by the formula:

$$F_{\text{GI}}(x) = \Phi\left(\sqrt{\frac{\gamma}{x}}\left(\frac{x}{\nu} - 1\right)\right) + \exp\left(\frac{2\gamma}{\nu}\right) \Phi\left(-\sqrt{\frac{\gamma}{x}}\left(\frac{x}{\nu} + 1\right)\right)$$

What happens when the drift of the Brownian motion is zero? Let us return to our introductory example. The question can now be stated in the following way: after how much time do we obtain a total return of 4%, when the average monthly return is zero? Again, we can immediately give the answer: the average period of time (the expectation of this period) is infinite, which is not to say that there will not be times when we pass through 4%. The distribution of the passage time through 4% is exactly a totally asymmetric 1/2-stable distribution (skewed to the right), i.e. with $\alpha = 1/2$ and $\beta = +1$.

Let us summarize what we have said so far. The inverse Gaussian distribution does not exist in the form of (5.28) unless the drift of the Brownian motion is non-zero.

If the drift $\mu = 0$, then $\nu = \infty$ and in the limit, as $\nu \to +\infty$, the density (5.28) is that of a 1/2-stable distribution and given by:

$$f_{\text{GI}}(x; \infty, \gamma) = \sqrt{\frac{\gamma}{2\pi x^3}} \exp\left(-\frac{\gamma}{2x}\right)$$

5.1.3.3 Series Representations

To display the densities of stable distributions graphically in all cases other than normal, Cauchy, or inverse Gaussian distributions, it is possible to use the following series representations.

For $\sigma = 1, \mu = 0$, we put: $a = -\frac{2}{\pi}\mathrm{arctg}\left(\beta\,\mathrm{tg}(\frac{\pi\alpha}{2})\right)$. For all $x > 0$ and all $\alpha \in\,]0,1[$, we have:

$$f_{\alpha,\beta}(x) = \frac{(1+\beta^2(\mathrm{tg}(\frac{\pi\alpha}{2}))^2)^{1/2\alpha}}{\pi x}\sum_{k=1}^{\infty}\frac{\Gamma(k\alpha+1)}{k!}(-x^{-\alpha})^k \sin\left(\frac{k\pi(a-\alpha)}{2}\right)$$

And for all $x > 0$ and all $\alpha \in\,]1,2[$:

$$f_{\alpha,\beta}(x) = \frac{(1+\beta^2(\mathrm{tg}(\frac{\pi\alpha}{2}))^2)^{1/2\alpha}}{\pi x}\sum_{k=1}^{\infty}\frac{\Gamma(k/\alpha+1)}{k!}(-x)^k \sin\left(\frac{k\pi(a-\alpha)}{2\alpha}\right)$$

The values the densities take for negative x are obtained using the identity:

$$f_{\alpha,\beta}(-x) = f_{\alpha,-\beta}(x)$$

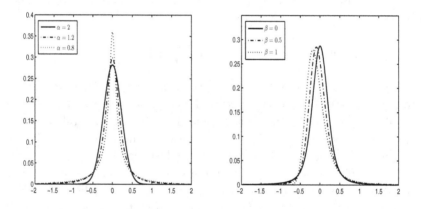

Figure 5.3 Densities of stable distributions in terms of α and β

Figure 5.3 shows examples of densities of stable distributions.

5.1.4 Simulating Paths of Stable Processes

Several simulation methods of stable random variables can be found in the research literature (for example Brown and Tukey (1946), Kanter (1975),

Samorodnitsky and Taqqu (1994), and Janicki and Weron (1993)). We present an intuitive approach to the simulation of such random variables.

5.1.4.1 Construction Principle

In general, a simulator for stable distributions is built from a combination of uniform and exponential distributions. We must therefore, in a first step, obtain a generator of exponentially distributed random variables from a generator of uniformly distributed pseudo-random numbers. Then, in a second step, we apply the formulas of Chambers, Mallows, and Stuck (1976).

The uniform distribution on $[0, 1]$ is the basic building block for the constructions that are to follow. This stems from the fact that it is possible to view values between 0 and 1 as the values taken by the distribution function F_X of any given random variable X. Since $U \in [0, 1]$, we can write $U = F_X(x)$, and then $x = F_X^{-1}(U)$ can be taken as a sample of the random variable X with distribution function F_X. In other words, we obtain the desired values for x (which represent random samples drawn for a given random variable X) by inverting the distribution function. Either we have F_X^{-1} available in closed form, or we use algorithms to approximate this function or algorithms that are specific to the distribution in question.

A generic simulation procedure is therefore given as follows. In a first step, we simulate a random variable U which is uniformly distributed on $[0, 1]$. In a second step, we calculate $x = F_X^{-1}(U)$, where F_X is the distribution function of X.

Let us apply this method to the exponential distribution. The exponential distribution with parameter $a > 0$ is the continuous distribution with density function:

$$f(x) = \begin{cases} a\, e^{-ax} & \text{if } x \geq 0 \\ 0 & \text{if } x < 0 \end{cases}$$

and distribution function:

$$F(x) = \begin{cases} 1 - e^{-ax} & \text{if } x \geq 0 \\ 0 & \text{if } x < 0 \end{cases}$$

For $x \geq 0$, we have $x = F_X^{-1}(U)$, where F_X is the distribution function of the exponential distribution. Assume this exponential distribution is given with the parameter a.

This leads to: $U = 1 - e^{-ax}$, hence $-ax = \ln(1 - U) \stackrel{d}{=} \ln U$, since $U \in [0, 1]$, and hence the values of x we are looking for:

$$x = -\frac{1}{a} \ln U$$

Algorithm 1 (Simulation from an Exponential Distribution).

To obtain a sample of n independent draws from an exponential distribution with parameter a, we must:

- *draw n samples u_i of a uniform random variable U on $[0, 1]$,*
- *and then calculate $x_i = -\frac{1}{a} \ln u_i$ for $i = 1, \cdots, n$.*

The x_i have an exponential distribution with parameter a.

5.1.4.2 Application to Stable Distributions

The distribution function of a Cauchy distribution is given by:

$$F(x) = \frac{1}{\pi} \text{Arctg} \, x + \frac{1}{2}$$

We must therefore shift the values taken by the uniform distribution. If U is uniformly distributed on $[0, 1]$, then $\theta = \pi (U - 1/2)$ is uniformly distributed on $[-\pi/2, \pi/2]$. Applying the method of inverting the distribution function, we obtain $\theta = \text{Arctg} \, x$, hence $x = \text{tg} \, \theta$. It follows that $\text{tg} \, \theta$ has a Cauchy distribution.

Algorithm 2 (Simulation from a Cauchy Distribution).

To obtain a sample of n independent draws from a standard Cauchy distribution, we must:

- *draw n samples u_i of a uniform random variable U on $[0, 1]$,*
- *and then calculate $x_i = \text{tg} \, \pi(u_i - 1/2)$ for $i = 1, \cdots, n$.*

The x_i have a standard Cauchy distribution.

For the normal distribution, the initial idea is to simulate two normal random variables simultaneously, instead of just one, or in other words a point (X, Y) with bivariate normal distribution given by the density:

$$f(x, y) = \frac{1}{2\pi} \exp\left(-\frac{x^2 + y^2}{2}\right)$$

and then to transform (X, Y) into polar coordinates (R, θ), where R and θ are assumed to be independent:

$$R^2 = X^2 + Y^2 \qquad \sin \theta = Y/R$$

In this situation, θ is uniformly distributed on $[0, 2\pi]$ and R^2 is exponentially distributed with parameter $1/2$. We find ourselves again in the previous setting of the exponential distribution. In this case, the random variables

$X_1 = R\cos\theta$ and $X_2 = R\sin\theta$ are independent standard normal random variables. R has the density function $f_R(r) = r\exp(-r^2/2)$. Using the method of inverting the distribution function, we have:

$$R = F_R^{-1}(U) = \sqrt{-2\log U}$$

It is therefore sufficient to simulate θ and R independently using the two independent uniform distributions U_1 and U_2, and then to transform these polar coordinates into Cartesian coordinates (X,Y). Hence the well-known simulation algorithm of Box and Muller (1958).

Algorithm 3 (Box–Muller Algorithm).
To obtain a sample of n independent draws from a standard normal distribution, we must:

- *draw n samples $u_{1,i}$ and $u_{2,i}$ ($i = 1,\cdots,n$) each from two uniform random variables U_1 and U_2 on $[0,1]$,*
- *and then calculate:*

$$\begin{cases} x_{1,i} = \cos(2\pi u_{2,i})\sqrt{-2\log u_{1,i}} \\ x_{2,i} = \sin(2\pi u_{2,i})\sqrt{-2\log u_{1,i}} \end{cases}$$

The $x_{1,i}$ and $x_{2,i}$, for all $i = 1,\cdots,n$, are independent and have a standard normal distribution. Either series may be chosen for the sample.

To simulate an $\mathcal{N}(m,\sigma)$-distributed random variable, it is enough to simulate either $Y = \sigma X_1 + m$ or $Y = \sigma X_2 + m$.

To simulate from a stable distribution, it is possible to take the preceding formula as a departure point and to generalize it to the case $\alpha < 2$.

Let us begin by giving the following equation for the normal distribution:

$$X = \sin(2\pi U_2)\sqrt{-2\log U_1}$$

or, to simplify the notation, multiplying by a factor of $\sqrt{2}$ and putting $W = -\log U_1$ and $\varphi = 2\pi U_2$:

$$X\sqrt{2} = 2\sin\varphi\sqrt{W}$$

which we transform into:

$$X\sqrt{2} = \frac{2\sin\varphi\cos\varphi}{\cos\varphi}\sqrt{W} = \frac{\sin 2\varphi}{\sqrt{\cos\varphi}}\left[\frac{\cos\varphi}{W}\right]^{-1/2}$$

or:

$$X\sqrt{2} = \frac{\sin 2\varphi}{(\cos\varphi)^{1/2}}\left[\frac{\cos((1-2)\varphi)}{W}\right]^{\frac{1-2}{2}}$$

The random variable $X\sqrt{2}$ has an $\mathcal{N}(0,\sqrt{2})$ normal distribution, which is a standard symmetric $\mathcal{S}(2,0,1,0)$ distribution. We can obtain a standard symmetric alpha-stable distribution by replacing 2 with α in the preceding equation, which leads to:

$$Y = \frac{\sin \alpha\varphi}{(\cos \varphi)^{1/\alpha}} \left[\frac{\cos((1-\alpha)\varphi)}{W} \right]^{\frac{1-\alpha}{\alpha}}$$

which has a standard symmetric stable distribution $\mathcal{S}(\alpha,0,1,0)$.

Algorithm 4 (Simulation from a ss-Stable Distribution).

To obtain a sample of n independent draws from a standard symmetric stable distribution, we must:

- *draw n samples $u_{1,i}$ and $u_{2,i}$ ($i=1,\cdots,n$) each from two uniform random variables U_1 and U_2 on $[0,1]$,*
- *and then calculate:*

$$x_i = \frac{\sin(\alpha\, 2\pi u_{1,i})}{(\cos 2\pi u_{1,i})^{1/\alpha}} \left[\frac{\cos[(1-\alpha)\, 2\pi u_{1,i}]}{\log u_{2,i}} \right]^{\frac{1-\alpha}{\alpha}}$$

The x_i are independent and have a standard symmetric stable distribution.

Since we have:

$$\mathcal{S}(\alpha,\beta,\gamma,\mu) \stackrel{d}{=} \mu + \gamma\mathcal{S}(\alpha,\beta,1,0) + \frac{2\beta\gamma\ln(\gamma)}{\pi}\mathbf{1}_{\alpha=1} \qquad (5.29)$$

it is enough to apply this relationship for $\beta = 0$ in order to simulate from any symmetric stable distribution $\mathcal{S}(\alpha,0,\gamma,\mu)$ by using a standard symmetric stable distribution $\mathcal{S}(\alpha,0,1,0)$.

Another method is that of Chambers, Mallows, and Stuck (1976), which we describe below.

The modified Chambers–Mallows–Stuck algorithm simulates random variables $X \sim \mathcal{S}(\alpha,\beta,1,0)$, with $\alpha \in (0,1) \cup (1,2]$ and $\beta \in [-1,1]$, and can be described as follows:

Algorithm 5 (Chambers–Mallows–Stuck Algorithm).

To obtain a sample of n independent draws from a standard stable distribution, we must:

- *draw n samples u_i of a uniform random variable U on $(-\frac{\pi}{2},\frac{\pi}{2})$,*
- *draw n samples w_i of an exponential random variable W with parameter 1,*

- and then calculate:

$$x_i = \mathcal{D}_{\alpha,\beta} + \frac{\sin(\alpha(u_i + C_{\alpha,\beta}))}{(\cos u_i)^{\frac{1}{\alpha}}} \left(\frac{\cos(u_i - \alpha(u_i + C_{\alpha,\beta}))}{w_i} \right)^{\frac{1-\alpha}{\alpha}} \quad (5.30)$$

where:

$$C_{\alpha,\beta} = \frac{\arctg\left(\beta \tg(\frac{\pi\alpha}{2})\right)}{1 - |1-\alpha|},$$

$$\mathcal{D}_{\alpha,\beta} = \left(\cos\left(\arctg\left(\beta \tg(\frac{\pi\alpha}{2})\right)\right)\right)^{-\frac{1}{\alpha}}$$

The x_i have a standard stable distribution $\mathcal{S}(\alpha, \beta, 1, 0)$.

Equation (5.29) can then be used to simulate a stable random variable with any other parameters.

Figure 5.4 Paths of stable processes

Figure 5.4 illustrates graphically two paths of stable processes generated by this method, using the following parameters: $\alpha = 1,4$, $\sigma = 1$, $\mu = 0$ and $\beta = \pm 0,8$.

5.2 Stable Financial Models

We present some simple illustrations of the use of stable distributions in finance.

5.2.1 With Pure Stable Distributions

We begin by using stable distributions just as they are given. Later on, we will study an example of a *tempered* stable distribution.

5.2.1.1 The Mandelbrot Model

Let us start with the classic equation for the stochastic process a share price follows defined for any time t:

$$S(t) = S(0)\, e^{X(t)}$$

Mandelbrot (1963) assumed that X was an alpha-stable process, or more precisely a Lévy process whose increments had symmetric stable marginal distributions. This model is therefore described by the characteristic exponent:

$$\Psi_{X_t}(u) = t\,(i\mu u - \gamma^\alpha |u|^\alpha)$$

5.2.1.2 The Carr and Wu Model

Carr and Wu (2003) have suggested using stable Lévy processes that have only downward jumps, i.e. with $\beta = -1$. They show that while the moments of the process of the returns X_t are infinite, the moments of the process of the stock prices remain finite. In Carr and Wu's model, $X_t = (r + \pi)t + \gamma L_t^{\alpha,-1}$ where r is the risk-free return, π a parameter to make the discounted stock price a martingale, γ a scale parameter characterizing the price fluctuations of the stock, and where $L_T^{\alpha,-1}$ has a totally asymmetric standard stable distribution with exponent $\alpha \in [1, 2]$ and skewness parameter $\beta = -1$.

To compute the moments, we must begin by determining the value of π. We have:

$$\mathbb{E}_Q[S_t] = S_0\, e^{rt}$$

hence:

$$\mathbb{E}_Q\left[e^{\gamma L_t^{\alpha,-1}}\right] = e^{-\pi t}$$

We can use a classic symmetry argument (see Samorodnitsky and Taqqu (1994)) to give this expectation in another form:

$$\mathbb{E}_Q \left[e^{\gamma L_t^{\alpha,-1}} \right] = \mathbb{E}_Q \left[e^{-\gamma L_t^{\alpha,1}} \right]$$

Now, the two-sided Laplace transform of L is defined only for $\beta = 1$ as:

$$\mathbb{E}_Q \left[e^{-\gamma L_t^{\alpha,1}} \right] = e^{-t\gamma^\alpha}$$

from which we can deduce the value of π:

$$\pi = \frac{\gamma'^\alpha}{\cos \frac{\pi \alpha}{2}}$$

with:

$$\gamma' = \gamma \left(\cos \frac{\pi \alpha}{2} \right)^{1/\alpha}$$

Knowing π, and using the two-sided Laplace transform again, we can deduce the equation for the moments of all orders of the share price:

$$\mathbb{E}\left[S_t^k \right] = S_0^k \exp\left(krt + t(k - k^\alpha) \gamma^\alpha \right)$$

In particular, they are all finite.

5.2.2 With Tempered Stable Distributions

To remedy the inconvenience of not having any moments for those Lévy processes whose measure is given by a power function built on the model:

$$\nu(x) = \begin{cases} \dfrac{C_-}{|x|^{1+\alpha}} & \text{if } x < 0 \\[6pt] \dfrac{C_+}{x^{1+\alpha}} & \text{if } x > 0 \end{cases}$$

it has been suggested to weight this measure by an exponential quantity in order to reduce large fluctuations, and therefore recover the moments. This idea corresponds to a set of Lévy processes whose marginal distributions are stable distributions truncated, or "tempered", by exponential functions, hence their name *tempered stable processes*. The distribution tails of these stable models, tempered by the truncation, are semi-light.

5.2.2.1 The Koponen Model

The first model to be built on this principle was Koponen's (1995), later taken up by Bouchaud and Potters (2003) and again by Boyarchenko and Levendorskii (2002), in which the Lévy measure is given by:

$$\nu(x) = \begin{cases} \frac{C_-}{|x|^{1+\alpha}} e^{-a_-\,|x|} & \text{if } x < 0 \\ \frac{C_+}{x^{1+\alpha}} e^{-a_+\,x} & \text{if } x > 0 \end{cases} \tag{5.31}$$

Observe that the parameterization is asymmetric in the decay rate of large jumps, but symmetric in the parameter α, which is the same for up and down jumps.

The CGMY model of Carr, Geman, Madan, and Yor (2002) (see Section 5.2.2.2 below) and the Variance Gamma model of Madan, Carr, and Chang (1998) (see Chapter 7, p. 166) are special cases of this model.

5.2.2.2 The CGMY Model

A special case of this model was brought into focus by Carr, Geman, Madan, and Yor (2002). It is called the CGMY model, and in it the coefficients C_- and C_+ have been symmetrized.

This model has the following Lévy measure:

$$\nu(x) = \begin{cases} \frac{C}{|x|^{1+\alpha}} e^{-a_-\,|x|} & \text{if } x < 0 \\ \frac{C}{x^{1+\alpha}} e^{-a_+\,x} & \text{if } x > 0 \end{cases} \tag{5.32}$$

As before, interpreting the parameters exposes two types of risk: size and shape. The parameter C is a measure of size (affecting all moments), whereas α, a_- and a_+ determine the shape (force of the jumps for α and asymmetry for a_- and a_+). When $a_- > a_+$, the distribution is curtailed on the left and stretched to the right (the left part is squashed by the exponential factor), and *vice versa*. Finally, for asymmetrical distributions, a change in α will result in a change in the asymmetry. In other words, α has an indirect leverage effect on the asymmetry.

The parameter α determines the behavior of the Lévy measure in a neighborhood of zero, as shown in Table 5.3.

The activity determines the frequency of the jumps, which increases as their size decreases: as x goes to zero, this frequency goes to infinity (more and more small jumps), hence the infinite value. The asterisk $^{(*)}$ in the table indicates that the infinite variation is caused by a very large number of very

Table 5.3 Values of α and type of CGMY process

α	Activity	Variation	Quadratic Variation
$\alpha < 0$	finite	finite	finite
$0 \leq \alpha < 1$	infinite	finite	finite
$1 < \alpha < 2$	infinite	infinite(*)	finite
$\alpha = 2$ (Brownian)		infinite(**)	finite

small jumps. The double asterisk (**) corresponds to Brownian motion: no activity by definition, irregularity of the continuous path (infinite variation), and finite quadratic variation.

The characteristic function can be obtained from the Lévy measure as follows. Starting with:

$$\Psi_{X_t}(u) = t \int_{-\infty}^{+\infty} (e^{iux} - 1)\nu(x)dx$$

gives:

$$\Psi_{X_t}(u) = Ct \int_{-\infty}^{0} (e^{iux} - 1)\frac{e^{a_- x}}{(-x)^{1+\alpha}}dx + Ct \int_{0}^{+\infty} (e^{iux} - 1)\frac{e^{-a_+ x}}{x^{1+\alpha}}dx$$

or:

$$\frac{\Psi_{X_t}(u)}{Ct} = \int_{-\infty}^{0} (e^{iux} - 1)\frac{e^{a_- x}}{(-x)^{1+\alpha}}dx + \int_{0}^{+\infty} (e^{iux} - 1)\frac{e^{-a_+ x}}{x^{1+\alpha}}dx$$

$$= \int_{0}^{+\infty} (e^{-iux} - 1)\frac{e^{-a_- x}}{x^{1+\alpha}}dx + \int_{0}^{+\infty} (e^{iux} - 1)\frac{e^{-a_+ x}}{x^{1+\alpha}}dx$$

which expands to:

$$\frac{\Psi_{X_t}(u)}{Ct} = \int_{0}^{+\infty} \frac{e^{-(a_- + iu)x}}{x^{1+\alpha}}dx - \int_{0}^{+\infty} \frac{e^{-a_- x}}{x^{1+\alpha}}dx$$
$$+ \int_{0}^{+\infty} \frac{e^{-(a_+ - iu)x}}{x^{1+\alpha}}dx - \int_{0}^{+\infty} \frac{e^{-a_+ x}}{x^{1+\alpha}}dx \qquad (5.33)$$

We must therefore be able to evaluate, in all generality, an integral of type:

$$\int_{0}^{+\infty} \frac{e^{-zx}}{x^{1+\alpha}}dx$$

where z is a complex number. Making the change of variable $y = zx$, we obtain:

$$\int_{0}^{+\infty} \frac{e^{-zx}}{x^{1+\alpha}}dx = \int_{0}^{+\infty} \frac{e^{-y}}{\left(\frac{y}{z}\right)^{1+\alpha}}d\left(\frac{y}{z}\right) = z^{\alpha} \int_{0}^{+\infty} \frac{e^{-y}}{y^{1+\alpha}}dy$$

Table 5.4 Comparison of four Lévy measures

Model	Lévy measure for: $x < 0$	$x > 0$
Mandelbrot (1963)	$\dfrac{C_-}{\|x\|^{1+\alpha}}$	$\dfrac{C_+}{x^{1+\alpha}}$
Koponen (1995)	$\dfrac{C_-}{\|x\|^{1+\alpha}} e^{-a_-\|x\|}$	$\dfrac{C_+}{x^{1+\alpha}} e^{-a_+ x}$
Carr et al. (2002)	$\dfrac{C}{\|x\|^{1+\alpha}} e^{-a_-\|x\|}$	$\dfrac{C}{x^{1+\alpha}} e^{-a_+ x}$
Madan et al. (1998)	$\dfrac{C}{\|x\|} e^{-a_-\|x\|}$	$\dfrac{C}{x} e^{-a_+ x}$

in which Euler's Gamma-function appears for the value $-\alpha$. Hence:

$$\int_0^{+\infty} \frac{e^{-zx}}{x^{1+\alpha}} dx = z^\alpha \Gamma(-\alpha)$$

Using this in Eq. (5.33) gives:

$$\frac{\Psi_{X_t}(u)}{Ct} = (a_- + iu)^\alpha \Gamma(-\alpha) - a_-^\alpha \Gamma(-\alpha) + (a_+ - iu)^\alpha \Gamma(-\alpha) - a_+^\alpha \Gamma(-\alpha)$$

or, in the following simple form:

$$\Psi_{X_t}(u) = Ct\, \Gamma(-\alpha) \left\{ (a_+ - iu)^\alpha + (a_- - iu)^\alpha - (a_+^\alpha + a_-^\alpha) \right\} \qquad (5.34)$$

The marginal distributions of the process can be obtained by numerical inversion using the so-called Fast Fourier Transform, or FFT, of the characteristic function just given.

In the case where the CGMY process is of finite variation and pure (i.e. without a diffusive part), its first moments are given as follows:

$$\mu = tC \left(\frac{1}{a_+^{1-\alpha}} - \frac{1}{a_-^{1-\alpha}} \right) \Gamma(1-\alpha) \qquad (5.35)$$

$$m_2 = tC \left(\frac{1}{a_+^{2-\alpha}} + \frac{1}{a_-^{2-\alpha}} \right) \Gamma(2-\alpha) \qquad (5.36)$$

$$m_3 = tC \left(\frac{1}{a_+^{3-\alpha}} - \frac{1}{a_-^{3-\alpha}} \right) \Gamma(3-\alpha) \qquad (5.37)$$

$$m_4 = tC \left(\frac{1}{a_+^{4-\alpha}} + \frac{1}{a_-^{4-\alpha}} \right) \Gamma(4-\alpha) + 3\, m_2^2 \qquad (5.38)$$

Finally, we note that the Variance Gamma model of Madan, Carr, and Chang (1998) is a special case of the CGMY model in which the coefficient $Y = 1 + \alpha$ is set to 1.

Table 5.4 gives an overview of the Lévy measures that can be derived from a stable Lévy measure.

Chapter 6

Laplace Distributions and Processes

The preceding chapter presented the class of stable distributions, as well as those Lévy processes that can be constructed from them. This chapter now turns to Laplace's first law of errors (1774), also known as the double exponential distribution, and the processes related to it. The Laplace distribution is infinitely divisible: as we saw in Chapter 4, this is an essential condition for it to be used as the distribution underlying a Lévy process.

The intuitive approach that guides the choice of Laplace's first law is explained at the beginning of this chapter. Then the main properties and representations characterizing it are described. Starting from this basic law, a family of generalized Laplace distributions can be constructed. This generalization is made explicit, and the properties of the corresponding distributions are also illustrated. Then the various processes that can be constructed using simple or generalized Laplace distributions are presented. This allows a reinterpretation of several financial models that use these particular cases of Laplace processes. The chapter concludes with a study of the existing link between Laplace distributions and hyperbolic distributions. Readers who are interested in further technical details can consult the book by Kotz, Kozubowski and Podgórski (2001).

6.1 The First Laplace Distribution

Laplace's first law of errors (see Laplace (1774)) precedes the normal distribution, or second law of errors (see Laplace (1778)), by four years. The normal distribution is also called the Laplace–Gauss distribution because of its rediscovery by Gauss at the beginning of the 19th century, or indeed Gaussian distribution because of the numerous applications this author found for it in various fields of the physical sciences. The first law of errors relies on the use of the standard decreasing exponential function, whereas the second law of errors is based on the use of the decreasing exponential function of the square of the variable.

6.1.1 *The Intuitive Approach*

Before we examine the construction and the use of Laplace's first law, we present the debate that puts it into contrast with Laplace's second law.

6.1.1.1 *The Choice of a Law of Errors*

The general question posed when faced with the choice of a law of errors is whether this distribution adequately represents the "middle" of the observations. In his dissertation from 1774, *Sur la probabilité des causes par les événements*, Laplace puts the question as follows: "to determine the middle that should be taken between three given observations of the same phenomenon". This question sends us back to the history of the mathematical study of errors, which spans a period from the Enlightenment to the first decades of the 19th century. The mathematical study of errors begins around 1750 and ends around 1830 with what is called the "synthesis of Laplace–Gauss".

Armatte (1995) has shown how this history is shaped by four key moments which led to the synthesis of 1830: firstly, outlining the problem and its stakes in the realms of astronomy and geodesy, realms characterized by the transition from natural philosophy to Newtonian science; next, elaborating a meaningful conceptual scheme, preliminary to any mathematical modeling of observation errors, which divides errors into "systematic errors" and "accidental errors", or a "grand separation of errors"; then, moving to the mathematical modeling of errors, which gives rise to three distinct problems, the middle error, the distribution of the errors, and the solving of systems of equations; and finally the solution of Laplace–Gauss, which directly links a certain middle (the arithmetic mean), a distribution of errors (the normal distribution), and a particular optimization method (that of least squares). The solution of Laplace–Gauss therefore establishes itself as the only solution, and becomes a kind of second nature in the treatment of statistical observations.

6.1.1.2 The Supremacy of the Second Law of Errors

Gauss' work (1809), which notably treats the theory of errors, consolidated statistics for a long time, hindering any questioning of the possibility of replacing the normal distribution by another distribution in statistical calculations. For example, Adolphe Quetelet's renamed synthesis on averages (1835) builds on the existence of a Gaussian distribution of errors. For Quetelet, this normal distribution is the proof that the average of a studied phenomenon is indeed pertinent and not just an artifact of a calculation. The mathematical developments of the 19th century could have allowed a challenge to the Gaussian supremacy. But this did not happen. For example, in a series of contributions to the *Académie des sciences* on the modeling of errors, Cauchy (1853) drew attention to distributions without a mean. But this debate was authoritatively closed by Bienaymé (1867), who confirmed the use of the Gaussian law and refused to consider any other possibility. Nevertheless, as Paul Lévy (1924) observed, "one has not understood a thing about the theory of errors if one has not understood *first* why accidental errors obey the Gaussian distribution; *then*, the calculation methods to be used in practice are obtained without difficulty" (our italics). What strikes us as important in Lévy's text is the insistence on the necessity of understanding that the errors are not simply to be taken for granted, but must be understood through an effort of reflection and abstraction. Only once this effort has been made (to understand why the errors obey such a distribution) is it possible to undertake any series of calculations.

6.1.1.3 Discussion of the Construction Procedure

It was Keynes who launched one of the first discussions on this construction effort, and therefore on the legitimacy of choosing Laplace's second law of errors over his first one.

In an article published in 1911 in the *Journal of the Royal Statistical Society*, Keynes addresses the question of the distribution of errors and the relevance of averages in statistical calculations, insisting on the arbitrary character of this choice (see p. 322):

"The object of the following investigation is to discover what laws of error, if we assume them, correspond to each of the simple types of average, and to discover this by means of a systematic method."

Keynes then asks the question (see p. 328):

"[...]the most probable value of the quantity is equal to the median of the measurements, what is the law of error?"

The answer is Laplace's first law. In 1923, in the same issue of the *Journal of the American Statistical Association*, Wilson (1923) and Crum (1923) analyze empirical distributions stemming from economic data. Crum concludes his article with the suggestion that (see p. 614):

"Good many series of data from the economic sources probably may be better treated by the median than by the mean."

This remark ends a century of complete domination of the mean and the normal distribution in statistical works: from now on, it is possible to question the predominance of the pair mean/median.

6.1.1.4 Two Different Universes of Thought

Although close in their origin and mathematical form, Laplace's two laws actually describe two very different intellectual universes, one founded on the use and relevance of medians (the first law), the other on those of means (the second law). Intuitively, Laplace's first law says that the frequency with which an error occurs is inversely related to the size of that error (an exponential function of the absolute value of the error), whereas Laplace's second law says that the frequency with which it occurs is inversely related to the square of the error (an exponential function of the square of the error). Taking logarithms, it follows that Laplace's first law says that the logarithm of the frequency of an error is a linear function of its magnitude, whereas the second law says that this function is a parabola.

6.1.2 Representations of the Laplace Distribution

Here we study the various forms taken by the Laplace distribution.

6.1.2.1 Density Functions

The Laplace distribution is obtained by taking the union of two exponential distributions. Let $\mathcal{E}(\lambda)$ denote the exponential distribution with parameter λ and density function:

$$f_{\mathcal{E}(\lambda)}(x) = \lambda e^{-\lambda x} \qquad x \geq 0$$

The standard exponential distribution is defined by $\lambda = 1$; where necessary we denote it by $\mathcal{E}(1)$:

$$f_{\mathcal{E}(1)}(x) = e^{-x} \qquad x \geq 0$$

Let us begin with the standard case.

Let $\mathcal{E}_1(1)$ and $\mathcal{E}_2(1)$ be two standard exponential distributions (both with parameter 1) defined on the support $[0, +\infty[$. The Laplace distribution in its original form will be denoted by $\mathcal{L}^*(0,1)$, where the asterisk indicates and recalls that this is the original form, in contrast to another more common form that we will present later on. The original form is the union of a standard exponential distribution and its opposite defined on the support $]-\infty, 0]$. Let $W_1 \sim \mathcal{E}_1(1)$ and $W_2 \sim \mathcal{E}_2(1)$ be two random variables having independent standard exponential distributions (both with parameter 1).

A random variable X that has a Laplace $\mathcal{L}^*(0,1)$ distribution satisfies the following relationship:

$$X \stackrel{d}{=} W_1 - W_2 \qquad (6.1)$$

where the symbol $\stackrel{d}{=}$ denotes equality in distribution.

To facilitate an intuitive understanding in the following of the existing relationship between the various probability distributions and their parameters, it will be useful to show these parameters explicitly in certain cases. For this, we introduce a specific notation:

Notation 1 (Equality in distribution: intuitive form).
If the random variable X_1 has the distribution ℓ_1 and the random variable X_2 has the distribution ℓ_2, then $\ell_1 = \ell_2$ is equivalent to $X_1 \stackrel{d}{=} X_2$, where $\stackrel{d}{=}$ denotes equality in distribution.

Using this notation, Eq. (6.1) can be written in the following way:

$$\mathcal{L}^*(0,1) = \mathcal{E}_1(1) - \mathcal{E}_2(1) \qquad (6.2)$$

This convenient notation makes evident the relationships between the parameters, which will be useful in cases less trivial than the present example. This does not mean in any way that the density function of $\mathcal{L}^*(0,1)$ can be written as the sum of the density functions of $\mathcal{E}_1(1)$ and $-\mathcal{E}_2(1)$.

To obtain the density function of the classic Laplace distribution we use Eq. (6.1), i.e. we add the two initial densities and multiply their sum by

1/2 to retain a value of 1 for the integral of the density function over the whole real line:

$$f_{\mathcal{L}^*(0,1)}(x) = \frac{1}{2}\left(\mathbf{1}_{]-\infty,0]}\, e^x + \mathbf{1}_{[0,+\infty[}\, e^{-x}\right) = \begin{cases} \frac{1}{2} e^{-x} & \text{if } x \geq 0 \\ \frac{1}{2} e^{x} & \text{if } x < 0 \end{cases}$$

or:

$$f_{\mathcal{L}^*(0,1)}(x) = \frac{1}{2} e^{-|x|} \tag{6.3}$$

The form (6.3) is the classic form of the standard Laplace distribution. The value $x = 0$ is at the same time the mode, the median, and the mean of the distribution. The classic Laplace distribution is symmetric ($S = 0$), and its kurtosis is given by $K = 6$.

We now proceed to the non-standardized form of the Laplace distribution. There are two different parameterizations for this non-standardized form, both representing Laplace's first law. The first parameterization uses a scale parameter, denoted by s, and the second one uses the standard deviation, denoted by c. In both cases, the localization parameter is denoted by θ; it corresponds to the mean and the mode of the distribution. To distinguish the two parameterizations, we introduce the following notation.

Notation 2 (Laplace distribution).
We denote by $\mathcal{L}^*(\theta, s)$ the Laplace distribution with parameters θ (mean, mode) and s (scale), and by $\mathcal{L}(\theta, c)$ the Laplace distribution with parameters θ (mean, mode) and c (standard deviation).

We now present these two parameterizations and give for each one the density function, the characteristic function, and the corresponding Lévy measure.

Parameterization in terms of the scale This non-standardized form of the Laplace distribution uses a scale parameter s and a localization parameter θ. The density function is then takes the form:

$$f_{\mathcal{L}^*(\theta,s)}(x) = \begin{cases} \frac{1}{2s} e^{-\frac{1}{s}(x-\theta)} & \text{if } x \geq \theta \\ \frac{1}{2s} e^{\frac{1}{s}(x-\theta)} & \text{if } x < \theta \end{cases}$$

or:

$$f_{\mathcal{L}^*(\theta,s)}(x) = \frac{1}{2s} e^{-\frac{1}{s}|x-\theta|} \tag{6.4}$$

The localization parameter θ is the mean, the median, and the mode of the distribution. The variance is $2s^2$. The values $\theta = 0$ and $s = 1$ give the classic standard form.

Let X_1 and X_2 be two independent random variables that have standard exponential distributions whose parameter has the value $1/s$. If X is a random variable that has an $\mathcal{L}^*(\theta, s)$ Laplace distribution, then in analogy to (6.1) we have:

$$X \stackrel{d}{=} \theta + X_1 - X_2 \tag{6.5}$$

In the same way, the analog of (6.2) is:

$$\mathcal{L}^*(\theta, s) = \theta + \mathcal{E}_1\left(\frac{1}{s}\right) - \mathcal{E}_2\left(\frac{1}{s}\right) \tag{6.6}$$

Let us give two new forms of Eqs (6.5) and (6.6). Recall the relationship existing between a standard exponential distribution $\mathcal{E}(1)$ and an exponential distribution $\mathcal{E}(\lambda)$ with parameter λ:

$$\mathcal{E}(\lambda) = \frac{1}{\lambda}\mathcal{E}(1) \tag{6.7}$$

In fact, let us assume that $X \sim \mathcal{E}(\lambda)$ and $W \sim \mathcal{E}(1)$. Beginning with:

$$\Pr(X < a) = \int_0^a \lambda e^{-\lambda t} dt = \int_0^a e^{-\lambda t} d(\lambda t)$$

we put $u = \lambda t$, or:

$$\Pr(X < a) = \int_0^{\lambda a} e^{-u} du = \Pr(W < \lambda a) = \Pr(W/\lambda < a)$$

hence:

$$X \stackrel{d}{=} \frac{1}{\lambda} W$$

which proves that Eq. (6.7) holds. We therefore obtain:

$$\mathcal{E}\left(\frac{1}{s}\right) = s\mathcal{E}(1)$$

which allows us to write Eqs (6.5) and (6.6) in the following way:

$$X \stackrel{d}{=} \theta + s(W_1 - W_2) \tag{6.8}$$

and:

$$\mathcal{L}^*(\theta, s) = \theta + s(\mathcal{E}_1(1) - \mathcal{E}_2(1)) \tag{6.9}$$

To conclude these developments, we show in Fig. 6.1 several examples of Laplace distributions with density functions given by Eq. (6.4).

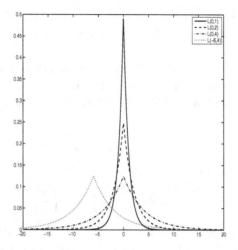

Figure 6.1 Laplace distributions according to their parameters θ and s

Parameterization in terms of the standard deviation The standard deviation of a Laplace random variable, whose density is given by (6.4), is equal to $s\sqrt{2}$, and therefore a second parameterization consists in setting $c = s\sqrt{2}$. The variance of the standard Laplace distribution is equal to 2 in the first case, and 1 in the second. The second form of Laplace's first law of errors is obtained by replacing s with $\frac{c}{\sqrt{2}}$ in the first form. We obtain:

$$f_{\mathcal{L}(\theta,c)}(x) = \begin{cases} \frac{1}{c\sqrt{2}} e^{-\frac{\sqrt{2}}{c}(x-\theta)} & \text{if } x \geq \theta \\ \frac{1}{c\sqrt{2}} e^{\frac{\sqrt{2}}{c}(x-\theta)} & \text{if } x < \theta \end{cases} \tag{6.10}$$

or:

$$f_{\mathcal{L}(\theta,c)}(x) = \frac{1}{c\sqrt{2}} e^{-\frac{\sqrt{2}}{c}|x-\theta|} \tag{6.11}$$

and for the standard form ($\theta = 0$, $c = 1$):

$$f_{\mathcal{L}(0,1)}(x) = \frac{1}{\sqrt{2}} e^{-\sqrt{2}|x|} \tag{6.12}$$

By analogy with Eqs (6.1), (6.5), and (6.8), we have:

$$X \stackrel{d}{=} \theta + \frac{c}{\sqrt{2}}(W_1 - W_2) \tag{6.13}$$

Table 6.1 The two parameterizations of Laplace's first law

Notation	First parameterization $\mathcal{L}^*(\theta, s)$	Second parameterization $\mathcal{L}(\theta, c)$
Standardized	$\frac{1}{2} e^{-\|x\|}$	$\frac{1}{\sqrt{2}} e^{-\sqrt{2}\|x\|}$
General	$\frac{1}{2s} e^{-\frac{1}{s}\|x-\theta\|}$	$\frac{1}{c\sqrt{2}} e^{-\frac{\sqrt{2}}{c}\|x-\theta\|}$
In distributions	$\theta + s(W_1 - W_2)$	$\theta + \frac{c}{\sqrt{2}}(W_1 - W_2)$
Intuitive (1)	$\theta + s(\mathcal{E}_1(1) - \mathcal{E}_2(1))$	$\theta + \frac{c}{\sqrt{2}}(\mathcal{E}_1(1) - \mathcal{E}_2(1))$
Intuitive (2)	$\theta + \mathcal{E}_1(\frac{1}{s}) - \mathcal{E}_2(\frac{1}{s})$	$\theta + \mathcal{E}_1(\frac{\sqrt{2}}{c}) - \mathcal{E}_2(\frac{\sqrt{2}}{c})$

and, in the same way, by analogy with Eqs (6.2), (6.6), and (6.9):

$$\mathcal{L}(\theta, c) = \theta + \frac{c}{\sqrt{2}} (\mathcal{E}_1(1) - \mathcal{E}_2(1)) \tag{6.14}$$

Finally, Table 6.1 compares the two approaches to Laplace's first law of errors.

6.1.2.2 *Characteristic Function*

To obtain the characteristic function of the Laplace distribution, we make use of the fact that this distribution is the difference between two exponential distributions. Let us begin with the classic case of the $\mathcal{L}^*(0,1)$ standard Laplace distribution. The computation of the characteristic function of the $\mathcal{L}^*(0,1)$ Laplace distribution can be carried out by starting from the relationship (6.1), which we recall here:

$$X \stackrel{d}{=} W_1 - W_2$$

from which it follows, since W_1 and W_2 are independent:

$$\mathbb{E}\left[e^{iuX}\right] = \mathbb{E}\left[e^{iu(W_1-W_2)}\right] = \mathbb{E}\left[e^{iuW_1} e^{-iuW_2}\right] = \mathbb{E}\left[e^{iuW_1}\right] \times \mathbb{E}\left[e^{-iuW_2}\right]$$

The characteristic function of the exponential distribution is given for positive values of x by:

$$\int_{\mathbf{R}^+} e^{iux} e^{-x} dx = \int_0^{+\infty} e^{(iu-1)x} dx = \left[\frac{e^{(iu-1)x}}{iu-1}\right]_0^{+\infty} = 0 - \frac{1}{iu-1} = \frac{1}{1-iu}$$

and for negative values by:

$$\int_{\mathbf{R}^-} e^{iux} e^{x} dx = \int_{-\infty}^{0} e^{(iu+1)x} dx = \left[\frac{e^{(iu+1)x}}{iu+1}\right]_{-\infty}^{0} = \frac{1}{iu+1} - 0 = \frac{1}{1+iu}$$

The characteristic function of the standard Laplace distribution is therefore:
$$\Phi_{\mathcal{L}^*(0,1)}(u) = \frac{1}{1-iu} \times \frac{1}{1+iu} = \frac{1}{1+u^2} \qquad (6.15)$$
This is the Cauchy distribution.

Let us now proceed to the non-standardized form by choosing for example the parameterization $\mathcal{L}(\theta, c)$. The argument is exactly the same. We start from the relationship (6.13), which we recall here:
$$X \stackrel{d}{=} \theta + \frac{c}{\sqrt{2}}(W_1 - W_2)$$
and obtain the characteristic function as before:
$$\mathbb{E}\left[e^{iuX}\right] = \mathbb{E}\left[e^{iu\left(\theta + \frac{c}{\sqrt{2}}(W_1 - W_2)\right)}\right]$$
or:
$$\mathbb{E}\left[e^{iu\left(\theta + \frac{c}{\sqrt{2}}(W_1 - W_2)\right)}\right] = \mathbb{E}\left[e^{i\theta u} e^{iu\frac{c}{\sqrt{2}}W_1} e^{-iu\frac{c}{\sqrt{2}}W_2}\right]$$
and:
$$\mathbb{E}\left[e^{iu\left(\theta + \frac{c}{\sqrt{2}}(W_1 - W_2)\right)}\right] = \mathbb{E}\left[e^{i\theta u}\right] \times \mathbb{E}\left[e^{iu\frac{c}{\sqrt{2}}W_1}\right] \times \mathbb{E}\left[e^{-iu\frac{c}{\sqrt{2}}W_2}\right]$$

The characteristic function of an exponential distribution with parameter λ is given for positive values of x by:
$$\lambda \int_{\mathbf{R}^+} e^{iux} e^{-\lambda x} dx = \lambda \int_0^{+\infty} e^{(iu-\lambda)x} dx = \lambda \left[\frac{e^{(iu-\lambda)x}}{iu - \lambda}\right]_0^{+\infty} = 0 - \frac{\lambda}{iu - \lambda}$$
or:
$$\lambda \int_{\mathbf{R}^+} e^{iux} e^{-\lambda x} dx = \frac{1}{1 - \frac{iu}{\lambda}}$$
and for negative values by:
$$\lambda \int_{\mathbf{R}^-} e^{iux} e^{\lambda x} dx = \lambda \int_{-\infty}^0 e^{(iu+\lambda)x} dx = \lambda \left[\frac{e^{(iu+\lambda)x}}{iu + \lambda}\right]_{-\infty}^0 = \frac{\lambda}{iu + \lambda} - 0$$
or:
$$\lambda \int_{\mathbf{R}^-} e^{iux} e^{\lambda x} dx = \frac{1}{1 + \frac{iu}{\lambda}}$$

The characteristic function of the non-standardized Laplace distribution in the parameterization $\mathcal{L}(\theta, c)$ is therefore:
$$\Phi_{\mathcal{L}(\theta,c)}(u) = e^{i\theta u} \times \frac{1}{1 - iu\frac{c}{\sqrt{2}}} \times \frac{1}{1 + iu\frac{c}{\sqrt{2}}} = \frac{e^{i\theta u}}{1 + \frac{1}{2}c^2 u^2} \qquad (6.16)$$

In the same way, the characteristic function of the non-standardized Laplace distribution in the parameterization $\mathcal{L}^*(\theta, s)$ is:
$$\Phi_{\mathcal{L}^*(\theta,s)}(u) = e^{i\theta u} \times \frac{1}{1 - ius} \times \frac{1}{1 + ius} = \frac{e^{i\theta u}}{1 + s^2 u^2} \qquad (6.17)$$

6.1.2.3 Infinitely Divisible Distribution

The form of the characteristic function allows us to show the infinite divisibility of Laplace distributions in a simple way. In fact, if we choose for example the parameterization in terms of the standard deviation of Eq. (6.16), we can write:

$$\Phi_{\mathcal{L}(\theta,c)}(u) = \left[e^{i\theta u/n} \times \left(\frac{1}{1 - iu\frac{c}{\sqrt{2}}} \right)^{1/n} \times \left(\frac{1}{1 + iu\frac{c}{\sqrt{2}}} \right)^{1/n} \right]^n \quad (6.18)$$

for any integer $n \geq 1$. It must therefore be verified that the function inside the brackets is the characteristic function of an existing probability distribution. Among the three terms that constitute this function, the first is the characteristic function of the constant θ/n, and the second and third are the characteristic functions of gamma distributions (see for example Eq. (6.35) given later on). It follows that if X_1 is a random variable with a $\mathcal{L}(\theta, c)$ Laplace distribution, then we have:

$$\Phi_{X_1}(u) = \left(\Phi_{X_{1/n}}(u) \right)^n \quad (6.19)$$

for any integer $n \geq 1$, where $X_{1/n}$ is written as the sum of three independent components (a constant and the difference between two gamma distributed random variables). In other words:

$$\Phi_{X_n}(u) = \left(\Phi_{X_1}(u) \right)^n$$

which allows us to come back to property (4.2) of infinitely divisible distributions. This property can be generalized to the case of continuous time t (positive real) as:

$$\Phi_{X_t}(u) = \left(\Phi_{X_1}(u) \right)^t \quad (6.20)$$

6.1.2.4 Lévy–Khintchine Formula

Since the Laplace distribution $\mathcal{L}(\theta, c)$ is infinitely divisible, it has – as we have seen in Chapter 4 – a unique representation given by the Lévy–Khintchine formula, where θ represents a shift term (i.e. a drift term in a dynamic context) and where the Gaussian component is zero.

If X_1 is a random variable with an $\mathcal{L}(\theta, c)$ Laplace distribution, then we verify that that its Lévy–Khintchine representation is:

$$\Phi_{X_1}(u) = \exp\left(\Psi_{X_1}(u) \right)$$

where:

$$\Psi_{X_1}(u) = i\theta u + \int_{\mathbf{R}^*} \left(e^{iux} - 1 \right) \nu(dx) \quad (6.21)$$

with associated Lévy measure:

$$\nu(x) = \begin{cases} \frac{1}{|x|} e^{\frac{\sqrt{2}}{c} x} & \text{if } x < 0 \\ \frac{1}{x} e^{-\frac{\sqrt{2}}{c} x} & \text{if } x \geq 0 \end{cases}$$

or:

$$\nu(x) = \frac{1}{|x|} e^{-\frac{\sqrt{2}}{c} |x|} \tag{6.22}$$

6.1.2.5 The Laplace Distribution as a Mixture Distribution

An interesting interpretation of the Laplace distribution can be given by considering it as a mixture of other probability distributions. A mixture distribution is obtained by an operation that allows the construction of complex distributions starting from the characteristics of simpler distributions. In a mixture distribution, one or more parameters of a given initial distribution are themselves random variables that have a different probability distribution.

To see how this is done, let f_θ be a probability density that depends on a parameter θ. If θ is assumed to be random and has the distribution function $G(\theta)$, then the density function resulting from the mixture is:

$$f(x) = \int_\Theta f_\theta(x) dG(\theta)$$

where Θ is the domain where the random variable θ is defined.

We now write the Laplace density as being obtained from a mixture distribution. Using formula 7.4.3 of Abramowitz and Stegun (1965), we have:

$$\frac{1}{2} e^{-|x|} = \int_0^{+\infty} \frac{1}{2\sqrt{\pi w}} e^{-\frac{1}{2}\left(\frac{x^2}{2w} + 2w\right)} dw$$

which can be rewritten as:

$$\frac{1}{2} e^{-|x|} = \int_{-\infty}^{+\infty} \frac{1}{\sqrt{2w}\sqrt{2\pi}} e^{-\frac{x^2}{2 \times 2w}} e^{-w} \mathbf{1}_{[0,+\infty[} dw$$

Putting $f_w(x) = \frac{1}{\sqrt{2w}\sqrt{2\pi}} e^{-\frac{x^2}{2 \times 2w}}$ and $g(w) = e^{-w} \mathbf{1}_{[0,+\infty[}$, we obtain:

$$\frac{1}{2} e^{-|x|} = \int_{-\infty}^{+\infty} f_w(x) g(w) dw = \int_{-\infty}^{+\infty} f_w(x) dG(w)$$

where in effect the Laplace distribution can be interpreted as a mixture of a normal and an exponential distribution, i.e. a normal distribution whose variance is random and has an exponential distribution.

Let us now set $\sigma^2 = 2W$, where W is a standard exponential random variable and Z a standard normal one.

If X is a Laplace distributed random variable, we can write:

$$X \stackrel{d}{=} \sqrt{2W}\, Z \tag{6.23}$$

It follows immediately that for a non-standardized Laplace random variable we have:

$$X \stackrel{d}{=} \theta + s\sqrt{2W}\, Z \tag{6.24}$$

This is the representation of the Laplace distribution as a mixture distribution.

If the variance is random and has an exponential distribution, this is the same as assuming – from a financial point of view – that the volatility is stochastic, with fluctuations governed by this exponential distribution.

Using the intuitive notation introduced above, Eq. (6.23) becomes:

$$\mathcal{L}^*(0,1) = \mathcal{N}\left[\mathcal{E}(1)\right] \tag{6.25}$$

In the same way, Eq. (6.24) becomes:

$$\mathcal{L}^*(\theta, s) = \theta + s\, \mathcal{N}\left[\mathcal{E}(1)\right] \tag{6.26}$$

In the parameterization in terms of the standard deviation, we have:

$$\mathcal{L}(\theta, c) = \theta + \frac{c}{\sqrt{2}}\, \mathcal{N}\left[\mathcal{E}(1)\right]$$

6.1.3 Laplace Motion

Since the Laplace distribution $\mathcal{L}(\theta, c)$ is infinitely divisible, it can – as we have seen in Chapter 4 – serve as an underlying distribution for the construction of a Lévy process. We will call the family of Lévy processes constructed from Laplace's first law "Laplace motions". The variance gamma process (see p. 166) is the best known example of such a motion.

In common with Brownian motions, Laplace motions have stationary and independent increments, and all moments are defined. However, they completely differ from Brownian motions in the fact that their paths are discontinuous everywhere; they are *pure jump processes* without any diffusive part. An important consequence of these discontinuities is the non-interchangeability between the space and time dimensions.

6.1.3.1 Definitions

Let τ be the unit of time of the process, c the scale parameter, and θ the localization parameter, where we use the notation in terms of the moments of the Laplace distribution given in Eq. (6.11), which we recall here:

$$f_{\mathcal{L}(\theta,c)} = \frac{1}{c\sqrt{2}} e^{-\frac{\sqrt{2}}{c}|x-\theta|}$$

A random process L is called a Laplace motion if $L(0) = 0$ and its increments are independent and distributed as:

$$L(t+\tau) - L(t) \sim \mathcal{L}(\theta, c) \tag{6.27}$$

where $\mathcal{L}(\theta, c)$ is a Laplace distribution with mean θ and standard deviation c. By construction, the increments of this process are stationary. Equation (6.27) is therefore equivalent to:

$$L(\tau) \sim \mathcal{L}(\theta, c) \tag{6.28}$$

The moments of Laplace motion at a date t are:

$$\mathbb{E}[L(t)] = \frac{\theta t}{\tau} \qquad \text{var } L(t) = \frac{c^2 t}{\tau}$$

where the quantity t/τ which appears in the two equations can be interpreted as the number of times (6.28) must be added to obtain $L(t)$.

At date $t = 1$, we have $\mathbb{E}[L(1)] = \theta/\tau$ and $\text{var } L(1) = c^2/\tau$. Imagine that we were looking for the equivalent of Brownian motion (Wiener process) in what we might call a "standard Laplace motion" with a normalized variance of 1. We see that there is no unique equivalent since there exist infinite possibilities for which the ratio c^2/τ is equal to 1, i.e. where $c^2 = \tau$.

For a variance equal to 1, there therefore exist infinitely many "standard" Laplace motions with paths that can take vastly differing forms.

In the construction of Brownian motion, the choice of time scale τ for the process is of little importance, since the increments will all have a Gaussian distribution.

In contrast, in the construction of Laplace motion, the choice of time scale τ is critical: the distributions of the increments of a Laplace motion depend on the size of the time step. The reader may think of a gamma random variable (we will give a detailed treatment later on), which is the sum of two exponential variables and provides an example where the distribution of the sum is not of the same type as the distribution of the summands.

In particular, Laplace motion is not self-similar, i.e., we cannot write $L(at) \stackrel{d}{=} a^H L(t)$ for all times t and all H, which can easily be verified using characteristic functions.

By analogy with Brownian motion, which is usually denoted by $B(t; \mu, \sigma)$, we will denote Laplace motion, defined using Eq. (6.27), by $L^\tau(t; \theta, c)$.

6.1.3.2 Characteristic Function

In order to obtain the characteristic function of Laplace motion, we start from Eq. (6.16) for the characteristic function of a Laplace distribution $\mathcal{L}(\theta, c)$, which we recall below:

$$\Phi_{\mathcal{L}(\theta,c)}(u) = \frac{e^{i\theta u}}{1 + \frac{1}{2}c^2 u^2}$$

Then we introduce the parameter τ for a unit of time at a date t and write the number t/τ of units of time at this date, using (6.20), which gives:

$$\Phi_{L_t}(u) = (\Phi_{L_1}(u))^{\frac{t}{\tau}} = \left(\frac{e^{i\theta u}}{1 + \frac{1}{2}c^2 u^2}\right)^{\frac{t}{\tau}} \quad (6.29)$$

The question we must address is which form the distribution $L(t)$ takes, knowing that the increments of the process L are Laplace distributed.

In case $t/\tau = n$ is an integer, (6.29) can be written as:

$$\Phi_{X_n}(u) = \left(\frac{e^{i\theta u}}{1 + \frac{1}{2}c^2 u^2}\right)^n \quad (6.30)$$

The characteristic function (6.30) of the random variable X_n is the characteristic function of the distribution given by the sum of n i.i.d. Laplace random variables, or in other words the summed Laplace distribution of order n. This is the distribution at a time horizon n of a risky market (in discrete time), whose marginal distributions are $\mathcal{L}(\theta, c)$ Laplace distributions.

In fact, let $L_i \sim \mathcal{L}(\theta, c), i = 1, \cdots, n$ be independent copies of L, and $X_n = L_1 + \cdots + L_n$. Then, by independence:

$$\Phi_{X_n}(u) = \Phi_{\sum_{i=1}^n L_i}(u) = \prod_{i=1}^n \Phi_{L_i}(u) = (\Phi_{L_1}(u))^n \quad (6.31)$$

The Laplace distribution does not have the property of self-similarity under addition of the stable distributions that we have seen in the preceding chapter, and which give simple examples of change of scale relationships: it is not stable under addition. This means that the sum of Laplace distributions is not a Laplace distribution, but some other distribution whose type still needs to be determined. To achieve this, we must understand how to pass from one time-scale to another.

6.1.3.3 Preliminary Results

Since the Laplace distribution is a union of exponential distributions, the sum of Laplace distributions will be a union of sums of exponential distributions. We begin by recalling two preliminary results concerning the sums of i.i.d. distributions and the relationships that hold between exponential and gamma distributions.

Sum of Exponential Distributions Take n independent and identically distributed random variables X_i. Starting with:

$$\mathbb{E}\left[e^{iu(X_1+\cdots+X_n)}\right] = \mathbb{E}\left[e^{iuX_1} \times \cdots \times e^{iuX_n}\right]$$

we obtain, since the X_i are independent:

$$\mathbb{E}\left[e^{iuX_1} \times \cdots \times e^{iuX_n}\right] = \mathbb{E}\left[e^{iuX_1}\right] \times \cdots \times \mathbb{E}\left[e^{iuX_n}\right]$$

Since they are identically distributed:

$$\mathbb{E}\left[e^{iuX_1}\right] \times \cdots \times \mathbb{E}\left[e^{iuX_n}\right] = \underbrace{\mathbb{E}\left[e^{iuX_1}\right] \times \cdots \times \mathbb{E}\left[e^{iuX_1}\right]}_{n \text{ times}} = \left(\mathbb{E}\left[e^{iuX_1}\right]\right)^n$$

We therefore obtain the following result:

$$\mathbb{E}\left[e^{iu \sum_{i=1}^n X_i}\right] = \left(\mathbb{E}\left[e^{iuX_1}\right]\right)^n \tag{6.32}$$

The sum of n independent X_i random variables all having exponential distribution with parameter λ is a random variable G that has a gamma distribution with parameters (n, λ). From Eq. (6.32) we have:

$$\left(\mathbb{E}\left[e^{iuX_1}\right]\right)^n = \mathbb{E}\left[e^{iuG}\right] \tag{6.33}$$

where G is a $\mathcal{G}(n, \lambda)$ gamma distributed random variable.

The Three Different Methods of Notation for the Gamma Distribution Gamma distributions form a two-parameter family of distributions built on Euler's gamma function defined in (2.36). The definition is as follows:

Definition 6.1 (Gamma Distribution in Terms of Parameters).
If a random variable X has a gamma distribution with parameters (α, λ), denoted by $\mathcal{G}(\alpha, \lambda)$, then it has the density function given by:

$$f_{\mathcal{G}(\alpha,\lambda)}(x) = \frac{\lambda^\alpha}{\Gamma(\alpha)} x^{\alpha-1} e^{-\lambda x} \tag{6.34}$$

for all $x \geq 0$. Its characteristic function is given by:

$$\Phi_{\mathcal{G}(\alpha,\lambda)}(u) = \left(1 - \frac{iu}{\lambda}\right)^{-\alpha} = \left(\frac{1}{1-\frac{iu}{\lambda}}\right)^\alpha \tag{6.35}$$

The parameter α is called the shape parameter of the distribution and λ is called the scale parameter. To be precise:

(1) Shape parameter. The distribution takes one of two different shapes, depending on whether α is greater or lesser than 1. If $\alpha > 1$, then as α goes to infinity, the gamma distribution converges towards a normal distribution with mean $\mu = (\alpha-1)/\lambda$ and variance $\sigma^2 = (\alpha-1)/\lambda^2$. The moments of a gamma distribution are $\mathbb{E}[X] = \alpha/\lambda$ and $\text{var}(X) = \alpha/\lambda^2$. The skewness is $S = 2/\sqrt{\alpha}$ and the kurtosis $K = 3 + 6/\alpha$. For $\alpha = 1$, we have again an exponential distribution with parameter λ:

$$\mathcal{G}(1, \lambda) = \mathcal{E}(\lambda) \tag{6.36}$$

(2) Scale parameter. If X is a random variable that has a gamma distribution with parameters (α, λ), then λX has a gamma distribution with parameters $(\alpha, 1)$. More generally, the scale law for the gamma distribution is given by the equation:

$$k\,\mathcal{G}(\alpha, \lambda) = \mathcal{G}\left(\alpha, \frac{\lambda}{k}\right) \tag{6.37}$$

For $\lambda = 1$, we obtain a standard gamma distribution $\mathcal{G}(\alpha, 1)$:

$$f_{\mathcal{G}(\alpha,1)}(x) = \frac{1}{\Gamma(\alpha)} x^{\alpha-1} e^{-x} \tag{6.38}$$

for all $x \geq 0$.

Let us consider the characteristic function (6.35). We see that:

$$\Phi_{\mathcal{G}(\alpha,\lambda)}(u) = \left(\Phi_{\mathcal{E}(\lambda)}(u)\right)^{\alpha} \tag{6.39}$$

If $\alpha = n$ is an integer, then the gamma distribution is called Erlang distribution and represents the sum of n exponential distributions with parameter λ. In intuitive notation:

$$\mathcal{G}(n, \lambda) = \sum_{i=1}^{n} \mathcal{E}(\lambda) \tag{6.40}$$

Let us recall that, with the exception of the Gaussian distribution, whose parameters are equal to its moments, all distributions can be written either in terms of their parameters or their moments. We now introduce this second notation. Let $m = \alpha/\lambda$ and $v = \alpha/\lambda^2$ be the expectation and the variance of a gamma distribution with parameters α and λ. It follows from these equations that the parameters α and λ can be expressed in terms of the moments m and v as: $\alpha = m^2/v$ and $\lambda = m/v$. Replacing α and λ with the values given in terms of m and v in (6.34) and (6.35), we obtain the second notation of the gamma distribution.

Table 6.2 Useful relationships for gamma and exponential distributions (w.r.t. the parameters)

$\lambda \mathcal{E}(\lambda) = \mathcal{E}(1)$ $\lambda \mathcal{G}(\alpha, \lambda) = \mathcal{G}(\alpha, 1)$
$\mathcal{G}(1, \lambda) = \mathcal{E}(\lambda)$ In case $n \in \mathbb{N}$ $\mathcal{G}(n, \lambda) = \sum_{i=1}^{n} \mathcal{E}(\lambda)$

Definition 6.2 (Gamma Distribution in Terms of Moments).
If a random variable X has a gamma distribution with parameters (α, λ), whose moments are $m = \alpha/\lambda$ and $v = \alpha/\lambda^2$, then it has its density function given by:

$$f_{\mathcal{G}^*(m,v)}(x) = \frac{\left(\frac{m}{v}\right)^{\frac{m^2}{v}}}{\Gamma\left(\frac{m^2}{v}\right)} x^{\frac{m^2}{v}-1} e^{-\frac{m}{v}x} \qquad (6.41)$$

and characteristic function:

$$\Phi_{\mathcal{G}^*(m,v)}(u) = \left(\frac{1}{1 - iu\frac{v}{m}}\right)^{\frac{m^2}{v}} \qquad (6.42)$$

It is also possible to normalize the gamma distribution by setting its mean to 1. This is the same as making the convention $\alpha = \lambda = 1/v$. We use formulas (6.41) and (6.42) with $m = 1$:

Definition 6.3 (Gamma Distribution: Mean Equal to 1).
If a random variable X has a gamma distribution with mean $m = 1$, then its parameters are $\alpha = \lambda = 1/v$, where v is the variance. It has the density function:

$$f_{\mathcal{G}^*(1,v)}(x) = \frac{\left(\frac{1}{v}\right)^{\frac{1}{v}}}{\Gamma\left(\frac{1}{v}\right)} x^{\frac{1}{v}-1} e^{-\frac{1}{v}x} \qquad (6.43)$$

and characteristic function:

$$\Phi_{\mathcal{G}^*(1,v)}(u) = \left(\frac{1}{1 - iuv}\right)^{\frac{1}{v}} \qquad (6.44)$$

This third form of notation is used when the distribution serves as a random clock in the context of time-changes, which are presented in Chapter 7.

6.1.3.4 Calculating the Characteristic Function

Equipped with these preliminary results, we can now compute the characteristic function of the sum of Laplace distributions. From (6.30) in its factorized form we obtain:

$$\Phi_{X_n}(u) = e^{iun\theta} \left(\frac{1}{1 - iu\frac{c}{\sqrt{2}}} \right)^n \left(\frac{1}{1 + iu\frac{c}{\sqrt{2}}} \right)^n \quad (6.45)$$

which allows us to write:

$$\Phi_{X_n}(u) = \mathbb{E}\left[e^{iun\theta}\right] \times \left(\mathbb{E}\left[e^{iu\frac{c}{\sqrt{2}}W_1}\right]\right)^n \times \left(\mathbb{E}\left[e^{-iu\frac{c}{\sqrt{2}}W_2}\right]\right)^n \quad (6.46)$$

where W_1 and W_2 are two independent random variables having standard exponential distributions with parameter 1.

We use the first preliminary result (6.32):

$$\Phi_{X_n}(u) = \mathbb{E}\left[e^{iun\theta}\right] \times \mathbb{E}\left[e^{iu\frac{c}{\sqrt{2}}\sum_{i=1}^n W_i}\right] \times \mathbb{E}\left[e^{-iu\frac{c}{\sqrt{2}}\sum_{i=1}^n W_i}\right] \quad (6.47)$$

Then we use the second preliminary result (6.33):

$$\Phi_{X_n}(u) = \mathbb{E}\left[e^{iun\theta}\right] \times \mathbb{E}\left[e^{iu\frac{c}{\sqrt{2}}G_1}\right] \times \mathbb{E}\left[e^{-iu\frac{c}{\sqrt{2}}G_2}\right] \quad (6.48)$$

where G_1 and G_2 are two independent random variables having standard gamma distributions with parameters $(n, 1)$.

We therefore obtain the relationship:

$$X_n \stackrel{d}{=} n\theta + \frac{c}{\sqrt{2}}(G_1 - G_2) \quad (6.49)$$

which shows that the sum of Laplace distributions can be interpreted as the difference of two gamma distributions.

Let us write (6.49) using intuitive notation in terms of parameters:

$$\sum_{i=1}^n \mathcal{L}(\theta, c) = n\theta + \frac{c}{\sqrt{2}}(\mathcal{G}_1(n, 1) - \mathcal{G}_2(n, 1)) \quad (6.50)$$

where $\mathcal{G}_1(n, 1)$ and $\mathcal{G}_2(n, 1)$ are two independent random variables having standard gamma distributions with parameters $(n, 1)$. If we want to bring these parameters into the gamma distribution, this gives:

$$\sum_{i=1}^n \mathcal{L}(\theta, c) = n\theta + \mathcal{G}_1\left(n, \frac{\sqrt{2}}{c}\right) - \mathcal{G}_2\left(n, \frac{\sqrt{2}}{c}\right) \quad (6.51)$$

where $\mathcal{G}_1(n, \lambda)$ and $\mathcal{G}_2(n, \lambda)$ are two independent random variables having gamma distributions with parameters (n, λ), where $\lambda = \sqrt{2}/c$.

Alternatively, we can start from Eq. (6.14), which we recall here:

$$\mathcal{L}(\theta, c) = \theta + \frac{c}{\sqrt{2}} (\mathcal{E}_1(1) - \mathcal{E}_2(1))$$

Taking sums gives:

$$\sum_{i=1}^{n} \mathcal{L}(\theta, c) = \sum_{i=1}^{n} \left(\theta + \frac{c}{\sqrt{2}} (\mathcal{E}_1(1) - \mathcal{E}_2(1)) \right)$$

$$= n\theta + \frac{c}{\sqrt{2}} \left(\sum_{i=1}^{n} \mathcal{E}_1(1) - \sum_{i=1}^{n} \mathcal{E}_2(1) \right)$$

or:

$$\sum_{i=1}^{n} \mathcal{L}(\theta, c) = n\theta + \frac{c}{\sqrt{2}} (\mathcal{G}_1(n, 1) - \mathcal{G}_2(n, 1)) \qquad (6.52)$$

We can proceed in the same manner for the standard form and start from Eq. (6.2), which we recall here:

$$\mathcal{L}^*(0, 1) = \mathcal{E}_1(1) - \mathcal{E}_2(1)$$

Taking sums gives:

$$\sum_{i=1}^{n} \mathcal{L}^*(0, 1) = \sum_{i=1}^{n} (\mathcal{E}_1(1) - \mathcal{E}_2(1))$$

We have:

$$\sum_{i=1}^{n} \mathcal{L}^*(0, 1) = \sum_{i=1}^{n} \mathcal{E}_1(1) - \sum_{i=1}^{n} \mathcal{E}_2(1) = \mathcal{G}_1(n, 1) - \mathcal{G}_2(n, 1) \qquad (6.53)$$

where $\mathcal{G}_1(n, 1)$ and $\mathcal{G}_2(n, 1)$ are two independent random variables having gamma distributions with parameters $(n, 1)$.

6.1.3.5 Laplace Motion: The Difference of Two Gamma Processes

From what we have seen, it follows that it is possible to represent Laplace motion as the difference of two independent gamma processes that correspond to increases and decreases in the total return of a financial asset.

Gamma Processes The definition of a gamma process is based on the gamma distribution. It is one of the most widely used stochastic processes in the area of systems reliability for modeling degradation over time. We now give its definition. Let $G = (G(t), t \geq 0)$ be a stochastic process. We say that G is a gamma process if: (1) $G(0) = 0$; (2) it has independent increments; and (3) its increments have a gamma distribution.

More precisely, let $\eta = (\eta(t), t \geq 0)$ be a strictly increasing real function with $\eta(0) = 0$. For all $t \geq 0$, the density of $G(t)$ is given by:

$$f_{G_t}(x) = \frac{\lambda^{\eta_t}}{\Gamma(\eta_t)} x^{\eta_t - 1} e^{-\lambda x} \qquad (6.54)$$

where $\Gamma(.)$ is Euler's gamma function.

The case most often encountered is that where the function η is linear, i.e. $\eta_t = \alpha t$. In this case, it follows from the three properties above that a gamma process also has stationary increments, which makes it a Lévy process. Its density is given by:

$$f_{G_t}(x) = \frac{\lambda^{\alpha t}}{\Gamma(\alpha t)} x^{\alpha t - 1} e^{-\lambda x} \qquad (6.55)$$

In this case, $G(t)$ has a gamma distribution with parameters $(\alpha t, \lambda)$ and moments

$$\mathbb{E}[G(t)] = \alpha t / \lambda \stackrel{\text{def}}{=} mt \qquad \text{var}(G(t)) = \alpha t / \lambda^2 \stackrel{\text{def}}{=} vt$$

The passing of time affects only the shape parameter (αt) of the distribution. Therefore, the shape of the distribution changes over time. The effect on the moments is the following: the more the time-span increases, the more the mean and variance increase and the skewness and kurtosis decrease (we see here again the convergence towards the Gaussian distribution as the shape parameter increases in value).

The Lévy measure of a gamma process is:

$$\nu(x) = \begin{cases} \frac{\alpha}{|x|} e^{-\lambda |x|} & \text{if } x < 0 \\ \frac{\alpha}{x} e^{-\lambda x} & \text{if } x > 0 \end{cases}$$

or:

$$\nu(x) = \frac{\alpha}{|x|} e^{-\lambda x} \qquad (6.56)$$

We see that the Lévy measure tends to infinity as x tends to zero, which corresponds to an infinite arrival rate of very small jumps. Since the integral of the Lévy measure over the real line is infinite, we say that the gamma process has infinite activity.

Figure 6.2 shows a path of a gamma process that has been simulated as the sum of increments with density given by (6.55), and where we have taken $m = 1$ (i.e. $\alpha = \lambda = 1/v$).

Finally, Eq. (6.49) becomes:

$$L(t) \stackrel{d}{=} \frac{\theta t}{\tau} + \frac{c}{\sqrt{2}} (G_1(t) - G_2(t)) \qquad (6.57)$$

where the two random variables $G_1(t)$ and $G_2(t)$ have a gamma distribution with parameters $(\alpha t, 1)$.

Table 6.3 Overview of gamma distributions and processes

	In terms of parameters: shape, scale Γ dist.	In terms of moments: mean, variance Γ dist.	Γ dist.: $\alpha = \lambda = 1/v$
	Standard Γ dist.		
$f_X(x)$	$\mathcal{G}(\alpha, 1)$	$\mathcal{G}(\alpha, \lambda)$	$\mathcal{G}(m, v)$	$\mathcal{G}(1, v)$
	$\frac{1}{\Gamma(\alpha)} x^{\alpha-1} e^{-x}$	$\frac{\lambda^\alpha}{\Gamma(\alpha)} x^{\alpha-1} e^{-\lambda x}$	$\frac{\left(\frac{m}{v}\right)^{\frac{m^2}{v}}}{\Gamma\left(\frac{m^2}{v}\right)} x^{\frac{m^2}{v}-1} e^{-\frac{m}{v} x}$	$\frac{\left(\frac{1}{v}\right)^{\frac{1}{v}}}{\Gamma\left(\frac{1}{v}\right)} x^{\frac{1}{v}-1} e^{-\frac{x}{v}}$
$\Phi_X(u)$	$\left(\frac{1}{1-iu}\right)^\alpha$	$\left(\frac{1}{1-\frac{iu}{\lambda}}\right)^\alpha$	$\left(\frac{1}{1-iu\frac{v}{m}}\right)^{\frac{m^2}{v}}$	$\left(\frac{1}{1-iuv}\right)^{\frac{1}{v}}$
$\mathbb{E}[X], \mathrm{var}(X)$	α, α	$\frac{\alpha}{\lambda} = m, \frac{\alpha}{\lambda^2} = v$	$\frac{\alpha}{\lambda} = m, \frac{\alpha}{\lambda^2} = v$	$1, v$
	Standard Γ process	Γ process	Γ process	Γ process: $\alpha = \lambda$
$G(t)$	$\mathcal{G}(t, 1)$	$\mathcal{G}(\alpha t, \lambda)$	$\mathcal{G}(mt, vt)$	$\mathcal{G}(t, vt)$
$\Phi_{G_t}(u)$	$\left(\frac{1}{1-iu}\right)^t$	$\left(\frac{1}{1-\frac{iu}{\lambda}}\right)^{\alpha t}$	$\left(\frac{1}{1-iu\frac{v}{m}}\right)^{\frac{m^2 t}{v}}$	$\left(\frac{1}{1-iuv}\right)^{\frac{t}{v}}$
$\mathbb{E}[G(t)], \mathrm{var}(X)$	t, t	$\frac{\alpha t}{\lambda} = mt, \frac{\alpha t}{\lambda^2} = vt$	$\frac{\alpha t}{\lambda} = mt, \frac{\alpha t}{\lambda^2} = vt$	t, vt
$\nu(x)$	$\frac{1}{|x|} e^{-|x|}$	$\frac{\alpha}{|x|} e^{-\lambda|x|}$	$\frac{m^2}{v} \frac{1}{|x|} e^{-\frac{m}{v}|x|}$	$\frac{1}{v} \frac{1}{|x|} e^{-\frac{|x|}{v}}$

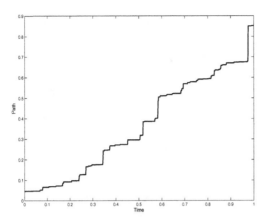

Figure 6.2   Gamma process trajectory

6.1.3.6   *The Sum of Laplace Distributions as a Mixture Distribution*

As performed above for the Laplace distribution, we will now write the sum of Laplace distributions as a mixture distribution.

Let the sum $X_n = L_1 + \cdots + L_n$ of n Laplace random variables be given, written in the form of a mixture distribution: for all i, $L_i = \sqrt{2W_i} Z_i$, where W_i and Z_i are independent standard exponential and normal distributions. Starting from:

$$X_n = \sum_{i=1}^{n} \sqrt{2W_i} Z_i \tag{6.58}$$

we want to compute the mean and variance of X_n conditional on the values taken by the W_i. For the conditional expectation, we have:

$$\mathbb{E}[X_n \mid W_1 \cdots W_n] = \sum_{i=1}^{n} \mathbb{E}[\sqrt{2W_i}\, Z_i \mid W_1 \cdots W_n] = \sum_{i=1}^{n} \sqrt{2W_i}\, \mathbb{E}[Z_i] = 0$$

For the conditional variance, we have:

$$\operatorname{var}(X_n \mid W_1 \cdots W_n) = \mathbb{E}[(X_n - \mathbb{E}[X_n \mid W_1 \cdots W_n])^2 \mid W_1 \cdots W_n]$$

or:

$$\operatorname{var}(X_n \mid W_1 \cdots W_n) = \mathbb{E}[X_n^2 \mid W_1 \cdots W_n]$$

We use (6.58) to replace X_n^2 with its given value:

$$\operatorname{var}(X_n \mid W_1 \cdots W_n) = \mathbb{E}\left[\left(\sum_{i=1}^{n} \sqrt{2W_i} Z_i\right)^2 \mid W_1 \cdots W_n\right]$$

or:

$$\text{var}(X_n \mid W_1 \cdots W_n) = \sum_{i=1}^{n} \mathbb{E}\left[2W_i Z_i^2 \mid W_1 \cdots W_n\right]$$
$$+ \sum_{i \neq j} \mathbb{E}\left[2\sqrt{W_i}\sqrt{W_j} Z_i Z_j \mid W_1 \cdots W_n\right]$$

hence:

$$\text{var}(X_n \mid W_1 \cdots W_n) = 2\sum_{i=1}^{n} W_i \mathbb{E}[Z_i^2 \mid W_1 \cdots W_n]$$
$$+ 2\sum_{i \neq j} \sqrt{W_i}\sqrt{W_j} \mathbb{E}\left[Z_i Z_j \mid W_1 \cdots W_n\right]$$

Since $\mathbb{E}[Z_i Z_j \mid W_1 \cdots W_n] = \mathbb{E}[Z_i Z_j] = 0$ and $\mathbb{E}[Z_i^2 \mid W_1 \cdots W_n] = \mathbb{E}[Z_i^2] = 1$, we finally obtain:

$$\text{var}(X_n \mid W_1 \cdots W_n) = 2\sum_{i=1}^{n} W_i \qquad (6.59)$$

Since it is the sum of n independent random variables having the same exponential distribution, the conditional variance of X_n has a gamma distribution. Noting that, conditioned on the W_i, X_n has a Gaussian distribution, we can write it as follows:

$$X_n = \sqrt{\text{var}(X_n \mid W_1 \cdots W_n)}\, Z \qquad (6.60)$$

where Z is a standard Gaussian random variable. This equation clearly displays X_n as a mixture of a normal and a gamma distribution.

Consequently, if the Laplace distribution can be represented as a normal distribution whose variance is an exponential random variable, then it can be seen from Eqs (6.59) and (6.60) that the sum of Laplace distributions can be represented as a normal distribution whose variance is a gamma random variable. This representation is a variant of that given in (6.50) as a difference of gamma distributions.

Using the intuitive notation introduced above, we write in analogy to (6.25):

$$\sum_{i=1}^{n} \mathcal{L}^*(0,1) = \mathcal{N}\left[\mathcal{G}(n,1)\right] \qquad (6.61)$$

To conclude, we show in Table 6.4 a comparison of Laplace's two laws.

Table 6.4 The two universes corresponding to Laplace's two laws

Date	1774	1778
Best middle	median	mean
Minimization	distances	squares
Usual form	$\dfrac{e^{-\frac{\|x-\theta\|}{s}}}{2s}$	$\dfrac{e^{-\frac{1}{2}\left(\frac{x-\mu}{\sigma}\right)^2}}{\sigma\sqrt{2\pi}}$
$\Phi_X(u)$ in terms of parameters	$\dfrac{e^{i\theta u}}{1+s^2 u^2}$	$e^{i\mu u - \frac{1}{2}\sigma^2 u^2}$
Expectation	θ	μ
Variance	$2s^2$	σ^2
Asymmetry coefficients	0	0
Kurtosis K	6	3
Standardized form in terms of parameters	$\frac{1}{2} e^{-\|x\|}$	$\dfrac{1}{\sqrt{2\pi}} e^{-\frac{x^2}{2}}$
$\Phi_X(u)$ standardized in terms of parameters	$\dfrac{1}{1+u^2}$	$e^{-\frac{u^2}{2}}$
Factorization	$\dfrac{1}{1+iu}\dfrac{1}{1-iu}$	
Exponential representation	$W_1 - W_2$	
Parameterization $c^2 = 2s^2$	$\dfrac{e^{-\sqrt{2}\frac{\|x-\theta\|}{c}}}{c\sqrt{2}}$	
$\Phi_X(u)$ in terms of moments	$\dfrac{e^{i\theta u}}{1+\frac{1}{2}c^2 u^2}$	$e^{i\mu u - \frac{1}{2}\sigma^2 u^2}$
Standardized form in terms of moments	$\dfrac{e^{-\sqrt{2}\|x\|}}{\sqrt{2}}$	$\dfrac{e^{-\frac{x^2}{2}}}{\sqrt{2\pi}}$
$\Phi_X(u)$ standardized in terms of moments	$\dfrac{1}{1+\frac{u^2}{2}}$	$e^{-\frac{u^2}{2}}$
Lévy measure	$\dfrac{e^{-\|x\|/s}}{\|x\|}$	
Infinitely divisible	yes	yes
Stability under addition	no	yes

6.2 The Asymmetrization of the Laplace Distribution

The Laplace distribution is symmetric. In many applications however, it is useful to obtain an asymmetrized version from it, or in other words to asymmetrize the Laplace distribution. A convenient way of achieving this asymmetrization is to introduce an asymmetry parameter β such that β and $1/\beta$ weight the positive and negative parts of the distribution, respectively. This parameter β determines and quantifies the probability mass lying on each side of the distribution.

6.2.1 *Construction of the Asymmetrization*

First of all, let us explain the principle of asymmetrizing the Laplace distribution.

6.2.1.1 The Principle of Asymmetrization

Let us begin by considering the standardized form. We construct a density of the type:

$$f(x) = C \begin{cases} \exp(-\beta x) & \text{if } x \geq 0 \\ \exp\left(\frac{1}{\beta} x\right) & \text{if } x < 0 \end{cases}$$

where C is a normalization constant for the density. To find C, we make use of the fact that $\int \beta e^{-\beta x} dx = 1$, from which we have $\int e^{-\beta x} dx = 1/\beta$ and $\int e^{-(1/\beta)x} dx = \beta$. It follows that $C = 1/(\beta + 1/\beta) = \beta/(\beta^2 + 1)$, and we obtain the density function of an asymmetrized and standardized Laplace distribution:

$$f(x) = \frac{\beta}{1 + \beta^2} \begin{cases} \exp(-\beta x) & \text{if } x \geq 0 \\ \exp\left(\frac{1}{\beta} x\right) & \text{if } x < 0 \end{cases} \tag{6.62}$$

We note that when $\beta = 1$, then $C = 1/2$, and we find the simple Laplace distribution again. The same argument leads to the general asymmetrized form of the Laplace distribution:

$$f(x) = \frac{\sqrt{2}}{c} \frac{\beta}{1 + \beta^2} \begin{cases} \exp\left(-\frac{\sqrt{2}\beta}{c}(x - \theta)\right) & \text{if } x \geq \theta \\ \exp\left(-\frac{\sqrt{2}}{c\beta}(-x + \theta)\right) & \text{if } x < \theta \end{cases} \tag{6.63}$$

The skewness and kurtosis both depend on the parameter β and are given by:

- Skewness (Pearson–Fisher γ_1):

$$S = 2 \frac{1/\beta^3 - \beta^3}{(1/\beta^2 + \beta^2)^{3/2}}$$

S takes values in the interval $[-2, +2]$ for β ranging from 0 to infinity.
- Kurtosis (Pearson–Fisher γ_2):

$$K = 9 - \frac{12}{(1/\beta^2 + \beta^2)^2}$$

K takes values in the interval $[6, 9]$ for β ranging from 0 to infinity.

Note that K is minimal and equal to 6 for $\beta = 1$, which corresponds to the standard Laplace distribution.

Notation 3 (Generalized Laplace Distribution).
Starting from a simple Laplace distribution $\mathcal{L}(\theta, c)$, we denote by $\mathcal{L}''(\theta, c, \beta)$ the Laplace distribution asymmetrized using the parameter β, and by $\mathcal{L}'(\theta, c, \delta)$ the Laplace distribution asymmetrized using the parameter δ.

Asymmetrization using the parameter δ will be introduced in Section 6.2.1.3.

6.2.1.2 *Characteristic Function*

As before, the characteristic function of a generalized Laplace distribution is obtained by starting from the equation for X given by:

$$X \stackrel{d}{=} \theta + Y_1 - Y_2$$

where Y_1 and Y_2 are two exponential distributions with parameters $\lambda_1 = \beta\frac{\sqrt{2}}{c}$ and $\lambda_2 = \frac{1}{\beta}\frac{\sqrt{2}}{c}$. In other words:

$$X \stackrel{d}{=} \theta + \frac{c}{\sqrt{2}}\left(\frac{1}{\beta} W_1 - \beta W_2\right) \qquad (6.64)$$

In intuitive notation:

$$\mathcal{L}''(\theta, c, \beta) = \theta + \frac{c}{\sqrt{2}}\left(\frac{1}{\beta}\mathcal{E}_1(1) - \beta\mathcal{E}_2(1)\right) \qquad (6.65)$$

where $\mathcal{E}_1(1)$ and $\mathcal{E}_2(1)$ are two standard exponential distributions. We recall that it follows from the properties of the scale parameter of exponential distributions that:

$$k\,\mathcal{E}(1) = \mathcal{E}\left(\frac{1}{k}\right)$$

and therefore:

$$\frac{c}{\beta\sqrt{2}}\mathcal{E}_1(1) = \mathcal{E}_1\left(\frac{\beta\sqrt{2}}{c}\right)$$

In other words:

$$\mathcal{L}''(\theta, c, \beta) = \theta + \mathcal{E}\left(\beta\frac{\sqrt{2}}{c}\right) - \mathcal{E}\left(\frac{1}{\beta}\frac{\sqrt{2}}{c}\right)$$

The characteristic function $\Phi_X(u)$ can be computed by starting from (6.64).

As before, we move from the sum of random variables to the product of characteristic functions:

$$\mathbb{E}\left[e^{iuX}\right] = \mathbb{E}\left[e^{iu\left(\theta + \frac{c}{\sqrt{2}}\left(\frac{1}{\beta}W_1 - \beta W_2\right)\right)}\right]$$

$$= \mathbb{E}\left[e^{i\theta u}\right] \times \mathbb{E}\left[e^{iu\frac{1}{\beta}\frac{c}{\sqrt{2}}W_1}\right] \times \mathbb{E}\left[e^{-iu\beta\frac{c}{\sqrt{2}}W_2}\right]$$

Therefore:
$$\Phi_X(u) = \frac{e^{i\theta u}}{\left(1 - iu\frac{1}{\beta}\frac{c}{\sqrt{2}}\right)\left(1 + iu\beta\frac{c}{\sqrt{2}}\right)} \qquad (6.66)$$

6.2.1.3 Trace of the Asymmetry

We will reformulate Eq. (6.66) in order to introduce a new and interesting parameter. We begin by rewriting Eq. (6.66) as:

$$\Phi_X(u) = \frac{e^{i\theta u}}{1 + \frac{1}{2}c^2 u^2 - i\left(\frac{1}{\beta} - \beta\right)\frac{c}{\sqrt{2}}u} \qquad (6.67)$$

We introduce a new parameter δ which depends on the asymmetry parameter β and is defined by:

$$\delta = \left(\frac{1}{\beta} - \beta\right) \times \frac{c}{\sqrt{2}} \qquad (6.68)$$

Using this new parameter in formula (6.67), we finally obtain the equation:

$$\Phi_X(u) = \frac{e^{i\theta u}}{1 + \frac{1}{2}c^2 u^2 - i\delta u} \qquad (6.69)$$

In the case of generalized distributions, one of the traces of asymmetry is an emerging difference between the mean and the mode. This property is satisfied here: the parameter δ represents this difference. The introduction of asymmetry via the parameter β displaces the mean from the mode: the asymmetrization moves the mean of the distribution away from the mode θ by an amount δ. The value of the new mean is given by $\mu = \theta + \delta$.

This displacement results from the asymmetry between upward or downward movements. The new mean therefore has a component that depends on the mode of the generalized Laplace distribution (the mean of the simple Laplace distribution) and a component that depends on the asymmetry between upward or downward movements.

The new mean resulting from the asymmetrization is therefore:

$$\mu = \theta + \left(\frac{1}{\beta} - \beta\right) \times s \qquad (6.70)$$

The value of this difference depends directly on the asymmetry parameter β, for a size given by the scale parameter s, according to the relationship (6.68). For example, if $s = 1$ and $\beta = 0.5$, then $\delta = 1.5$, i.e. the mean μ is 1.5 greater than the mode θ. If $s = 1$ and $\beta = 10$, then $\delta = -9.9$, i.e. the

Table 6.5 Characteristic functions of the generalized Laplace distribution

Shape	In terms of δ	In terms of β	In terms of β, factorized
With s	$\dfrac{e^{i\theta u}}{1+s^2 u^2 - i\delta u}$	$\dfrac{e^{i\theta u}}{1+s^2 u^2 - i\left(\frac{1}{\beta}-\beta\right)su}$	$\dfrac{e^{i\theta u}}{\left(1-iu\frac{1}{\beta}s\right)(1+iu\beta s)}$
With c	$\dfrac{e^{i\theta u}}{1+\frac{1}{2}c^2 u^2 - i\delta u}$	$\dfrac{e^{i\theta u}}{1+\frac{1}{2}c^2 u^2 - i\left(\frac{1}{\beta}-\beta\right)\frac{c}{\sqrt{2}}u}$	$\dfrac{e^{i\theta u}}{\left(1-iu\frac{1}{\beta}\frac{c}{\sqrt{2}}\right)\left(1+iu\beta\frac{c}{\sqrt{2}}\right)}$

Table 6.6 The various notations used for Laplace distributions

Laplace distribution Form	Simple	Generalized Notation in terms of δ	Generalized Notation in terms of β
With s	$\mathcal{L}^*(\theta, s)$	N/A	N/A
With c	$\mathcal{L}(\theta, c)$	$\mathcal{L}'(\theta, c, \delta)$	$\mathcal{L}''(\theta, c, \beta)$

mean μ is 9.9 less than the mode θ. Fixing s, we see that the shift increases as β increases. Large values of β crush the right tail of the distribution and draw out its left tail. For $\beta = 1$, $\delta = 0$, and $\mu = \theta$, we find the simple Laplace distribution with mean θ.

We denote by $\mathcal{L}'(\theta, c, \delta)$ the generalized Laplace distribution in terms of the parameter δ, starting from the simple Laplace distribution $\mathcal{L}(\theta, c)$. Table 6.6 summarizes the various notations we use to write Laplace distributions.

If we view this from the perspective of a stochastic process instead of a probability distribution, then this means that large values of β translate into the appearance of a small number of significant losses, while there will be a large number of small gains, hence a skewness $S < 0$.

6.2.1.4 Infinitely Divisible Distribution and Lévy Measure

As we did for simple Laplace distributions, we verify that generalized Laplace distributions are infinitely divisible. In fact, it is clear from the characteristic function (6.69) that:

$$\Phi_X(u) = \frac{e^{i\theta u}}{1 + \frac{1}{2}c^2 u^2 - i\delta u} = \left[e^{i\theta u/n} \times \left(\frac{1}{1 + \frac{1}{2}c^2 u^2 - i\delta u} \right)^{1/n} \right]^n \quad (6.71)$$

for any integer $n \geq 1$.

Being an infinitely divisible distribution, the generalized Laplace distribution has a unique representation by the Lévy–Khintchine formula. It can be shown that θ represents a drift term, the diffusion component σ is zero, and that the Lévy measure of a Laplace distribution $\mathcal{L}(\theta, c, \beta)$ is given by:

$$\nu(x) = \begin{cases} \frac{1}{|x|} \exp\left(-\frac{\sqrt{2}\beta}{c} x\right) & \text{if } x < 0 \\ \frac{1}{x} \exp\left(-\frac{\sqrt{2}}{\beta c} |x|\right) & \text{if } x > 0 \end{cases} \quad (6.72)$$

Figure 6.3 shows several examples of density functions of asymmetrized Laplace distributions.

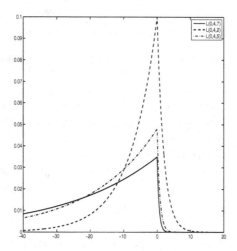

Figure 6.3 Densities of asymmetrized Laplace distributions for different values of β

6.2.2 Laplace Processes

We now come to the construction of Laplace processes.

6.2.2.1 Definitions

Just as we constructed Laplace motion from simple Laplace distributions, we can construct a new stochastic process from generalized Laplace distributions. We use the generalized distribution $\mathcal{L}'(\theta, c, \delta)$ as the underlying distribution of the Lévy process. This new stochastic process X will be called the Laplace process: it is an asymmetric Laplace motion with scale parameter c for time-unit parameter τ with $X(0) = 0$. Its increments are

independent and stationary with:
$$X(t+\tau) - X(t) \sim \mathcal{L}'(\theta, c, \delta) \tag{6.73}$$
where $\mathcal{L}'(\theta, c, \delta)$ is a generalized Laplace distribution with scale parameter c and difference mean–mode δ.

The moments of a Laplace process at date t are:
$$\mathbb{E}[X(t)] = (\theta + \delta)\frac{t}{\tau} \qquad \operatorname{var} X(t) = \frac{t}{\tau}c^2$$

When $\delta = 0$, we have again:
$$\mathbb{E}[X(t)] = \theta\frac{t}{\tau} \qquad \operatorname{var} X(t) = \frac{t}{\tau}c^2$$

6.2.2.2 Sum of Generalized Laplace Distributions

We follow the same route for generalized Laplace distributions as we did for simple Laplace distributions. The distribution of X_n is a generalized Laplace distribution summed to order n. From the characteristic function (6.69), we obtain:
$$\Phi_{X_n}(u) = \left(\frac{e^{i\theta u}}{1 + \frac{1}{2}c^2 u^2 - i\delta u}\right)^n \tag{6.74}$$

or in factorized form:
$$\Phi_{X_n}(u) = e^{iun\theta} \left(\frac{1}{1 - iu\frac{1}{\beta}\frac{c}{\sqrt{2}}}\right)^n \left(\frac{1}{1 + iu\beta\frac{c}{\sqrt{2}}}\right)^n$$

or:
$$\mathbb{E}\left[e^{iuX_n}\right] = \mathbb{E}\left[e^{iun\theta}\right] \times \left(\mathbb{E}\left[e^{iu\frac{1}{\beta}\frac{c}{\sqrt{2}}W_1}\right]\right)^n \times \left(\mathbb{E}\left[e^{-iu\beta\frac{c}{\sqrt{2}}W_2}\right]\right)^n$$

An argument identical to that made for simple Laplace distributions leads to:
$$\Phi_{X_n}(u) = \mathbb{E}\left[e^{iun\theta}\right] \times \mathbb{E}\left[e^{iu\frac{1}{\beta}\frac{c}{\sqrt{2}}\mathcal{G}_1}\right] \times \mathbb{E}\left[e^{-iu\beta\frac{c}{\sqrt{2}}\mathcal{G}_2}\right]$$

where $\mathcal{G}_1$ and $\mathcal{G}_2$ are two independent random variables with identical standard gamma distribution. We therefore obtain the equality in distribution:
$$X_n \stackrel{d}{=} n\theta + \frac{c}{\sqrt{2}}\left(\frac{1}{\beta}\mathcal{G}_1 - \beta\mathcal{G}_2\right) \tag{6.75}$$

or, in intuitive notation in terms of parameters:
$$\sum_{i=1}^n \mathcal{L}'(\theta, c, \delta) = n\theta + \frac{c}{\sqrt{2}}\left(\frac{1}{\beta}\mathcal{G}_1(n,1) - \beta\mathcal{G}_2(n,1)\right) \tag{6.76}$$

Table 6.7 The sum of generalized Laplace distributions

	L_i	$X_n = \sum_{i=1}^n L_i$
class.	$\theta + \frac{c}{\sqrt{2}}(\mathcal{E}_1(1) - \mathcal{E}_2(1))$	$n\theta + \frac{c}{\sqrt{2}}(\mathcal{G}_1(n,1) - \mathcal{G}_2(n,1))$
asym.	$\theta + \frac{c}{\sqrt{2}}\left(\frac{1}{\beta}\mathcal{E}_1(1) - \beta\mathcal{E}_2(1)\right)$	$n\theta + \frac{c}{\sqrt{2}}\left(\frac{1}{\beta}\mathcal{G}_1(n,1) - \beta\mathcal{G}_2(n,1)\right)$
class.	$\theta + \mathcal{E}_1\left(\frac{\sqrt{2}}{c}\right) - \mathcal{E}_2\left(\frac{\sqrt{2}}{c}\right)$	$n\theta + \mathcal{G}_1\left(n, \frac{\sqrt{2}}{c}\right) - \mathcal{G}_2\left(n, \frac{\sqrt{2}}{c}\right)$
asym.	$\theta + \mathcal{E}\left(\beta\frac{\sqrt{2}}{c}\right) - \mathcal{E}\left(\frac{1}{\beta}\frac{\sqrt{2}}{c}\right)$	$n\theta + \mathcal{G}\left(n, \beta\frac{\sqrt{2}}{c}\right) - \mathcal{G}\left(n, \frac{1}{\beta}\frac{\sqrt{2}}{c}\right)$

The general notation gives:

$$\sum_{i=1}^n \mathcal{L}'(\theta, c, \delta) = n\theta + \mathcal{G}_1\left(n, \beta\frac{\sqrt{2}}{c}\right) - \mathcal{G}_2\left(n, \frac{1}{\beta}\frac{\sqrt{2}}{c}\right) \qquad (6.77)$$

Setting $\lambda_1 = \beta\frac{\sqrt{2}}{c}$ and $\lambda_2 = \frac{1}{\beta}\frac{\sqrt{2}}{c}$, we have:

$$\sum_{i=1}^n \mathcal{L}'(\theta, c, \delta) = n\theta + \mathcal{G}(n, \lambda_1) - \mathcal{G}(n, \lambda_2) \qquad (6.78)$$

In the standardized case with $\theta = 0$ and $c = \sqrt{2}$ (unit variance):

$$\sum_{i=1}^n \mathcal{L}'(0, \sqrt{2}, \delta) = \frac{1}{\beta}\mathcal{G}_1(n,1) - \beta\mathcal{G}_2(n,1)$$

Table 6.7 summarizes these equivalences.

6.2.2.3 The Laplace Process as Difference of two Gamma Processes

The relationship (6.75) can be generalized to continuous time as:

$$X_t \stackrel{d}{=} \theta t + \frac{c}{\sqrt{2}}(G_1(t) - G_2(t)) \qquad (6.79)$$

where $G_1(t)$ and $G_2(t)$ are two gamma random variables.

Figure 6.4 shows a Laplace process given by the sum of a Laplace motion and a drift term (shown individually with dotted lines).

6.3 The Laplace Distribution as the Limit of Hyperbolic Distributions

Here we study the link between Laplace's first law and hyperbolic distributions.

Table 6.8 Overview of Laplace distributions and processes

Standard distribution	Simple distribution	Generalized distribution
$\mathcal{L}(0,1)$	$\mathcal{L}(\theta,c)$	$\mathcal{L}'(\theta,c,\delta)$
$X \stackrel{d}{=} \frac{1}{\sqrt{2}}(W_1 - W_2)$	$X \stackrel{d}{=} \theta + \frac{c}{\sqrt{2}}(W_1 - W_2)$	$X \stackrel{d}{=} \theta + \frac{c}{\sqrt{2}}\left(\frac{1}{\beta}W_1 - \beta W_2\right)$
		$\delta = \frac{c}{\sqrt{2}}\left(\frac{1}{\beta} - \beta\right)$
		$\Phi_X(u) = \frac{e^{iu\theta}}{1 + \frac{1}{2}c^2u^2 - i\delta u}$
$\mathbb{E}[X] = 0 \quad W_i \sim \mathcal{E}(1)$	$\mathbb{E}[X] = \theta$	$\mathbb{E}[X] = \theta + \delta$
$\mathrm{var}(X) = 1 \quad f_{W_i}(x) = e^{-x}$	$\mathrm{var}(X) = c^2$	$\mathrm{var}(X) = c^2 + \delta^2$
Sum of standard distributions	Sum of simple distributions	Sum of generalized distributions
$\sum_{i=1}^n \mathcal{L}(0,1)$	$\sum_{i=1}^n \mathcal{L}(\theta,c)$	$\sum_{i=1}^n \mathcal{L}'(\theta,c,\delta)$
$X_n \stackrel{d}{=} \frac{1}{\sqrt{2}}(G_1 - G_2)$	$X_n \stackrel{d}{=} n\theta + \frac{c}{\sqrt{2}}(G_1 - G_2)$	$X_n \stackrel{d}{=} n\theta + \frac{c}{\sqrt{2}}\left(\frac{1}{\beta}G_1 - \beta G_2\right)$
		$\Phi_{X_n}(u) = \frac{e^{iun\theta}}{\left(1 + \frac{1}{2}c^2u^2 - i\delta u\right)^n}$
$\mathbb{E}[X_n] = 0 \quad G_i \sim \mathcal{G}(n,1)$	$\mathbb{E}[X_n] = n\theta$	$\mathbb{E}[X_n] = n(\theta + \delta)$
$\mathrm{var}(X_n) = n \quad f_{G_i}(x) = \frac{1}{\Gamma(n)}x^{n-1}e^{-x}$	$\mathrm{var}(X_n) = nc^2$	$\mathrm{var}(X_n) = n(c^2 + \delta^2)$
Standard Laplace motion	Laplace motion	Laplace process
$A(t) - A(t-\tau) \sim \mathcal{L}(0,1)$	$L(t) - L(t-\tau) \sim \mathcal{L}(\theta,c)$	$X(t) - X(t-\tau) \sim \mathcal{L}'(\theta,c,\delta)$
$A(t) \stackrel{d}{=} \frac{1}{\sqrt{2}}(G_1(t) - G_2(t))$	$L_t \stackrel{d}{=} \theta\frac{t}{\tau} + \frac{c}{\sqrt{2}}(G_{1,t} - G_{2,t})$	$X(t) \stackrel{d}{=} \theta\frac{t}{\tau} + \frac{c}{\sqrt{2}}\left(\frac{1}{\beta}G_1(t) - \beta G_2(t)\right)$
		$\Phi_{X_t}(u) = \frac{e^{iu\theta\frac{t}{\tau}}}{\left(1 + \frac{1}{2}c^2u^2 - i\delta u\right)^{\frac{t}{\tau}}}$
$\mathbb{E}[A(t)] = 0 \quad G_i \sim \mathcal{G}(\frac{t}{\tau},1)$	$\mathbb{E}[L(t)] = \theta\frac{t}{\tau}$	$\mathbb{E}[X(t)] = \frac{t}{\tau}(\theta + \delta)$
$\mathrm{var}(A(t)) = \frac{t}{\tau} \quad \Phi_{G_t}(x) = \left(\frac{1}{1-iu}\right)^{\frac{t}{\tau}}$	$\mathrm{var}(L(t)) = \frac{t}{\tau}c^2$	$\mathrm{var}(X(t)) = \frac{t}{\tau}(c^2 + \delta^2)$

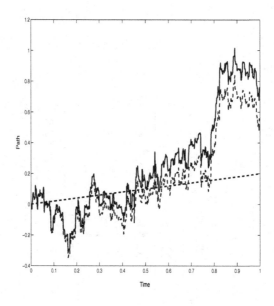

Figure 6.4   Laplace process

6.3.1 *Motivation for Hyperbolic Distributions*

The difficulties encountered when working with alpha-stable distributions led to alpha-stable Lévy processes being abandoned rather quickly in the 1970s. Research in finance then led to the development of mixed processes of jump-diffusion type, which are, as we have seen, simple non-stable Lévy processes with finite activity and variation. The finite activity also limited financial modeling. Then, in the 1990s, with the use of hyperbolic distributions, non-stable Lévy processes were rediscovered without the limitation imposed by finite activity.

These so-called hyperbolic distributions originated from the work of geophysicists, in particular the study of blown sand carried out in Denmark and Germany, and led straight to the works of Barndorff-Nielsen in the 1970s. One of the arguments given from the beginning in favor of applying these distributions to finance was that they were not stable. In this vein, Eberlein and Keller (1995), write (see p. 282):

> [...] real stock-price paths change drastically if we look at them on different time scales.

Hyperbolic distributions are infinitely divisible and can therefore be used to construct Lévy processes by specifying the underlying marginal distribution (for the date $t = 1$). However, they are not stable. In other words, if the underlying distribution is hyperbolic at date $t = 1$, then this does not imply that it will remain this way at any other date t. We will often use a numerical computation to go from the scale $t = 1$ to any other given scale t.

The first hyperbolic distribution, introduced into finance by Barndorff-Nielsen (1997) in 1997, was the normal inverse Gaussian (NIG) distribution obtained by mixing normal and inverse Gaussian distributions. Then in 1998, the generalized hyperbolic distribution was systematically analyzed by Eberlein et al. (1998). Prause's thesis (1999) in turn gave a first synthesis of these distributions and their application to the valuation of financial derivatives. It seemed that they modeled both the skewness and leptokurtic features encountered in empirical distributions from the actual financial world rather well, without running into the (perceived) practical inconveniences of alpha-stable distributions.

6.3.2 *Construction of Hyperbolic Distributions*

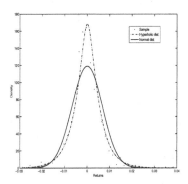 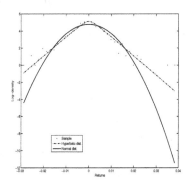

Figure 6.5 Empirical and theoretical densities of hourly returns of BNP stock in a classic chart (left) and a semi-logarithmic chart (right)

Let us try to develop an intuitive understanding of what motivates the name "hyperbolic" for these distributions. If we draw the graph of a Gaussian density in a semi-logarithmic chart (i.e. one in which the ordinate is given on a logarithmic scale), then density function (with e^{-x^2}) will have the shape of a parabola (since we have $-x^2$ now). This parabola is characterized visually by the rapid fall (rapid decline) of the distribution's tails. If we try to obtain more drawn-out distribution tails, we can slow down the fall by replacing the parabola by a hyperbola. In this case, the distribution's density function will be of hyperbolic form in a semi-logarithmic chart. This is the reason why distributions constructed in this manner are called hyperbolic. Their usefulness for the modeling of price movements on the stock markets stems from the slower decrease of their tails.

In Fig. 6.5 we show the densities of hourly returns of BNP stock between September 2, 2010 and February 3, 2011, both in a standard chart (left) and in a semi-logarithmic chart (right). We also show the Gaussian and hyperbolic fittings, i.e. the Gaussian parabola and the hyperbola fitted to BNP stock prices. The figure shows clearly that the hyperbola fits the given data better.

This intuition allows us to reach in a simple way the general, complicated form of these distributions. In fact, the equation defining a hyperbola (with vertex on top in order to obtain a density) is easily obtained as:

$$y(x) = -\alpha\sqrt{1+x^2} + \beta x \tag{6.80}$$

where α and β are two constants that determine the shape of the hyperbola. As x goes to $\pm\infty$, the asymptotes are the two lines with slope $-\alpha + \beta$ (for $x \to +\infty$) and $\alpha + \beta$ (for $x \to -\infty$). It follows directly from Eq. (6.80) that the general form of the density function of a hyperbolic distribution is given by:

$$f(x) = a \exp\left(-\alpha\sqrt{1+x^2} + \beta x\right) \tag{6.81}$$

where a is a normalization constant for the integral of the density function.

By considering the slopes of the two asymptotes, we obtain a direct interpretation of the parameters: $\alpha > 0$ and $0 < |\beta| < \alpha$ determine the shape of the distribution, affecting the fatness of the tails (for α) and the asymmetry (for β).

Introducing a localization parameter denoted by μ and a dispersion (or scale) parameter denoted by γ, we obtain a more general version of the preceding equation:

$$f(x) = a' \exp\left(-\alpha\sqrt{\gamma^2 + (x-\mu)^2} + \beta(x-\mu)\right) \tag{6.82}$$

We must now define the normalization constant. Since the density must integrate to 1, we compute its integral and deduce the value of the normalization constant from it. In this way we obtain the result that the density function of a hyperbolic distribution (H) is given by:

$$f_H(x) = \frac{\sqrt{\alpha^2 - \beta^2}}{2\alpha\gamma K_1\left(\gamma\sqrt{\alpha^2 - \beta^2}\right)} \exp\left(-\alpha\sqrt{\gamma^2 + (x-\mu)^2} + \beta(x-\mu)\right)$$
(6.83)

where $K_1(.)$ is a modified Bessel function of the third kind of order 1.

We see that, like alpha-stable distributions, hyperbolic distributions are defined by four parameters: the kurtosis, the asymmetry, the dispersion, and the localization of the distribution. In other words, here too we have completed the size of the random movements (their dispersion, or scale: γ) with their form (fatness of tails and asymmetry: α and β). The localization parameter keeps the same meaning in all cases. Although different in their asymptotic behavior, all hyperbolic distributions retain a two-fold notion of risk in their description of price movements on an exchange (size–form, or amplitude–structure).

The construction of the density function of a hyperbolic distribution is therefore carried out in two steps:

(1) by defining the hyperbolic part of the exponential function in its general form with four parameters $(\alpha, \beta, \gamma, \mu)$:

$$-\alpha\sqrt{\gamma^2 + (x-\mu)^2} + \beta(x-\mu)$$

(2) by adjusting the exponential of the hyperbola by a normalization constant that depends on the parameters: $a(\alpha, \beta, \gamma)$. This constant involves modified Bessel functions denoted by $K_\lambda(.)$.

We recall that Bessel functions are in fact solutions to second-order differential equations:

$$x^2 y'' + xy' + (x^2 - \lambda^2)y = 0 \qquad (6.84)$$
$$x^2 y'' + xy' - (x^2 + \lambda^2)y = 0 \qquad (6.85)$$

The finite solution of the first equation, denoted by $J_\lambda(x)$, is called Bessel function of the first kind of order λ, and that of the second equation, denoted by $Y_\lambda(x)$, is called Bessel function of the second kind of order λ.

If at present we consider the modified differential equation, then the solutions are called modified Bessel functions of the first and second kinds, denoted by $I_\lambda(x)$ and $K_\lambda(x)$. There are as many solutions as there are values for $\lambda \in \mathbb{R}$. We can therefore generalize the density (6.83) by choosing an order λ different from 1 for the modified Bessel function to obtain a generalized version of the hyperbolic distribution. The Bessel function of the second kind of order λ is defined for all $x > 0$ by:

$$K_\lambda(x) = \int_0^\infty y^{\lambda-1} e^{-\frac{1}{2}x(y+y^{-1})} dy$$

- For $\lambda = 1$, we have the case K_1 of the simple hyperbolic distribution.
- For $\lambda = -1/2$ we obtain the density of the normal inverse Gaussian distribution:

$$f_{\text{NIG}}(x) = \frac{\alpha\gamma}{\pi} \frac{K_1(\alpha\sqrt{\gamma^2 + (x-\mu)^2})}{\sqrt{\gamma^2 + (x-\mu)^2}} \exp\left(\gamma\sqrt{\alpha^2 - \beta^2} + \beta(x-\mu)\right) \tag{6.86}$$

The dispersion parameter γ takes values between 0 and infinity. The two limit cases correspond to Laplace's two laws.

Starting from the simple hyperbolic distribution with density (6.83), it can be shown that:

- when $\gamma \to +\infty$ and $\beta \to 0$, then $\gamma/\sqrt{\alpha^2 - \beta^2} \to \sigma^2$, and the simple hyperbolic density converges towards a Gaussian one with mean μ and standard deviation σ
- when $\gamma \to 0$, the simple hyperbolic density converges towards an asymmetric Laplace density:

$$f(x) = C \begin{cases} e^{-(\alpha-\beta)(|x|-\mu)} & \text{if } x \geq \mu \\ e^{-(\alpha+\beta)(|x|-\mu)} & \text{if } x < \mu \end{cases} \tag{6.87}$$

This shows Laplace's two laws as limit laws of hyperbolic functions.

- For $\lambda \neq 1$, we have the case of the generalized hyperbolic distribution (GH), whose density function is given by:

$$f_{\text{GH}}(x) = a\left(\gamma^2 + (x-\mu)^2\right)^{(\lambda-\frac{1}{2})/2} K_{\lambda-\frac{1}{2}}\left(\alpha\sqrt{\gamma^2 + (x-\mu)^2}\right) e^{\beta(x-\mu)} \tag{6.88}$$

with normalizing constant:

$$a = \frac{(\alpha^2 - \beta^2)^{\lambda/2}}{\sqrt{2\pi}\alpha^{\lambda-\frac{1}{2}}\gamma^\lambda K_\lambda\left(\gamma\sqrt{\alpha^2 - \beta^2}\right)}$$

This distribution has the following features:

- Characteristic function:

$$\mathbb{E}\left[e^{iuX_1}\right] = e^{i\mu u} \left(\frac{\alpha^2 - \beta^2}{\alpha^2 - (\beta+iu)^2}\right)^{\frac{\lambda}{2}} \frac{K_\lambda(\gamma\sqrt{\alpha^2 - (\beta+iu)^2})}{K_\lambda(\gamma\sqrt{\alpha^2 - \beta^2})} \tag{6.89}$$

- Moment-generating function:

$$\mathbb{E}\left[e^{uX_1}\right] = e^{\mu u} \left(\frac{\alpha^2 - \beta^2}{\alpha^2 - (\beta+u)^2}\right)^{\frac{\lambda}{2}} \frac{K_\lambda(\gamma\sqrt{\alpha^2 - (\beta+u)^2})}{K_\lambda(\gamma\sqrt{\alpha^2 - \beta^2})} \quad (6.90)$$

- Expectation:

$$\mathbb{E}[X] = \mu + \frac{\beta\gamma}{\sqrt{\alpha^2 - \beta^2}} \frac{K_{\lambda+1}(\gamma\sqrt{\alpha^2 - \beta^2})}{K_\lambda(\gamma\sqrt{\alpha^2 - \beta^2})} \quad (6.91)$$

- Lévy measure when λ is positive:

$$\nu(x) = \frac{\lambda e^{-\alpha|x|+\beta x}}{|x|} + \frac{e^{\beta x}}{|x|} \int_0^{+\infty} \frac{e^{-|x|\sqrt{2t+\alpha^2}}}{\pi^2 t(J_\lambda^2(\gamma\sqrt{2t}) + Y_\lambda^2(\gamma\sqrt{2t}))} dt \quad (6.92)$$

where J_λ and Y_λ are Bessel functions.

6.3.3 Hyperbolic Distributions as Mixture Distributions

Recall that the first hyperbolic distribution, used by Barndorff-Nielsen in 1997, was obtained by mixing a normal distribution with an inverse Gaussian distribution. An understanding just as intuitive as the previous one of hyperbolic distributions can therefore be obtained via the approach of mixture laws: fat tails are obtained by a mixture of simple probability distributions. Just as the simple hyperbolic distribution results from a mixture of a normal and an inverse Gaussian distribution, generalized hyperbolic distributions can be interpreted as the result of a mixture of a normal and a generalized inverse Gaussian distribution.

As in the preceding situations, the mixture is carried out by taking a normal distribution whose variance is random and has a generalized inverse Gaussian distribution. We then make the additional assumption that the mean m depends on the variance σ^2 via the relationship $m = \mu + \beta\sigma^2$. The only remaining step is to construct the resulting distribution.

The normal distribution whose expectation m is given as a function of the variance by the relationship $m = \mu + \beta\sigma^2$ has the density:

$$f_N(x) = \frac{1}{\sigma\sqrt{2\pi}} \exp\left(-\frac{1}{2}\left(\frac{x-\mu-\beta\sigma^2}{\sigma}\right)^2\right)$$

The inverse Gaussian distribution presented in Section 5.1.3.2 (p. 92) was generalized by introducing a third parameterization that makes use of modified Bessel functions. Let us rewrite the density (5.28) of an inverse Gaussian distribution in the following way:

$$f_{IG}(x) = \sqrt{\frac{\gamma}{2\pi}} x^{-\frac{1}{2}-1} e^{-\frac{1}{2}\frac{(x-\nu)^2}{\nu^2 x}\gamma}$$

We can generalize this form by replacing the value $-1/2$ with a parameter λ and by adjusting the normalization constant. Recall that:

$$e^{-\frac{1}{2}\frac{(x-\nu)^2}{\nu^2 x}}\gamma = k\, e^{-\frac{1}{2}\left(\frac{\chi}{x}+\psi x\right)}$$

Setting $\chi = \gamma$, $\psi = \gamma/\nu^2$, and where k is a constant, we can write the previous density as:

$$f_{\mathrm{IG}}(x) = a\, x^{-\frac{1}{2}-1}\, e^{-\frac{1}{2}\left(\frac{\chi}{x}+\psi x\right)}$$

with $a = k\sqrt{\frac{\gamma}{2\pi}}$. We then obtain the generalized inverse Gaussian distribution, whose density is given by:

$$f_{\mathrm{GIG}}(x) = a'\, x^{\lambda-1}\, e^{-\frac{1}{2}\left(\frac{\chi}{x}+\psi x\right)} \qquad (6.93)$$

where $a' = \dfrac{(\psi/\chi)^{\frac{\lambda}{2}}}{2K_\lambda(\sqrt{\psi\chi})}$ is the new normalization constant.

If σ is a random variable with generalized inverse Gaussian distribution, then we have for its density:

$$f_X(x) = \int_0^{+\infty} f_{\mathrm{N}(\mu+\beta y, y)}(x)\, f_{\mathrm{GIG}}(y)\, dy$$

where $f_{\mathrm{GIG}}(y)$ is the density (6.93).

Let us note two special cases:

(1) For $\lambda = -\frac{1}{2}$, this distribution becomes the inverse Gaussian distribution (IG).
(2) For $\chi = 0$, this distribution simplifies to the gamma distribution.

The mnemonic overview given below visually summarizes the construction of hyperbolic distributions as mixture distributions. For each line – each case of mixture – the interlocking of a generalized inverse Gaussian distribution GIG (with parameter given in parenthesis) inside a normal distribution N is shown and brought next to the resulting simple hyperbolic distribution H or generalized hyperbolic distribution GH.

$$\begin{cases} \mathrm{N[IG]} & = \mathrm{NIG} \\ \mathrm{N[GIG(1)]} & = \mathrm{GH}(1) & = \mathrm{H} \\ \mathrm{N[GIG}(\lambda)] & = \mathrm{GH}(\lambda) \\ \mathrm{N[GIG}(-\tfrac{1}{2})] & = \mathrm{GH}(-\tfrac{1}{2}) = \mathrm{NIG} \\ \mathrm{N[GIG}(-\tfrac{1}{2})] & = \mathrm{N[IG]} & = \mathrm{NIG} \end{cases} \qquad (6.94)$$

Let us summarize our results and draw an interesting conclusion regarding the interpretation of share price movements. We were able to construct infinitely divisible leptokurtic distributions by mixing distributions.

If we now think in terms of stochastic processes instead of distributions and recall that variance is linear in time, then we find a new interpretation for random variance: the random variance becomes a stochastic clock and the mixture distribution, in a static context, becomes a change of clock in a dynamic context.

Let us now replace the normal distribution by Brownian motion. We assume that the clock we use has a generalized inverse Gaussian distribution. We see that it is possible to construct a Lévy process by simply changing the way we measure time. Static relationships such as:

$$N[GIG(1)]$$

become dynamic relationships:

$$\text{Brownian motion}\,[GIG(1)\text{ clock}]$$

In more compact notation:

$$X[\Theta(t)]$$

where Θ is a clock that measures time according to a GIG(1) distribution.

This new representation in terms of time changes is very rich since it gives us a mechanism with which to deform time on stock markets. We will address this topic in the next chapter.

Chapter 7

The Time Change Framework

This chapter gives a detailed presentation of the time change framework introduced previously in Chapter 2, p. 22. The nature of time changes is recalled in the introduction. It is illustrated with a first example based on the Poisson process. The notion of subordination is then introduced and defined in full generality, before we can apply it to the time change of Brownian motion, in the case of a gamma clock. The situations where the daily variations of market activity are not independent – for instance when exchanged volumes display a certain degree of persistence – are modeled with a square root of time process.

The time change representation is useful in that is allows us to interpret any Lévy process in calendar time (a time devoid of social events) as a Brownian motion in the social time of the market.

7.1 Time Changes

The use of a time change within the modeling of market fluctuations dates back to Mandelbrot. As exposed in *Fractals and Scaling in Finance* (see Mandelbrot (1997), p. 39):

> The key step is to introduce an auxiliary quantity called trading time. The term is self-explanatory and embodies two observations. While price changes over fixed clock time intervals are long-tailed, price changes between successive transactions stay near-Gaussian over sometimes long periods between discontinuities. Following variations in the trading volume, the time intervals between successive transactions vary greatly. This suggests that trading time is related to volume.

The idea of a time that is intrinsic to the market can be related to the developments of economic historians and sociologists on the memory of social time, adopting adequate formal representations. We now set out this approach in detail.

7.1.1 Historical Survey

We begin by giving a very brief overview of the literature on time changes in finance. Two periods can be roughly distinguished within finance theory: that of the introduction of the idea of time change (in the 1970s) and that of its systematic use (in the 1990s).

The first finance works introducing the concept of a modification of the clock are those of Mandelbrot and Taylor (1967) and of Clark (1973). Mandelbrot and Taylor had used the process of the cumulative number of transactions (denoted hereafter by N) as a stochastic clock. They formulated the hypothesis that this process was an increasing alpha-stable motion of exponent $\alpha < 1$: the lags $\Delta N(t)$ between two consecutive quotes followed a Pareto law of exponent $\alpha/2$. In the time of transactions, the distribution of return variations was Gaussian. The conclusion of their article indicated that there is essentially no difference between a Paretian marginal distribution in physical time and a Gaussian marginal distribution in the social time of exchanges, as measured by transactions. They pointed out the possibility of considering these two analytical frameworks as identical. An alpha-stable motion in physical time was nothing more than a Brownian motion measured in a Paretian social time.

In his 1973 paper, Clark considered that, rather than measuring time through transactions, the speed of evolution of time should be rendered by exchanged volumes – more precisely by the total cumulative volume between dates 0 and t, yielding the process defined at any time t by $X(t) = \xi(V(t))$.

This strand of the literature does not seem to have been pursued. However, new research on time changes and their link to market activity were conducted in the 1990s by making use of the newly-available high-frequency data (Müller et al. (1990), Müller et al. (1997), Gourieroux, Jasiak, and Le Fol (1999), Engle and Russel (1994)). The article of Geman (2008) presents a historical survey of recent works on this topic.

The idea of using a Brownian motion in social time to explain non-Gaussian distributions has been explored systematically since then. The modification of physical time brings us back to Gaussian distributions (see Ané and Geman (2000)).

7.1.2 A First Modeling Example

We present a first example of modeling in social time based on the standard Poisson process defined in Chapter 4, p. 56. In this setting, the probability of observing long lags between two quotes decreases exponentially, or, expressed differently, very long lags between two quotes are highly unlikely.

Mandelbrot and Taylor had chosen a Pareto distribution for the social law. This implied on the contrary that very long lags between two quotes were possible – at least less unlikely than with an exponential law. Indeed, the slow Paretian rate of decay allows us to obtain long time intervals between transactions: long lags where nothing happens on the market. However, for the sake of simplicity, we limit ourselves here to the pedagogical example of the Poisson law.

7.1.2.1 The Social Time of Exchanges

We first construct a calendar, defined by the sequence of quote times:

$$0 = t_0 < t_1 < t_2 < \cdots < t_n = T$$

Then, we formulate a hypothesis concerning the social random clock. We assume in this example that the number of quotes having appeared until time t, denoted by $N(t)$, follows the marginal distribution attached to a standard Poisson process.

Let us show how this choice of stochastic clock allows us to interpret the evolution of Bouygues stock, as displayed in Table 2.3, p. 24. In this sequence of quotes, the average time between two quotes is ten seconds. A time unit must be chosen: let it be one hour, for instance. In this case,

Table 7.1 Empirical data: probabilities attached to the stochastic clock of Bouygues

Exponential distribution hypothesis on Bouygues	
Probability:	that the time between two quotes exceeds:
90.48%	1 second
60.65%	5 seconds
36.79%	10 seconds
13.53%	20 seconds
4.98%	30 seconds
0.25%	1 minute
0.0006%	2 minutes
0.000002%	3 minutes

the average lag is $\mathbb{E}[\Delta t_k] = 10/3,600 = 1/360$, yielding the value $\lambda = 360$, which can be interpreted in the following way: approximately 360 quotes can be observed per hour. Equipped with this result, we can immediately deduce the average number of quotes observed over any lag $t - s$. For example, when $t - s = 3$ hours, then $\lambda(t - s) = 360 \times 3 = 1,080$. On average, around $1,080$ quotes will be observed over three hours. If the time unit is now equal to one second, then $\mathbb{E}[\Delta t_k] = 10/1 = 10$, and therefore $\lambda = 1/10$. In this situation, a quote would appear every ten units of time, and so on.

Let us choose the second as time unit. In this case, assuming occurrences of quotes are driven by a Poisson process, and therefore by an exponential social clock, the probability of exceeding a duration τ between two quotes equals:

$$\Pr(\Delta t_k > \tau) = 1 - F(\tau) = e^{-0.10\,\tau} \qquad (7.1)$$

We display in Table 7.1 an application of formula (7.1), which represents the probability that the time between two quotes exceeds a given value. We see that the exponential distribution produces a very rapid decrease in the probabilities attached to the time lags between exchanges, a duration superior to three minutes being extremely rare. However, market regimes where there are no exchanges for three minutes are empirically possible. This illustrates the limits of modeling a stochastic clock by means of an exponential distribution. Consequently, this leads us to choose other probability distributions for clocks, as explored later.

We use these results in order to simulate the dynamics of a proxy of Bouygues stock in the time of exchanges driven by a social Poisson clock.

We compute the first fifteen lags ($k = 1, \cdots, 15$) choosing the following time unit: five seconds. With this unit, and knowing that the average time between two quotes is ten seconds, a quote appears on average every two time units, yielding the parameter of the Poisson distribution $\lambda = 1/2 = 0.5$. The first fifteen computed values of the time lag $\Delta t_k = t_k - t_{k-1}$ are given in the left part of Fig. 7.1, and are expressed in time units. The right side of the figure represents the number $N(t_k)$ of quotes that occur as a function of time. The interpretation is the same for Fig. 2.1 of Chapter 2. At each quote instant t_k, for $k = 1, \cdots, 15$, this number is increased by 1. The trajectory of the process N is thus stepped. Between two quotes, there are no exchanges and therefore the graph of the process N remains flat.

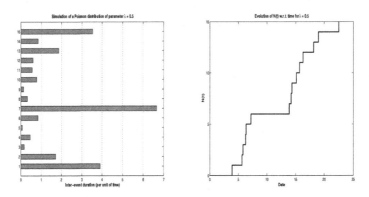

Figure 7.1 Simulation of the first fifteen quote instants by a Poisson process of parameter $\lambda = 0.5$ (time unit: five seconds)

This example illustrates the effect of the choice of probability distribution on the social clock, and therefore the importance of this distribution for the fine tuning of the clock defining the time of the market.

7.1.2.2 Market Capitalization in Social Time

Let us now study the size of quote jumps. At each quote time t_k, a jump of a certain size occurs; it is necessary to complete the previous description by modeling how this amplitude evolves in time.

Let us start from a very simple example: classic capitalization. This is a situation where the amplitude of jumps is a constant which depends on the non-random interest rate. Jumps occur at prefixed non-random dates

t_k. In this case, successive stock prices increase according to the formula:

$$S_n = S_0(1+i)^n \tag{7.2}$$

This is a very convenient starting point, but the situation described is not realistic. To generalize this model, we can assume an interest rate that is still non-random but that is not constant. In this case, we go from the value S_{k-1} to the value S_k using the elementary formula $S_k = S_{k-1}(1+i_k)$ where i_k is the (non-random) rate of increase over the period $[k-1, k]$. The trajectory of this process is then:

$$S_n = S_0 \underbrace{(1+i_1)(1+i_2)\cdots(1+i_n)}_{n \text{ jumps}} = S_0 \prod_{k=1}^{n}(1+i_k) \tag{7.3}$$

A simple natural generalization of the above model consists in assuming that interest rates are random. In this case, at each jump time t_k, the relative jump height denoted by φ_k is a random variable. At each new quote, the stock price goes from $S(t_{k-1})$ to $S(t_k)$ in the proportion given by the relative jump height φ_k such that $S(t_k)/S(t_{k-1}) = 1 + \varphi_k$. This coefficient corresponds to the periodic return defined in the natural computation space.

We then replace $(1+i_1)(1+i_2)\cdots(1+i_n)$ with a sequence of random variables $(1+\varphi_1)(1+\varphi_2)\cdots(1+\varphi_n)$. If the φ_k take only two values (up or down), then, as in a binomial model, we obtain a binomial tree. More generally, this modeling is that of a random walk (in discrete time) characterized by the following evolution of the stock price:

$$S_n = S_0 \underbrace{(1+\varphi_1)(1+\varphi_2)\cdots(1+\varphi_n)}_{n \text{ jumps}} = S_0 \prod_{k=1}^{n}(1+\varphi_k) \tag{7.4}$$

Until now, jump dates where non-random and prefixed. We now turn to a continuous time setting where jump dates occur at random times. As above, we denote by $N(t)$ the number of quote jumps that have occurred until time t. The stock price $S(t)$ results from these $N(t)$ jumps and is given by the formula:

$$S_t = S_0 \underbrace{(1+\varphi_{t_1})(1+\varphi_{t_2})\cdots(1+\varphi_{t_{N_t}})}_{N(t) \text{ jumps}} = S_0 \prod_{k=1}^{N(t)}(1+\varphi_{t_k}) \tag{7.5}$$

We now describe the evolution of the cumulative return associated with this representation. To do so, we introduce the return $X(t)$ by taking the

Table 7.2 Comparison of the two clocks: physical time and social time

	Random walk model	Compound Poisson model
	Values at fixed times	Values at random times
Discrete time	Continuous time	Continuous time
$\{S_k, k = 0, 1, \cdots, n\}$	$(S(t), t \geq 0)$	$(S(N(t)), t \geq 0)$
	Physical clock: t	Random clock: $N(t)$
i.i.d. random variables φ_k,	$Y_k = \ln(1 + \varphi_k)$ has the density f_Y	

logarithm of the formula (7.5):

$$\ln \frac{S_t}{S_0} = \ln \prod_{k=1}^{N(t)} (1 + \varphi_{t_k}) = \sum_{k=1}^{N(t)} \ln(1 + \varphi_{t_k})$$

The coefficient φ_{t_k} of the quote jumps is defined, as an interest rate, by a percentage. By analogy with interest rates, the logarithm of $1 + \varphi_{t_k}$ is the continuous equivalent of this coefficient. This corresponds to the periodic continuous return rate examined before. Let us write $Y_k = \ln(1 + \varphi_{t_k})$. Because, by definition $X(t) = \ln S(t) - \ln S(0)$, we have:

$$X(t) = \sum_{k=1}^{N(t)} Y_k \qquad (7.6)$$

If the random times of quote jumps are described by a standard Poisson process of parameter λ, and if quote jumps are independent random variables, the formula (7.6) defines a compound Poisson process. In this simple case, the distribution of amplitudes is stationary, and denoted by f_Y. As far as the Bouygues stock is concerned (see Table 2.3, p. 24) its parameters are given by $E[Y] = -0.038\%$ and $\sigma(Y) = 0.072\%$. The market clock is that defined in the social time of exchanges. Table 7.2 summarizes the similarities and differences between the two models in physical and social time.

7.1.2.3 Subordination of a Random Process

We now introduce a new expression of the random walk in order to consider explicitly the change of time referential. Let us start from the compound Poisson process:

$$X(t) = \sum_{k=1}^{N(t)} Y_k$$

This formula shows that the cumulative return is merely the sum of the price variations (quote jumps) taken at each quote instant (social time).

Suppose now that we look at the evolution of the jumps Y_k in physical time by calibrating the clock on a constant number of time units, and that we compute the cumulative return that results from this change of clock. Because the time lags between two jumps are not the same, the cumulative return corresponding to this new clock will be different. In a discrete-time model, if we denote this return at time n by ξ_n, it is defined as:

$$\xi_n = \sum_{k=1}^{n} Y_k \tag{7.7}$$

Let us come back to continuous time. Denoting by ξ the process expressed in physical calendar time, it is possible to express the random variable $X(t)$ as a function of ξ in the following way:

$$X(t) = \xi(N(t)) \tag{7.8}$$

This last formula shows that the value at time t of the stochastic process driving the cumulative return X is equal to the value at this date of the stochastic process ξ measured in the social time of the market.

The operation of measuring the evolution of any stochastic process $\xi(t)$ in the time of social events, but not in that of natural events, is called a random clock or time change. The random clock (here $N(t)$) is a non-decreasing stochastic process called the driving process of the process ξ. In a way, $\xi(.)$ is driven by $N(t)$, yielding the terminology:

> observed process at *physical time* = driven process at *social time*

When the time change is an increasing process with stationary and independent increments, i.e. an increasing Lévy process, it is called the subordinator of the process ξ. Any subordinator is a driving process, but the converse is not true.

Denote by $\Theta(t)$ a given subordinator. The process $X(t) = \xi(\Theta(t))$ is called the subordinated process, whereas this time change operation is called subordination. If ξ is any Lévy process independent from Θ, the subordinated process X remains a Lévy process. Conversely, it is possible to interpret any Lévy process $X(t)$ as a process $\xi(t)$ measured in units of social time $\Theta(t)$. The ultimate question is that of the choice of social time.

7.2 Subordinated Brownian Motions

We now consider a particular type of subordination, that of a Brownian motion whose clock is described by a subordinator. This allows us to present the "variance gamma" process, which will be used later on for the computation of VaR and other risk measures. This process was introduced by Madan and Seneta (1990) and Madan and Milne (1991) to generalize the Black–Scholes formula for option pricing. It was then generalized by Madan, Carr, and Chang (1998). This process was introduced, among other reasons, to explain a feature of the market: the volatility smile.

We first present the mechanics of subordination, followed by the construction of the time change. We apply these ideas by considering a gamma clock measuring the time of a Brownian motion.

7.2.1 *The Mechanics of Subordination*

We start by examining the idea of subordination, i.e. the way a time change can be achieved by using increasing Lévy processes.

7.2.1.1 *Subordinators*

A subordinator is a positive (strictly) increasing Lévy process, with finite variation and positive jumps and with no Gaussian component. The time change of a Lévy process by a subordinator is very convenient for constructing new Lévy processes. Conversely, it is not possible to apply a subordinator to a Lévy process in order to construct processes whose increments are not independent or stationary. See Section 7.3, p. 173.

Denote by Ψ_X the Fourier exponent and by ϕ_X the Laplace exponent of a Lévy process X. They are given:

$$\mathbb{E}\left[e^{iuX_t}\right] = \exp\left(t\,\Psi_X(u)\right)$$

and:

$$\mathbb{E}\left[e^{uX_t}\right] = \exp\left(t\,\phi_X(u)\right)$$

Let the process Z be defined by $Z(t) = X_2(X_1(t))$ where X_2 is a Lévy process and X_1 is a subordinator. We can write the characteristic function of Z at any time t as:

$$\mathbb{E}\left[e^{iuZ_t}\right] = \exp\left(t\,\phi_{X_1}(\Psi_{X_2}(u))\right)$$

By the tower property of conditional expectations, we obtain:

$$\mathbb{E}\left[e^{iuZ_t}\right] = \mathbb{E}\left[e^{iuX_2(X_1(t))}\right] = \mathbb{E}\left[\mathbb{E}\left[e^{iuX_2(X_1(t))}|X_1(t)\right]\right]$$

and then, by identification:

$$\mathbb{E}\left[e^{iuZ_t}\right] = \mathbb{E}\left[e^{X_1(t)\cdot\Psi_{X_2}(u)}\right] = e^{t\phi_{X_1}(\Psi_{X_2}(u))} \qquad (7.9)$$

Brownian motion in gamma time (also called Variance Gamma or VG process) can be obtained in this way. Indeed, Brownian motion is a particular example of a Lévy process, and the gamma process, representing the measure of time, is a particular case of a strictly increasing positive Lévy process.

7.2.1.2 Generalization

The preceding property can be generalized in the following way. Let Z be a process defined by $Z(t) = X_n \circ X_{n-1} \circ \cdots \circ X_2 \circ X_1(t)$ where X_n is a Lévy process and $X_{n-1}, ..., X_1$ are subordinators. The characteristic function of Z can be written at any time t as:

$$\mathbb{E}\left[e^{iuZ_t}\right] = e^{t\phi_{X_1}\circ\phi_{X_2}\circ\cdots\circ\phi_{X_{n-1}}\circ\Psi_{X_n}(u)}$$

Indeed:

$$\mathbb{E}\left[e^{iuZ_t}\right] = \mathbb{E}\left[e^{iuX_n\circ X_{n-1}\circ\cdots\circ X_2\circ X_1(t)}\right]$$

can be expressed, by means of the tower property of conditional expectations, as:

$$\mathbb{E}\left[e^{iuZ_t}\right] = \mathbb{E}\left[\mathbb{E}\left[e^{iuX_n\circ X_{n-1}\circ\cdots\circ X_2\circ X_1(t)}|X_1(t),...,X_{n-1}(t)\right]\right]$$

giving via identification:

$$\mathbb{E}\left[e^{iuZ_t}\right] = \mathbb{E}\left[e^{\psi_{X_n}(u)\,\cdot\,X_{n-1}\circ\cdots\circ X_2\circ X_1(t)}\right]$$

Then, again by the tower property of conditional expectations:

$$\mathbb{E}\left[e^{iuZ_t}\right] = \mathbb{E}\left[\mathbb{E}\left[e^{\psi_{X_n}(u)\,\cdot\,X_{n-1}\circ\cdots\circ X_2\circ X_1(t)}|X_1(t),\cdots,X_{n-2}(t)\right]\right]$$

so that:

$$\mathbb{E}\left[e^{iuZ_t}\right] = \mathbb{E}\left[e^{\phi_{X_{n-1}}\circ\psi_{X_n}(u)\,\cdot\,X_{n-2}\circ\cdots\circ X_2\circ X_1(t)}\right]$$

and then by induction:

$$\mathbb{E}\left[e^{iuZ_t}\right] = \cdots = \mathbb{E}\left[e^{\phi_{X_2}\circ\cdots\circ\phi_{X_{n-1}}\circ\Psi_{X_n}(u)\,\cdot\,X_1(t)}\right]$$

finally yielding:

$$\mathbb{E}\left[e^{iuZ_t}\right] = e^{t\phi_{X_1}\circ\phi_{X_2}\circ\cdots\circ\phi_{X_{n-1}}\circ\Psi_{X_n}(u)}$$

The spectrum of the processes constructed by subordination can thus be obtained very easily.

We will show that the processes constructed along these lines are Lévy processes. Therefore, we will argue that any process Z defined at any time t by $Z(t) = X_2(X_1(t))$, where X_2 is a Lévy process and X_1 is an independent subordinator, is a Lévy process.

To do so, we will verify the independence and the stationarity of the increments of Z. Consider the following times: $t_1 \leq t_2 < t_3 \leq t_4$ such that the time steps $t_4 - t_3$ and $t_2 - t_1$ are non-overlapping. We first check that $Z_{t_2} - Z_{t_1}$ and $Z_{t_4} - Z_{t_3}$ are independent. We start by computing:

$$\mathbb{E}\left[e^{i\alpha[Z_{t_2} - Z_{t_1}] + i\beta[Z_{t_4} - Z_{t_3}]}\right]$$

$$= \mathbb{E}\left[e^{i\alpha[X_2(X_1(t_2)) - X_2(X_1(t_1))] + i\beta[X_2(X_1(t_4)) - X_2(X_1(t_3))]}\right] \quad (7.10)$$

By the usual tower property of conditional expectations, we can rewrite the right-hand side of formula (7.10) as:

$$\mathbb{E}\left[\mathbb{E}\left[e^{i\alpha[X_2(X_1(t_2)) - X_2(X_1(t_1))] + i\beta[X_2(X_1(t_4)) - X_2(X_1(t_3))]} \mid X_1\right]\right]$$

Note first that $X_1(t_1) \leq X_1(t_2) < X_1(t_3) \leq X_1(t_4)$ due to the increasing nature of the process X_1. We deduce that when $X_1(t_1), X_1(t_2), X_1(t_3), X_1(t_4)$ are set constant, then $X_2(X_1(t_2)) - X_2(X_1(t_1))$ and $X_2(X_1(t_4)) - X_2(X_1(t_3))$ are independent, by means of the independence of the increments of the Lévy process X_2. The right-hand side of formula (7.10) can thus be decomposed as follows:

$$\mathbb{E}\left[\mathbb{E}\left[e^{i\alpha[X_2(X_1(t_2)) - X_2(X_1(t_1))]} \mid X_1\right] \mathbb{E}\left[e^{i\beta[X_2(X_1(t_4)) - X_2(X_1(t_3))]} \mid X_1\right]\right]$$

The stationarity of the increments of X_2 allows us to rewrite this expression in the following form:

$$\mathbb{E}\left[\mathbb{E}\left[e^{i\alpha[X_2(X_1(t_2) - X_1(t_1))]} \mid X_1\right] \mathbb{E}\left[e^{i\beta[X_2(X_1(t_4) - X_1(t_3))]} \mid X_1\right]\right]$$

Observe that $\mathbb{E}\left[e^{i\alpha[X_2(X_1(t_2) - X_1(t_1))]} \mid X_1\right]$ is a function of $X_1(t_2) - X_1(t_1)$, whereas $\mathbb{E}\left[e^{i\beta[X_2(X_1(t_4) - X_1(t_3))]} \mid X_1\right]$ is a function of $X_1(t_4) - X_1(t_3)$. By the independence property of the increments of X_1, we can write the former quantity as follows:

$$\mathbb{E}\left[\mathbb{E}\left[e^{i\alpha[X_2(X_1(t_2) - X_1(t_1))]} \mid X_1\right]\right] \mathbb{E}\left[\mathbb{E}\left[e^{i\beta[X_2(X_1(t_4) - X_1(t_3))]} \mid X_1\right]\right]$$

and, using the reverse of the tower property, we obtain:

$$\mathbb{E}\left[e^{i\alpha[X_2(X_1(t_2)) - X_2(X_1(t_1))]}\right] \mathbb{E}\left[e^{i\beta[X_2(X_1(t_4)) - X_2(X_1(t_3))]}\right]$$

so that we get:

$$\mathbb{E}\left[e^{i\alpha(Z_{t_2}-Z_{t_1})+i\beta(Z_{t_4}-Z_{t_3})}\right] = \mathbb{E}\left[e^{i\alpha(Z_{t_2}-Z_{t_1})}\right] \mathbb{E}\left[e^{i\beta(Z_{t_4}-Z_{t_3})}\right]$$

yielding the independence of the increments of Z. The stationarity of the increments of Z will be checked by considering:

$$\mathbb{E}\left[e^{iu(Z_{t+\varepsilon}-Z_{s+\varepsilon})}\right] = \mathbb{E}\left[e^{iu[X_2(X_1(t+\varepsilon))-X_2(X_1(s+\varepsilon))]}\right] \quad (7.11)$$

Note first that:

$$X_1(t+\varepsilon) = X_1(t+\varepsilon) - X_1(t) + X_1(t) \stackrel{d}{=} \hat{X}_1(\varepsilon) + X_1(t)$$

by the stationarity of the increments of X_1 and where $\hat{X}_1(\varepsilon)$ is a copy of $X_1(\varepsilon)$ independent from $X_1(t)$. Similarly, we have:

$$X_1(s+\varepsilon) \stackrel{d}{=} \hat{X}_1(\varepsilon) + X_1(s)$$

We can rewrite formula (7.11) as follows:

$$\mathbb{E}\left[e^{iu(Z_{t+\varepsilon}-Z_{s+\varepsilon})}\right] = \mathbb{E}\left[e^{iu\left(X_2[\hat{X}_1(\varepsilon)+X_1(t)]-X_2[\hat{X}_1(\varepsilon)+X_1(s)]\right)}\right]$$

So, by the usual tower property of conditional expectations, we get:

$$\mathbb{E}\left[e^{iu(Z_{t+\varepsilon}-Z_{s+\varepsilon})}\right] = \mathbb{E}\left[\mathbb{E}\left[e^{iu\left(X_2[\hat{X}_1(\varepsilon)+X_1(t)]-X_2[\hat{X}_1(\varepsilon)+X_1(s)]\right)}\big|X_1,\hat{X}_1\right]\right]$$

Then, by stationarity of the increments of X_2, we obtain:

$$\mathbb{E}\left[e^{iu(Z_{t+\varepsilon}-Z_{s+\varepsilon})}\right] = \mathbb{E}\left[\mathbb{E}\left[e^{iu(X_2[X_1(t)]-X_2[X_1(s)])}\big|X_1\right]\right]$$

so that we get:

$$\mathbb{E}\left[e^{iu(Z_{t+\varepsilon}-Z_{s+\varepsilon})}\right] = \mathbb{E}\left[e^{iu(X_2[X_1(t)]-X_2[X_1(s)])}\right]$$

We have thus proved:

$$\mathbb{E}\left[e^{iu(Z_{t+\varepsilon}-Z_{s+\varepsilon})}\right] = \mathbb{E}\left[e^{iu(Z_t-Z_s)}\right]$$

which is nothing more than the stationarity of the increments of Z.

7.2.2 Construction of a Time Change

We first give the expression of the characteristic function of a subordinated arithmetic Brownian motion.

7.2.2.1 Subordinator and Characteristic Function

Start from a classic arithmetic Brownian motion B defined at any time t by:

$$B(t) = \mu t + \sigma W(t)$$

where W is a standard Wiener process. Choose a subordinator θ, independent from W, for the stochastic clock of the arithmetic Brownian motion. We obtain a Brownian motion measured in this new referential and defined at any time t by:

$$X(t) = B(\theta(t)) = \mu\,\theta(t) + \sigma W(\theta(t)) \qquad (7.12)$$

This is a time-changed Brownian motion, called in this situation subordination Brownian motion. This type of process X will be used to model the process of cumulative returns.

To obtain the characteristic function of the above subordinated Brownian motion, we start from the expression:

$$\Phi_{X_t}(u) = \mathbb{E}\left[e^{iuX_t}\right] = \mathbb{E}\left[e^{iuB_{\theta_t}}\right]$$

We then use the tower property of the conditional expectation at time $\theta(t)$, knowing that, at that time and due to (7.12), $B(\theta(t))$ follows a Gaussian distribution of expectation $\mu\,\theta(t)$ and standard deviation $\sigma\sqrt{\theta(t)}$. This yields:

$$\Phi_{X_t}(u) = \mathbb{E}\left[\mathbb{E}\left[e^{iuB_{\theta_t}}|\theta_t\right]\right] = \mathbb{E}\left[e^{i\mu u\theta_t - \frac{\sigma^2}{2}u^2\theta_t}\right]$$

and thus:

$$\Phi_{X_t}(u) = \mathbb{E}\left[e^{(i\mu u - \frac{\sigma^2}{2}u^2)\theta_t}\right] \qquad (7.13)$$

The choice of θ determines the shape of the characteristic function.

Infinitely many possibilities of time changes exist: it is sufficient to define a stochastic clock, in order to move from calendar time (the time of natural events) to financial social time, and to apply it to any Lévy process.

As explained below, one of the consequences of a time change is the emergence of non-Gaussianity in physical time.

7.2.2.2 Non-Gaussianity in Calendar Time

What happens when the distributions of $X(t)$, for any t, are measured in calendar time? These marginal distributions are non-Gaussian. Let us explain why, by computing the modification of moments that results from a time change.

We present the computation of the moments of X_t in detail. Note first that, when θ_t is fixed and thus deterministic, W_{θ_t} follows a centered Gaussian law whose variance is equal to θ_t. From this, we deduce:

$$\mathbb{E}[W_{\theta_t}] = \mathbb{E}[\,\mathbb{E}[W_{\theta_t}|\theta_t]\,] = 0 \qquad (7.14)$$

by the definition of standard Brownian motion: the mean of the latter process in a non-random time is null. We can also compute the following covariance:

$$\mathrm{cov}(\theta_t, W_{\theta_t}) = \mathbb{E}[\theta_t\, W_{\theta_t}] - \mathbb{E}[\theta_t]\,\mathbb{E}[W_{\theta_t}] = \mathbb{E}[\theta_t\, W_{\theta_t}]$$

by using Eq. (7.14). Then, we obtain:

$$\mathbb{E}[\theta_t\, W_{\theta_t}] = \mathbb{E}[\,\mathbb{E}[\theta_t\, W_{\theta_t}|\theta_t]\,] = \mathbb{E}[\,\theta_t \times \mathbb{E}[W_{\theta_t}|\theta_t]\,] = 0$$

for the same reason as above. Consequently:

$$\mathrm{cov}(\theta_t, W_{\theta_t}) = \mathbb{E}[\theta_t\, W_{\theta_t}] = 0 \qquad (7.15)$$

and by extension, for any positive integer n:

$$\mathbb{E}[\theta_t^n\, W_{\theta_t}] = 0 \qquad (7.16)$$

Mean We can now compute the first moment of X_t:

$$\mathbb{E}[X_t] = \mathbb{E}[\mu\theta_t + \sigma W_{\theta_t}] = \mu\mathbb{E}[\theta_t] + \sigma\mathbb{E}[W_{\theta_t}]$$

so that, using Eq. (7.14):

$$\mathbb{E}[X_t] = \mu\,\mathbb{E}[\theta_t] \qquad (7.17)$$

Variance Let us consider the computation of the non-centered second moment of X_t:

$$\mathbb{E}[X_t^2] = \mathbb{E}\left[(\mu\theta_t + \sigma W_{\theta_t})^2\right] = \mathbb{E}\left[\mu^2\theta_t^2 + \sigma^2 W_{\theta_t}^2 + 2\mu\sigma\theta_t W_{\theta_t}\right]$$

which can be developed as:

$$\mathbb{E}[X_t^2] = \mu^2\,\mathbb{E}[\theta_t^2] + \sigma^2\,\mathbb{E}[\,\mathbb{E}[W_{\theta_t}^2|\theta_t]\,] + 2\mu\sigma\,\mathbb{E}[\theta_t W_{\theta_t}]$$

so that:

$$\mathbb{E}[X_t^2] = \mu^2\,\mathbb{E}[\theta_t^2] + \sigma^2\,\mathbb{E}[\theta_t] + 2\mu\sigma.0$$

using the fact that the non-centered moment of order 2 of W_{θ_t} is, conditionally on θ_t, equal to its centered moment, so to θ_t, and using Eq. (7.15). Finally, we can write:

$$\mathbb{E}[X_t^2] = \mu^2\,\mathbb{E}[\theta_t^2] + \sigma^2\,\mathbb{E}[\theta_t] \qquad (7.18)$$

Let us compute the variance of X_t, where we use Eqs (7.17) and (7.18):

$$\text{var}(X_t) = \mathbb{E}\left[X_t^2\right] - \mathbb{E}\left[X_t\right]^2 = \mu^2 \, \mathbb{E}\left[\theta_t^2\right] + \sigma^2 \, \mathbb{E}\left[\theta_t\right] - \mu^2 \, \mathbb{E}\left[\theta_t\right]^2$$

so that we obtain:

$$\text{var}(X_t) = \mu^2 \, \text{var}(\theta_t) + \sigma^2 \, \mathbb{E}\left[\theta_t\right] \tag{7.19}$$

In the particular and simple case where $\mu = 0$, the increase of variance that results from subordination is exactly equal to the expectation of the subordinator:

$$\text{var}^{\mu=0}(X_t) = \sigma^2 \, \mathbb{E}[\theta_t]$$

Note that if the subordinator has an infinite expectation, then the marginal variance of the resulting process is infinite. For example, if we use a Pareto clock of exponent $\alpha < 1$ (so that $\mathbb{E}[\theta_t] = \infty$), then Brownian motion measured in this Paretian social time becomes an alpha-stable motion with $\alpha < 2$ (and therefore $\text{var}(X_t) = \infty$).

Asymmetry Let us now come to the computation of the non-centered third moment of a subordinated Brownian motion:

$$\mathbb{E}\left[X_t^3\right] = \mathbb{E}\left[(\mu\theta_t + \sigma W_{\theta_t})^3\right]$$

so that, developing:

$$\mathbb{E}\left[X_t^3\right] = \mathbb{E}\left[\mu^3\theta_t^3 + 3\mu^2\theta_t^2\sigma W_{\theta_t} + 3\mu\theta_t\sigma^2 W_{\theta_t}^2 + \sigma^3 W_{\theta_t}^3\right]$$

If we use Eq. (7.16) with $n = 2$ for the second term and if we note that:

$$\mathbb{E}\left[\theta_t W_{\theta_t}^2\right] = \mathbb{E}\left[\mathbb{E}\left[\theta_t W_{\theta_t}^2 \mid \theta_t\right]\right] = \mathbb{E}\left[\theta_t \mathbb{E}\left[W_{\theta_t}^2 \mid \theta_t\right]\right] = \mathbb{E}\left[\theta_t . \theta_t\right]$$

and that:

$$\mathbb{E}\left[W_{\theta_t}^3\right] = \mathbb{E}\left[\mathbb{E}\left[W_{\theta_t}^3 \mid \theta_t\right]\right] = \mathbb{E}[0] = 0$$

Then, if we use the symmetry of $W_{\theta_t}^3$ conditionally on θ_t, we obtain:

$$\mathbb{E}\left[X_t^3\right] = \mu^3 \mathbb{E}\left[\theta_t^3\right] + 3\mu^2\sigma.0 + 3\mu\sigma^2 \mathbb{E}\left[\theta_t.\theta_t\right] + \sigma^3.0$$

So, finally, we get:

$$\mathbb{E}\left[X_t^3\right] = \mu^3 \, \mathbb{E}\left[\theta_t^3\right] + 3\,\mu\,\sigma^2 \, \mathbb{E}\left[\theta_t^2\right] \tag{7.20}$$

We can now compute the centered moment of order 3 and the skewness, using formulas (7.17), (7.18), and (7.20). Indeed:

$$m_3\left(X_t\right) = \mathbb{E}\left[[X_t - \mathbb{E}\left[X_t\right]]^3\right] = \mathbb{E}\left[[X_t - \mu\mathbb{E}\left[\theta_t\right]]^3\right]$$

can be developed into:

$$m_3(X_t) = \mathbb{E}[X_t^3] + \mathbb{E}[3X_t^2(-\mu\mathbb{E}[\theta_t])] + \mathbb{E}\left(3X_t(-\mu\mathbb{E}[\theta_t])^2\right) - \mu^3\mathbb{E}[\theta_t]^3$$

and, substituting the non-centered moments of orders 1, 2, and 3 of X_t:

$$m_3(X_t) = \mu^3\mathbb{E}[\theta_t^3] + 3\mu\sigma^2\mathbb{E}[\theta_t^2] - 3\mu^3\mathbb{E}[\theta_t^2]\mathbb{E}[\theta_t] - 3\mu\sigma^2\mathbb{E}[\theta_t]^2$$
$$+ 3\mu^3\mathbb{E}[\theta_t]^3 - \mu^3\mathbb{E}[\theta_t]^3$$

Thus, grouping terms, we get:

$$m_3(X_t) = \mu^3\mathbb{E}[\theta_t^3] - 3\mu^3\mathbb{E}[\theta_t^2]\mathbb{E}[\theta_t] + 3\mu\sigma^2\text{Var}(\theta_t) + 2\mu^3\mathbb{E}[\theta_t]^3 \quad (7.21)$$

We can also derive an expression for the skewness:

$$S(X_t) = \frac{\mu^3\mathbb{E}[\theta_t^3] - 3\mu^3\mathbb{E}[\theta_t^2]\mathbb{E}[\theta_t] + 3\mu\sigma^2\text{Var}(\theta_t) + 2\mu^3\mathbb{E}[\theta_t]^3}{(\mu^2\text{Var}(\theta_t) + \sigma^2\mathbb{E}[\theta_t])^{\frac{3}{2}}} \quad (7.22)$$

Note that, for $\mu = 0$, we obtain the following noteworthy identities:

$$S^{\mu=0}(X_t) = m_3^{\mu=0}(X_t) = \mathbb{E}^{\mu=0}[X_t^3] = 0$$

Leptokurticity The computation of the fourth non-centered moment is based on the observation that the kurtosis of a Gaussian distribution (typically W_{θ_t} conditionally on θ_t) is equal to three. We thus use the following relationship:

$$\frac{m_4(W_{\theta_t}|\theta_t)}{m_2(W_{\theta_t}|\theta_t)^2} = \frac{\mathbb{E}\left[[W_{\theta_t} - \mathbb{E}[W_{\theta_t}]]^4 |\theta_t\right]}{\mathbb{E}\left[[W_{\theta_t} - \mathbb{E}[W_{\theta_t}]]^2 |\theta_t\right]^2} = 3$$

or, because W_{θ_t} is centered conditionally on θ_t:

$$\mathbb{E}[W_{\theta_t}^4|\theta_t] = 3\,\mathbb{E}[W_{\theta_t}^2|\theta_t]^2 \quad (7.23)$$

Let us now state:

$$\mathbb{E}[X_t^4] = \mathbb{E}\left[(\mu\theta_t + \sigma W_{\theta_t})^4\right]$$

which can be expressed as:

$$\mathbb{E}[X_t^4] = \mathbb{E}\left[\mu^4\theta_t^4 + 4\mu^3\theta_t^3\sigma W_{\theta_t} + 6\mu^2\theta_t^2\sigma^2 W_{\theta_t}^2 + 4\mu\theta_t\sigma^3 W_{\theta_t}^3 + \sigma^4 W_{\theta_t}^4\right]$$

and can be developed as:

$$\mathbb{E}[X_t^4] = \mu^4\mathbb{E}[\theta_t^4] + 4\mu^3\sigma\mathbb{E}\left[\mathbb{E}[\theta_t^3 W_{\theta_t}|\theta_t]\right] + 6\mu^2\sigma^2\mathbb{E}\left[\mathbb{E}[\theta_t^2 W_{\theta_t}^2|\theta_t]\right]$$
$$+ 4\mu\sigma^3\mathbb{E}\left[\mathbb{E}[\theta_t W_{\theta_t}^3|\theta_t]\right] + \sigma^4\mathbb{E}\left[\mathbb{E}[W_{\theta_t}^4|\theta_t]\right]$$

Using Eq. (7.23), we can write:
$$\mathbb{E}\left[X_t^4\right] = \mu^4 \mathbb{E}\left[\theta_t^4\right] + 4\mu^3 \sigma \mathbb{E}\left[\theta_t^3 \mathbb{E}\left[W_{\theta_t}|\theta_t\right]\right] + 6\mu^2 \sigma^2 \mathbb{E}\left[\theta_t^2 \mathbb{E}\left[W_{\theta_t}^2|\theta_t\right]\right]$$
$$+ 4\mu\sigma^3 \mathbb{E}\left[\theta_t \mathbb{E}\left[W_{\theta_t}^3|\theta_t\right]\right] + \sigma^4 \mathbb{E}\left[3\ \mathbb{E}\left[W_{\theta_t}^2|\theta_t\right]^2\right]$$

yielding, by identification:
$$\mathbb{E}\left[X_t^4\right] = \mu^4 \mathbb{E}\left[\theta_t^4\right] + 4\mu^3 \sigma \mathbb{E}\left[\theta_t^3.0\right] + 6\mu^2 \sigma^2 \mathbb{E}\left[\theta_t^2.\theta_t\right]$$
$$+ 4\mu\sigma^3 \mathbb{E}\left[\theta_t.0\right] + 3\sigma^4 \mathbb{E}\left[\theta_t^2\right]$$

so that, finally, we get:
$$\mathbb{E}\left[X_t^4\right] = \mu^4\ \mathbb{E}\left[\theta_t^4\right] + 6\ \mu^2\ \sigma^2\ \mathbb{E}\left[\theta_t^3\right] + 3\ \sigma^4\ \mathbb{E}\left[\theta_t^2\right] \tag{7.24}$$

Equipped with the formulas (7.17), (7.18), (7.20), and (7.24), which give the expressions of the first four non-centered moments of X_t, we can write:
$$m_4(X_t) = \mathbb{E}\left[\left[X_t - \mathbb{E}\left[X_t\right]\right]^4\right] = \mathbb{E}\left[\left[X_t - \mu \mathbb{E}\left[\theta_t\right]\right]^4\right]$$

which is equivalent to:
$$m_4(X_t) = \mathbb{E}\left[X_t^4\right] - 4\mathbb{E}\left[X_t^3 \mu \mathbb{E}\left[\theta_t\right]\right] + 6\mathbb{E}\left[X_t^2 \mu^2 \mathbb{E}\left[\theta_t\right]^2\right]$$
$$- 4\mathbb{E}\left[X_t \mu^3 \mathbb{E}\left[\theta_t\right]^3\right] + \mu^4 \mathbb{E}\left[\theta_t\right]^4$$

and to:
$$m_4(X_t) = \mathbb{E}\left[X_t^4\right] - 4\mu \mathbb{E}\left[\theta_t\right] \mathbb{E}\left[X_t^3\right] + 6\mu^2 \mathbb{E}\left[\theta_t\right]^2 \mathbb{E}\left[X_t^2\right]$$
$$- 4\mu^3 \mathbb{E}\left[\theta_t\right]^3 \mathbb{E}\left[X_t\right] + \mu^4 \mathbb{E}\left[\theta_t\right]^4$$

under the form:
$$m_4(X_t) = \mu^4 \mathbb{E}\left[\theta_t^4\right] + 6\mu^2 \sigma^2 \mathbb{E}\left[\theta_t^3\right] + 3\sigma^4 \mathbb{E}\left[\theta_t^2\right]$$
$$- 4\mu \mathbb{E}\left[\theta_t\right]\left(\mu^3 \mathbb{E}\left[\theta_t^3\right] + 3\mu \sigma^2 \mathbb{E}\left[\theta_t^2\right]\right)$$
$$+ 6\mu^2 \mathbb{E}\left[\theta_t\right]^2 \left(\mu^2 \mathbb{E}\left[\theta_t^2\right] + \sigma^2 \mathbb{E}\left[\theta_t\right]\right) - 4\mu^3 \mathbb{E}\left[\theta_t\right]^3 \mu \mathbb{E}\left[\theta_t\right] + \mu^4 \mathbb{E}\left[\theta_t\right]^4$$

Therefore, we have the centered moment of order 4:
$$m_4(X_t) = \mu^4 \mathbb{E}\left[\theta_t^4\right] + 6\mu^2 \sigma^2 \mathbb{E}\left[\theta_t^3\right] + 3\sigma^4 \mathbb{E}\left[\theta_t^2\right]$$
$$- 4\mu^4 \mathbb{E}\left[\theta_t\right] \mathbb{E}\left[\theta_t^3\right] - 12\mu^2 \sigma^2 \mathbb{E}\left[\theta_t\right] \mathbb{E}\left[\theta_t^2\right]$$
$$+ 6\mu^4 \mathbb{E}\left[\theta_t\right]^2 \mathbb{E}\left[\theta_t^2\right] + 6\mu^2 \sigma^2 \mathbb{E}\left[\theta_t\right]^3 - 3\mu^4 \mathbb{E}\left[\theta_t\right]^4 \tag{7.25}$$

From there, we can deduce the expression of the kurtosis K as:
$$K(X_t) = \frac{m_4(X_t)}{\left(\mu^2 \text{Var}(\theta_t) + \sigma^2 \mathbb{E}\left[\theta_t\right]\right)^2} \tag{7.26}$$

Table 7.3 From social to physical time: moments in physical time of a Brownian motion in social time, subordinated by a social clock θ

Moment	Expression
$\mathbb{E}[X_t]$	$\mu \, \mathbb{E}[\theta_t]$
$\mathbb{E}[X_t^2]$	$\mu^2 \, \mathbb{E}[\theta_t^2] + \sigma^2 \, \mathbb{E}[\theta_t]$
$\mathrm{var}(X_t)$	$\mu^2 \, \mathrm{var}(\theta_t) + \sigma^2 \, \mathbb{E}[\theta_t]$
$\mathbb{E}[X_t^3]$	$\mu^3 \, \mathbb{E}[\theta_t^3] + 3\, \mu \, \sigma^2 \, \mathbb{E}[\theta_t^2]$
$m_3(X_t)$	$\mu^3 \mathbb{E}[\theta_t^3] - 3\mu^3 \mathbb{E}[\theta_t^2]\mathbb{E}[\theta_t] + 3\mu\sigma^2 \mathrm{Var}(\theta_t) + 2\mu^3 \mathbb{E}[\theta_t]^3$
$S(X_t)$	$\dfrac{\mu^3 \mathbb{E}[\theta_t^3] - 3\mu^3 \mathbb{E}[\theta_t^2]\mathbb{E}[\theta_t] + 3\mu\sigma^2 \mathrm{Var}(\theta_t) + 2\mu^3 \mathbb{E}[\theta_t]^3}{(\mu^2 \mathrm{Var}(\theta_t) + \sigma^2 \mathbb{E}[\theta_t])^{\frac{3}{2}}}$
$\mathbb{E}[X_t^4]$	$\mu^4 \, \mathbb{E}[\theta_t^4] + 6\,\mu^2\,\sigma^2\,\mathbb{E}[\theta_t^3] + 3\,\sigma^4\,\mathbb{E}[\theta_t^2]$
$m_4(X_t)$	$\mu^4 \mathbb{E}[\theta_t^4] + 6\mu^2\sigma^2 \mathbb{E}[\theta_t^3] + 3\sigma^4 \mathbb{E}[\theta_t^2]$
	$-4\mu^4 \mathbb{E}[\theta_t]\mathbb{E}[\theta_t^3] - 12\mu^2\sigma^2 \mathbb{E}[\theta_t]\mathbb{E}[\theta_t^2]$
	$+6\mu^4 \mathbb{E}[\theta_t]^2 \mathbb{E}[\theta_t^2] + 6\mu^2\sigma^2 \mathbb{E}[\theta_t]^3 - 3\mu^4 \mathbb{E}[\theta_t]^4$
$K(X_t)$	$\dfrac{m_4(X_t)}{(\mu^2 \mathrm{Var}(\theta_t) + \sigma^2 \mathbb{E}[\theta_t])^2}$

Note that if $\mu = 0$, then $m_4^{\mu=0}(X_t) = 3\sigma^4 \mathbb{E}[\theta_t^2]$ and:

$$K^{\mu=0}(X_t) = \frac{3\sigma^4 \mathbb{E}[\theta_t^2]}{(\sigma^2 \mathbb{E}[\theta_t])^2} = 3\, \frac{\mathbb{E}[\theta_t^2]}{\mathbb{E}[\theta_t]^2}$$

yielding the expression for the excess kurtosis $\hat{K}$:

$$\hat{K}^{\mu=0}(X_t) = 3\, \frac{\mathbb{E}[\theta_t^2]}{\mathbb{E}[\theta_t]^2} - 3 = 3\, \frac{\mathbb{E}[\theta_t^2] - \mathbb{E}[\theta_t]^2}{\mathbb{E}[\theta_t]^2} = 3\, \frac{\mathrm{Var}(\theta_t)}{\mathbb{E}[\theta_t]^2}$$

Table 7.3 groups together the results obtained for the centered and non-centered moments of a Brownian motion subordinated by a stochastic clock θ. Let us now study an example of time change.

7.2.3 Brownian Motion in Gamma Time

We now present the characteristics of a Brownian motion measured in a social time that follows the gamma distribution. With this clock, the inter-quote time lags are gamma-distributed. Brownian motion measured in gamma time is called Variance Gamma process in the finance literature. We shall show precisely how this process can be constructed and what its financial characteristics are, stressing the relationship with the representation of Laplace processes studied in the previous chapter.

7.2.3.1 Choice of Gamma Time

We set out in Chapter 2 the merits of time change using the example of an exponential clock (Section 7.1.2.1, p. 149). The gamma distribution is a simple generalization of the exponential distribution. The gamma process (introduced in Chapter 6, p. 124) is a strictly increasing Lévy process whose increments follow a gamma distribution. We can thus consider it as a possible subordinator.

As in the preceding chapter, we denote by $\mathcal{G}(.)$ the gamma distribution, and by G the gamma process whose increments follow this distribution.

In the finance research literature, the gamma process is usually described through its moments but not its parameters. The parameter v expresses the variance of the distribution density of the random clock, permitting a direct interpretation. This is the reason we consider the characteristic function of Brownian motion in gamma time with this type of parameterization. Furthermore, we assume by convention $m = 1$: the expectation of the time change is set to 1 by definition.

Thus, the gamma clock satisfies:

$$\theta(t) \sim G(t; 1, v) \tag{7.27}$$

The gamma process $G(t; 1, v)$ has the probability density at time t:

$$f_{G(t)}(x) = \frac{\left(\frac{1}{v}\right)^{\frac{t}{v}}}{\Gamma\left(\frac{t}{v}\right)} x^{\frac{t}{v}-1} e^{-\frac{x}{v}} \tag{7.28}$$

Its moments are:

$$\mathbb{E}[G(t)] = t \qquad \text{var}(G(t)) = vt$$

Its characteristic function is:

$$\Phi_{G(t)}(u) = \left(\frac{1}{1 - iuv}\right)^{\frac{t}{v}}$$

Denote by $B(t;\mu,\sigma)$ a classic arithmetic Brownian motion such that $B(t;\mu,\sigma) = \mu t + \sigma W(t)$; and by $G(t;1,v)$ the gamma process used as social clock (written along the moment-style parameterization). The resulting process is:

$$X(t) = B\left(G(t;1,v);\mu,\sigma\right) = \mu G(t;1,v) + \sigma W(G(t;1,v))$$

the so-called Variance Gamma process.

7.2.3.2 Characteristic Function

To compute the characteristic function of $X(t)$, we need, following (7.13), to assess:

$$\Phi_{X_t}(u) = \mathbb{E}\left[e^{\left(i\mu u - \frac{\sigma^2}{2}u^2\right)G(t)}\right] = \mathbb{E}\left[e^{sG(t)}\right]$$

with $s = i\mu u - \sigma^2 u^2/2$.

We use the Laplace transform of a gamma distribution. If X follows the distribution $\mathcal{G}(\alpha,\lambda)$, then its Laplace transform is:

$$\mathbb{E}[e^{sX}] = \left(1 - \frac{s}{\lambda}\right)^{-\alpha} \tag{7.29}$$

Here, we have $\alpha = \lambda = 1/v$, so that:

$$\mathbb{E}[e^{sX}] = (1 - sv)^{-1/v}$$

Because $s = i\mu u - \sigma^2 u^2/2$, we obtain:

$$\Phi_{X_1}(u) = \left(1 - \left(i\mu u - \frac{1}{2}\sigma^2 u^2\right)v\right)^{-1/v}$$

The previous expression represents the characteristic function of a Brownian motion in gamma time when the time unit is $t=1$. At date t (elevating to the power of t), the characteristic function becomes:

$$\Phi_{X_t}(u) = \mathbb{E}\left[e^{iuX_t}\right] = \left(\frac{1}{1 - i\mu v u + \frac{1}{2}v\sigma^2 u^2}\right)^{\frac{t}{v}} \tag{7.30}$$

Set $\delta = \mu v$, $c^2 = \sigma^2 v$, and $v = \tau$. The preceding expression can then be written as:

$$\Phi_{X_t}(u) = \left(\frac{1}{1 - i\delta u + \frac{1}{2}c^2 u^2}\right)^{\frac{t}{\tau}} \tag{7.31}$$

We recognize the characteristic function of a Laplace process with $\theta = 0$. The Variance Gamma (VG) process, see Madan, Carr, and Chang (1998),

Table 7.4 Equivalences

Symmetrical Variance Gamma	Madan and Seneta Madan and Milne	1990 1991	Laplace motion without drift
Variance Gamma (VG)	Madan, Carr, and Chang	1998	Laplace process without drift
Variance Gamma with Drift (VGD)	Le Courtois and Walter	2011	Laplace process

is thus a Laplace process without drift, constructed based on the centered generalized Laplace distribution $\mathcal{L}'(0, c, \delta)$.

When $\mu = 0$, we have:
$$X(t) = \sigma W(G(t; 1, v))$$

This is the symmetrical Variance Gamma process introduced by Madan and Seneta (1990) and Madan and Milne (1991). Because $\delta = 0$, we obtain:

$$\Phi_{X_t}(u) = \left(\frac{1}{1 + \frac{1}{2}c^2 u^2}\right)^{\frac{t}{\tau}}$$

We recognize the characteristic function of the Laplace motion with $\theta = 0$. The symmetrical Variance Gamma process is thus a Laplace motion without drift, constructed from the distribution of the centered simple Laplace distribution $\mathcal{L}(0, c)$.

By extension (see Le Courtois and Walter (2011)), if we add a drift $\theta \neq 0$ to the Variance Gamma process to construct a Variance Gamma process with Drift (VGD), we in fact obtain a Laplace process. These equivalences are summarized in Table 7.4.

The asymmetry parameter δ, which results from the asymmetrization of the simple Laplace distribution, and which induces an asymmetry effect on the time t distribution, is a function of the expectation μ of Brownian motion multiplied by the variance v of the gamma process. It is possible to construe $v = \tau$ as a time length, and we see that μv represents the value of the Brownian drift over a lag corresponding to the variance. The higher the variance v of the gamma process, the more important the effect on the drift. This will produce an asymmetry effect due precisely to this drift. In the first symmetrical Variance Gamma process of Madan and Seneta (1990), μ (and therefore δ) was zero. Only leptokurticity could be rendered, but not asymmetry.

The variance c^2 of the Laplace distribution is equal to the variance σ^2 of Brownian motion multiplied by the variance v of the gamma process.

The parameter σ corresponds to the degree of agitation of Brownian motion. The parameter v acts directly on the shape of the distribution and permits the emergence of leptokurticity, introducing another dimension to risk. A strong agitation of the time change process creates an important leptokurticity of the process resulting from the time change.

To finish, note from (7.30) that:

$$\mathbb{E}\left[e^{iuX_t}\right] = \left(1 - \frac{v}{t}\left(i\mu tu - \frac{1}{2}\sigma^2 tu^2\right)\right)^{-\frac{t}{v}}$$

So that:

$$\mathbb{E}\left[e^{iuX_t}\right] = \mathbb{E}\left[e^{iuX_1}\right]^t$$

A Variance Gamma process remains a Variance Gamma process at all scales.

7.2.3.3 Distribution Density

To obtain the distribution density of $X(t)$, we can consider Brownian motion in gamma time at a date t as having a Gaussian density conditioned by the realization of a time change given by the gamma distribution at time t (knowing the passage of time is related to the parameter v), i.e. a probability mixture: a Gaussian random variable whose variance follows the gamma distribution. This interpretation as a mixture of distributions allows us to write directly the density of the marginal distribution.

Denoting, as above, the distribution density of the gamma process at time t by $f_{G_t}(x)$, we have:

$$f_{X_t}(x) = \int_0^{+\infty} \frac{1}{\sigma\sqrt{2\pi y}} \exp\left(-\frac{1}{2y}\left(\frac{x-\mu y}{\sigma}\right)^2\right) f_{G_t}(y)\,dy$$

where $f_{G_t}(y)$ is given by (7.28). Replacing $f_{G_t}(y)$ by its expression, we obtain:

$$f_{X_t}(x) = \int_0^{+\infty} \frac{1}{\sigma\sqrt{2\pi y}} \exp\left(-\frac{1}{2y}\left(\frac{x-\mu y}{\sigma}\right)^2\right) \frac{y^{\frac{t}{v}-1} e^{-\frac{y}{v}}}{v^{\frac{t}{v}} \Gamma\left(\frac{1}{v}\right)}\,dy \quad (7.32)$$

which can be cast in the analytical form:

$$f(x) = \sqrt{\frac{\sigma^2 v}{2\pi}} \frac{(\mu^2 v + 2\sigma^2)^{\frac{1}{4}-\frac{\mu}{2v}}}{\Gamma\left(\frac{t}{v}\right)} |x|^{\frac{t}{v}-\frac{1}{2}} e^{\frac{\mu x}{\sigma^2}} K_{\frac{t}{v}-\frac{1}{2}}\left(\frac{1}{\sigma}\sqrt{\frac{2}{v}+\frac{\mu^2}{\sigma^2}}|x|\right)$$

7.2.3.4 Difference of Two Gamma Processes

Because the Laplace process without drift – Brownian motion in gamma time – is a finite variation process, it is possible to represent it as the difference of two gamma processes in calendar time. We now seek an equivalence between the time change and the calendar time representations.

The first gamma process defines negative jumps whereas the second one corresponds to positive jumps. Denote by G_+ and G_- the two gamma processes of moments (m_+, v_+) and (m_-, v_-) with $m_+ \neq 1$ and $m_- \neq 1$. The parameters m_+ and v_+, m_- and v_- respectively, represent the expectations and variances of the positive and negative jumps, and are a function of the clock v.

By definition, we have for all t:

$$X(t) \stackrel{d}{=} G_+(t) - G_-(t)$$

Recall that we are in the situation where $\theta = 0$. We position ourselves at date $t = 1$. By definition of (7.30):

$$\Phi_{X_1}(u) = \left(\frac{1}{1 - i\mu v u + \frac{1}{2}v\sigma^2 u^2}\right)^{\frac{1}{v}}$$

and by the starting hypothesis:

$$\Phi_{X_1}(u) = \left(\frac{1}{1 - iu\frac{v_+}{m_+}}\right)^{\frac{m_+^2}{v_+}} \left(\frac{1}{1 + iu\frac{v_-}{m_-}}\right)^{\frac{m_-^2}{v_-}}$$

We therefore need to solve the equation:

$$\left(\frac{1}{1 - i\mu v u + \frac{1}{2}v\sigma^2 u^2}\right)^{\frac{1}{v}} = \left(\frac{1}{1 - iu\frac{v_+}{m_+}}\right)^{\frac{m_+^2}{v_+}} \left(\frac{1}{1 + iu\frac{v_-}{m_-}}\right)^{\frac{m_-^2}{v_-}}$$

Set:

$$\frac{m_+^2}{v_+} = \frac{m_-^2}{v_-} = \frac{1}{v} \tag{7.33}$$

We obtain the new equation:

$$\frac{1}{1 - i\mu v u + \frac{1}{2}v\sigma^2 u^2} = \left(\frac{1}{1 - iu\frac{v_+}{m_+}}\right)\left(\frac{1}{1 + iu\frac{v_-}{m_-}}\right)$$

so that:

$$1 - i\mu v u + \frac{1}{2}v\sigma^2 u^2 = \left(1 - iu\frac{v_+}{m_+}\right)\left(1 - iu\frac{v_-}{m_-}\right)$$

yielding:

$$\begin{cases} \frac{v_+}{m_+} - \frac{v_-}{m_-} = \mu v = \delta \\ \frac{v_+}{m_+} \frac{v_-}{m_-} = \frac{\sigma^2 v}{2} = \frac{c^2}{2} \end{cases}$$

We obtain the two solutions:

$$\begin{cases} \frac{v_-}{m_-} = \frac{1}{2}\left(\sqrt{\mu^2 v^2 + 2\sigma^2 v} - \mu v\right) = \frac{1}{2}\left(\sqrt{\delta^2 + 2c^2} - \delta\right) \\ \frac{v_+}{m_+} = \frac{1}{2}\left(\sqrt{\mu^2 v^2 + 2\sigma^2 v} + \mu v\right) = \frac{1}{2}\left(\sqrt{\delta^2 + 2c^2} + \delta\right) \end{cases}$$

giving finally, for the expectation parameters:

$$\begin{cases} m_- = \frac{1}{2}\left(\sqrt{\mu^2 + \frac{2\sigma^2}{v}} - \mu\right) \\ m_+ = \frac{1}{2}\left(\sqrt{\mu^2 + \frac{2\sigma^2}{v}} + \mu\right) \end{cases}$$

and, using (7.33), we have for the variance parameters:

$$\begin{cases} v_- = v\, m_-^2 = v\left(\frac{1}{2}\sqrt{\mu^2 + \frac{2\sigma^2}{v}} - \frac{\mu}{2}\right)^2 \\ v_+ = v\, m_+^2 = v\left(\frac{1}{2}\sqrt{\mu^2 + \frac{2\sigma^2}{v}} + \frac{\mu}{2}\right)^2 \end{cases}$$

7.2.3.5 *Lévy Measure*

The Lévy measure of the Variance Gamma process is:

$$\nu(x) = \frac{\frac{1}{v}}{|x|} \exp\left(\frac{\mu x}{\sigma^2} - \frac{1}{\sigma}\sqrt{\frac{2}{v} + \frac{\mu^2}{\sigma^2}}|x|\right) \tag{7.34}$$

With the definition of the VG process as the difference of two gamma processes, the Lévy measure is given by:

$$\nu(x) = \begin{cases} \frac{\frac{m_-^2}{v_-}}{|x|} \exp\left(-\frac{m_-}{v_-}|x|\right) & \text{if } x < 0 \\ \frac{\frac{m_+^2}{v_+}}{x} \exp\left(-\frac{m_+}{v_+}x\right) & \text{if } x > 0 \end{cases}$$

so that:
$$\nu(x) = \begin{cases} \frac{\frac{1}{v}}{|x|} \exp\left(-\frac{m_-}{v_-}|x|\right) & \text{if } x < 0 \\ \frac{\frac{1}{v}}{x} \exp\left(-\frac{m_+}{v_+}x\right) & \text{if } x > 0 \end{cases} \qquad (7.35)$$

Denote $C = \frac{1}{v}$, $a_+ = \frac{m_+}{v_+}$ and $a_- = \frac{m_-}{v_-}$. We obtain:
$$\nu(x) = \begin{cases} \frac{C}{|x|} e^{-a_-|x|} & \text{if } x < 0 \\ \frac{C}{x} e^{-a_+ x} & \text{if } x > 0 \end{cases} \qquad (7.36)$$

Recalling that the Lévy measure of a CGMY process, as defined in (5.32), is:
$$\nu(x) = \begin{cases} \frac{C}{|x|^{1+\alpha}} e^{-a_-|x|} & \text{if } x < 0 \\ \frac{C}{x^{1+\alpha}} e^{-a_+ x} & \text{if } x > 0 \end{cases}$$

we note that a Brownian motion in gamma time is a particular case of a CGMY process where $\alpha = 0$.

With this new parameterization, the interpretation of the Lévy measure is the following. As $C = 1/v$, where v is the variance of the time change, C represents a measure of the level of activity of the process. Written differently, the calibration of activity (average number of jumps per unit of time) can be done directly with C. This is a measure of the size of risk. As with the CGMY process, the shape of risk is given by the parameters a_- and a_+ which control the decrease rate of the distribution tails and the asymmetry between stock price increases and decreases. When $a_- > a_+$, the distribution presents a thicker right tail.

In Fig. 7.2, we plot two simulations of trajectories of a Variance Gamma process: by subordinating a Brownian motion by a gamma process (top graph), and by computing the difference between two gamma processes (bottom graph). The VG trajectory in the top graph (plain curve) is obtained by simulating a Brownian motion (dotted curve) by a gamma process (dashed curve). The VG trajectory in the bottom graph (plain curve, zoomed three times) is obtained as the difference of two gamma processes (dashed and dotted curves).

7.2.3.6 Moments

Recall that the moment of order k of a random variable X is related to the k-th derivative at zero of the characteristic function $\Phi_X(u)$:
$$\delta_k = i^{-k} \Phi_X^{(k)}(0)$$

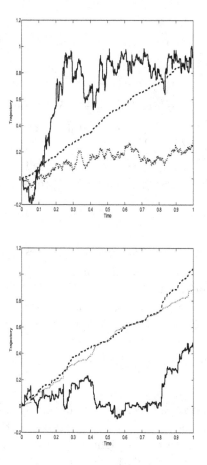

Figure 7.2   Two simulations of a VG process: by subordination of a Brownian motion by a gamma process (top graph), by difference between two gamma processes (bottom graph)

We obtain for the Variance Gamma process the following expressions:

$$\begin{cases} \mathbb{E}[X_t] = t\,\mu \\ m_2(X_t) = t\left(\mu^2 v + \sigma^2\right) \\ m_3(X_t) = t\left(2\mu^3 v^2 + 3\sigma^2 \mu v\right) \\ m_4(X_t) = t\left(3\sigma^4 v + 12\sigma^2 \mu^2 v^2 + 6\mu^4 v^3\right) + t^2\left(3\sigma^4 + 6\sigma^2 \mu^2 v + 3\mu^4 v^2\right) \end{cases}$$

When $\mu < 0$ (resp. $\mu > 0$), the probability of obtaining negative (resp. positive) values of x becomes larger than that of obtaining positive (resp.

negative) values of the same variable: an asymmetry appears and is consistently linked to a negative (resp. positive) asymmetry coefficient. Therefore, the parameter μ allows us to tune the asymmetry of the distribution.

When $\mu = 0$ (symmetrical VG process), the moments are:

$$\begin{cases} \mathbb{E}[X_t] = 0 \\ m_2(X_t) = t\sigma^2 \\ m_3(X_t) = 0 \\ m_4(X_t) = t\,3\sigma^4(1+v) \end{cases}$$

An interesting interpretation of the parameter v can be made. Indeed, starting from:

$$m_4(X_1) = 3\sigma^4(1+v)$$

we deduce the expression of the kurtosis of the symmetrical Variance Gamma process (let us denote it by K_{VGS}):

$$K(X_1) = \frac{m_4}{\sigma^4} = 3(1+v) = K_{\text{VGS}}$$

Because the normal value of K is 3 (denote it by K_{N}), we see that:

$$\frac{K_{\text{VGS}}}{K_{\text{N}}} = 1 + v$$

Written differently, the parameter v corresponds exactly to the excess kurtosis of the distribution that stems from a time change by a gamma clock of variance v. Consequently, if $v = 0.30$, this means that the observed distribution displays a coefficient K which is 30% larger than the normal value.

$\boxed{\delta = 0\text{: gamma clock variance} = \text{excess kurtosis in percent}}$

7.3 Time-Changed Laplace Process

One of the difficulties – and therefore one of the limits – linked to the use of Lévy process as social stochastic clocks, comes from their inability to take into account certain market stylized features, such as volatility clustering, which are well-documented in the finance research literature. A way to model this phenomenon is to introduce (see Carr, Geman, Madan, and Yor (2003)) a stochastic volatility inside a jump model, such as the CGMY model. In this situation, the time change creates return autocorrelations and therefore volatility persistence.

The developments that follow examine the construction of stochastic processes that are more general than Lévy processes, but which still have a characteristic function and thus allow for an easy computation of standard risk measures (VaR and Expected Shortfall).

7.3.1 Mean-Reverting Clock

We want to construct a time change allowing the introduction of non-independent or non-stationary increments.

The appeal of this approach lies in the fact that it goes beyond subordinators to use non-Lévy processes for time changes. This allows us to introduce a VaR term structure richer than that obtained with processes such as the Variance Gamma process with drift. In particular, this allows us to take into account a time decrease of kurtosis slower than that associated with Lévy processes.

7.3.1.1 Construction of an Auto-Regressive Process

Suppose that there exists a bijective link between the changes of time and the changes of the scale of fluctuations. In this case, any time change is a change of volatility and *vice versa*. To take into account volatility persistence, we see that it is necessary to introduce a mean-reverting feature inside time. We therefore want to construct a mean-reverting stochastic process. The most classic example of these processes is the Ornstein–Uhlenbeck process, from the names of Leonard Ornstein (1880–1941) and George Uhlenbeck (1900–1988). This is a stochastic process described by the stochastic differential equation:

$$dx_t = a(\mu - x_t)\,dt + \sigma\,dW_t$$

where θ, μ are σ are deterministic parameters and W is a Wiener process.

Just as Brownian motion represents the limit of a discrete random walk, the Ornstein–Uhlenbeck process is the limit of a process in discrete time. To describe this latter process, we can imagine an urn containing n balls that can be either blue or yellow. Each time a ball is randomly selected, it is replaced by a ball of the other color. If X_n is the number of blue balls (for instance), then $\frac{X_{[nt]} - n/2}{\sqrt{n}}$ converges to an Ornstein–Uhlenbeck process when n tends to infinity.

This process was applied in finance by Oldrich Vasicek in 1977 to model stochastic interest rates. In this case, x_t represents the short-term interest rate, which is supposed to be the only factor affecting bond prices. Over a long period, this value is attracted by the quantity μ. The parameter a represents the speed of adjustment of the short rate to the long-term mean value, and the parameter σ is the instantaneous volatility of the short rate. The fluctuations of the interest rate x_t around μ are due to the random quantity dW_t and occur with the speed a.

One of the weaknesses of the Vasicek model lies in the fact that it allows negative interest rates. This is the reason why several generalizations were developed subsequently, including that of Cox, Ingersoll, and Ross (CIR) in 1985. The CIR process (or "square root process") is continuous, positive (provided $ab > 0$, where a and b are defined hereafter), and described by the stochastic differential equation:

$$dx_t = a(b - x_t)dt + c\sqrt{x_t}\, dW_t \qquad (7.37)$$

Contrary to the Wiener process, whose drift term is constant, the CIR process has a drift which depends on the value of the process itself. This specificity allows, after integration, the construction of a change of time more complex than that stemming from the use of a subordinator.

7.3.1.2 Properties of the CIR Process

We show first that the CIR process is Markovian and that its increments are neither independent nor stationary. So a CIR process is not a Lévy process: indeed, the only possible continuous Lévy process is arithmetic Brownian motion. Note also that independent increment processes are Markovian but that the converse is not necessarily true.

To show that the increments of the CIR process are not stationary, we start from Itō's lemma which allows us to restate Eq. (7.37) as follows:

$$x_t = b + (x_0 - b)e^{-at} + ce^{-at}\int_0^t e^{au}\sqrt{x_u}dW_u \qquad (7.38)$$

Then, we readily have (with x_0 deterministic):

$$\mathbb{E}(x_t) = b + (x_0 - b)e^{-at}$$

because the last term of Eq. (7.38) is a martingale (being a continuous positive local martingale). Therefore:

$$\mathbb{E}(x_t - x_s) = (x_0 - b)\left(e^{-at} - e^{-as}\right)$$

and:

$$\mathbb{E}(x_{t-s} - x_0) = (x_0 - b)\left(e^{-a(t-s)} - 1\right)$$

so that:

$$\mathbb{E}(x_t - x_s) = e^{-as}\mathbb{E}(x_{t-s} - x_0)$$

This shows that the increments of x are not stationary, even if this type of non-stationarity is quite simple.

Then, we show that the increments of the CIR process are not independent. To do so, we compute for $0 \leq s \leq t$:

$$\mathrm{cov}(x_t - x_s, x_s - x_0) = \mathbb{E}\left[(x_t - x_s - \mathbb{E}[x_t - x_s])(x_s - x_0 - \mathbb{E}[x_s - x_0])\right]$$

This expression can be rewritten, after substitution, as:

$$\mathbb{E}\left[\left(ce^{-at}\int_0^t e^{au}\sqrt{x_u}dW_u - ce^{-as}\int_0^s e^{au}\sqrt{x_u}dW_u\right)ce^{-as}\int_0^s e^{au}\sqrt{x_u}dW_u\right]$$

which is equal to, using the isometry property:

$$c^2 e^{-a(t+s)}\int_0^s e^{2au}\mathbb{E}[x_u]du - c^2 e^{-2as}\int_0^s e^{2au}\mathbb{E}[x_u]du$$

or:

$$c^2\left(e^{-a(t+s)} - e^{-2as}\right)\int_0^s e^{2au}\mathbb{E}[x_u]du$$

and finally, by integration:

$$\mathrm{cov}(x_t - x_s, x_s - x_0) = c^2\left(e^{-a(t+s)} - e^{-2as}\right)$$
$$\times \left[\frac{b}{2a}\left(e^{2as} - 1\right) + \frac{x_0 - b}{a}\left(e^{as} - 1\right)\right]$$

The non-zero covariance between $x_t - x_s$ and $x_s - x_0$ shows that the increments of the process x are not independent.

To conclude, we show that the CIR process is Markovian. We express formula (7.37) as:

$$x_t = x_s + a\int_s^t (b - x_u)du + c\int_s^t \sqrt{x_u}dW_u$$

This means that the knowledge of x_t, knowing all the values of x between 0 and s, can be reduced to the knowledge of x_t knowing x_s. This is the Markov property.

7.3.1.3 The Time Change Process

We select the integrated CIR (ICIR) process for performing time changes. The ICIR process is continuous, positive, and increasing (contrary to the CIR process), which is a necessary condition for modeling time:

$$dX_t = x_t dt$$

or:

$$X_t = \int_0^t x_s ds$$

We show that the ICIR process is, similar to the CIR process, a Markovian process whose increments are neither independent nor stationary. We start by computing:

$$\mathbb{E}[X_t] = \mathbb{E}\left[\int_0^t x_s ds\right] = \int_0^t \mathbb{E}[x_s] ds = \int_0^t \mathbb{E}\left[b + (x_0 - b)e^{-as}\right] ds$$

from which we get:

$$\mathbb{E}[X_t] = bt + \frac{x_0 - b}{a}\left(1 - e^{-at}\right)$$

Then, we see that:

$$\mathbb{E}[X_t - X_s] = b(t - s) + \frac{x_0 - b}{a}\left(e^{-as} - e^{-at}\right)$$

is different from:

$$\mathbb{E}[X_{t-s} - X_0] = b(t - s) + \frac{x_0 - b}{a}\left(1 - e^{-a(t-s)}\right)$$

This shows the (again elementary) non-stationarity of the increments of this process.

The characteristic function of the ICIR process X (chosen such as $X_0 = 0$) is known and can be expressed as follows:

$$\mathbb{E}\left[e^{iuX_t}\right] = \frac{e^{\frac{ba^2}{c^2}t}}{\left(\cosh\left(\frac{d}{2}t\right) + \frac{a}{d}\sinh\left(\frac{d}{2}t\right)\right)^{\frac{2ab}{c^2}}} \qquad (7.39)$$

where $d = \sqrt{a^2 - 2iuc^2}$.

From this formula, we can obtain:

$$\mathbb{E}\left[e^{iuX_t}\right] \neq \mathbb{E}\left[e^{iuX_1}\right]^t$$

which confirms that the ICIR process is not a Lévy process.

We represent the spectrum of the ICIR process in the left panel of Fig. 7.3. Observe that this spectrum makes complete loops around the origin: this comes from the presence of a complex non-integer power function in the formula of the spectrum. If we try to obtain the corresponding density function by a fast Fourier transform, numerous irregularities or oscillations are produced, as illustrated in the central panel of Fig. 7.3. The proper treatment of the phase of the complex non-integer power function, which amounts to counting the number of loops done by the spectrum around the origin, allows us to obtain smooth density functions, as illustrated in the right-hand panel of Fig. 7.3.

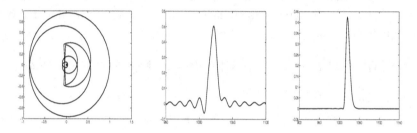

Figure 7.3 Spectrum, non-corrected and corrected densities of the ICIR process

Which variable should we choose to represent the social time of exchanges? Two main possibilities are offered: the number of transactions and the volume of exchanges. We consider, for the sake of illustration, the volume of exchanges, and conduct a calibration of the ICIR process to this quantity obtained on BNP stock. We use daily data between December 10, 2009 and June 22, 2010. The histogram in Fig. 7.4 represents empirical volumes, where the abscissa has been normalized and corresponds to the different possible volumes. The coordinate represents the number of draws in each of the possible volume layers. The plain curve displays the calibration of the histogram by an ICIR density, obtained after properly taking into account the above-mentioned phase problems.

7.3.2 The Laplace Process in ICIR Time

We can now model the evolution of the return X. We select a Laplace process without drift (a Variance Gamma process in the terminology of the finance literature), i.e. a process having the marginal Laplace distribution $\mathcal{L}'(0, c, \delta)$, and on which we apply an ICIR time change.

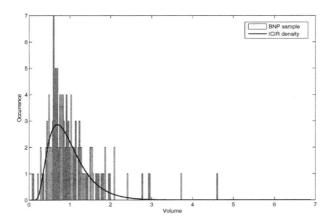

Figure 7.4 Calibration of the ICIR clock on daily exchanged volumes on the BNP share

Denote by T the ICIR clock and by Y the driftless Laplace process. The return in calendar time is the process defined at all time t by the equation $X(t) = Y(T(t))$ and having six parameters.

Because the process T is not a subordinator, in the sense that it is not a strictly increasing positive Lévy process, we cannot use formula (7.9) to compute the spectrum of the process X. We show first that, for Z defined for all t by $Z(t) = X_2(X_1(t))$ where X_2 is a Lévy process and X_1 is a strictly increasing positive stochastic process, then the characteristic function of Z can be written as:

$$\mathbb{E}\left[e^{iuZ_t}\right] = e^{\varphi_{X_1}(\psi_{X_2}(u),t)} \tag{7.40}$$

where φ_{X_1} is not necessarily proportional to time in the general case.

Formula (7.40) can be obtained by setting:

$$\mathbb{E}\left[e^{iuZ_t}\right] = \mathbb{E}\left[e^{iuX_2(X_1(t))}\right] = \mathbb{E}\left[\mathbb{E}\left[e^{iuX_2(X_1(t))}|X_1(t)\right]\right]$$

then, by identification:

$$\mathbb{E}\left[e^{iuZ_t}\right] = \mathbb{E}\left[e^{X_1(t)\cdot\psi_{X_2}(u)}\right] = e^{\varphi_{X_1}(\psi_{X_2}(u),t)}$$

For X_1 an ICIR process, we have the moment generating function:

$$\mathbb{E}\left[e^{uX_t}\right] = e^{\varphi_{X_1}(u,t)}$$

with:

$$\varphi_{X_1}(u,t) = \frac{ba^2}{c^2}t - \frac{2ab}{c^2}\ln\left(\cosh\left(\frac{\sqrt{a^2 - 2uc^2}}{2}t\right)\right.$$
$$\left. + \frac{a}{\sqrt{a^2 - 2uc^2}}\sinh\left(\frac{\sqrt{a^2 - 2uc^2}}{2}t\right)\right) \tag{7.41}$$

and where this expression is real only for $u < \frac{a^2}{2c^2}$.

Recall that the characteristic function of the driftless Laplace process can be written as:

$$\mathbb{E}\left[e^{iuX_2(t)}\right] = e^{t\psi_{X_2}(u)} = \left(\frac{1}{1 - i\mu vu + \frac{1}{2}\sigma^2 vu^2}\right)^{\frac{t}{v}}$$

where naturally:

$$\psi_{X_2}(u) = \frac{1}{v}\ln\left(\frac{1}{1 - i\mu vu + \frac{1}{2}\sigma^2 vu^2}\right) \tag{7.42}$$

The spectrum of the process $X(T(.))$ can finally be obtained by including formulas (7.41) and (7.42) within Eq. (7.40).

Chapter 8

Tail Distributions

The classic theory of extreme values distinguishes between small and large variations in stock exchange quantities. One of the difficulties often encountered in the practical application of this theory is the choice of a threshold beyond which a variation is considered large. Such an approach is distinct from the concepts developed in this book, where all admissible variations are taken into account. One of the main goals of this chapter is to show how the Lévy process approach can be reconciled with that of extreme values.

This chapter starts by recalling the approach related to the laws of maxima, before examining threshold exceedances. Next follows a study of the hierarchy of large values and of the concentration of results (gains or losses) on the largest values. Finally, we describe a statistical method that allows us to improve the estimation of the core parameter of the theory of extreme values: the exponent of the weight of distribution tails.

8.1 Largest Values Approach

The goal here is to characterize and model the largest values of a random quantity X. In this perspective, we introduce the following definitions:

- a threshold x_0 beyond which we consider the values taken by X. This threshold is necessarily high because we are examining the largest variations of X,
- a maximal value $x_{\max}$, called the right endpoint, which represents the physical limit that can be attained by X, independently from the mathematical properties of the process that generates X. This is a final point of the domain of definition of the distribution. This value $x_{\max}$ stems either from a purely exogenous argument ("it is impossible for this

reason that this value exceeds this quantity..."), or from an observation of the past values taken by the process, or again from the *a priori* choice of a value judged relevant for the considered risk measure. Formally, this value is defined as follows:

$$x_{\max} = \max\{x \in \mathbb{R}; F(x) < 1\} \leq \infty$$

where $F(x) = \Pr(X < x)$ is the c.d.f. of the random variable X.

We now consider the values of X that can exceed the threshold x_0.

Two computation methods are possible:

- either we consider the largest values of X, i.e. of the random variable $\{X|X > x_0\}$, whose maximum value is $x_{\max}$,
- or we concentrate on the exceedances $X - x_0$, i.e. on the exceedances of the random variable $\{X - x_0 \,|\, X > x_0\}$, whose maximum value is $x_{\max} - x_0$.

It is important not to mix up these two computation methods. For ease of reading and to better distinguish between the two methods, we choose to index by x_0 the computations related to exceedances.

8.1.1 The Laws of Maxima

The convergence of extremes represents for the maxima the equivalent of the convergence of means: another law of large numbers. If the distribution of means around their expectation tends to be Gaussian when variance is finite, the distribution of extremes also converges toward a particular limit. We now present this limit distribution.

8.1.1.1 The Fisher–Tippett Theorem

Let $\{X_1, X_2, \cdots, X_n\}$ be n i.i.d. random variables having the c.d.f. $F_X(x) = \Pr(X \leq x)$. Consider a sample consisting of the n realizations $\{x_1, \cdots, x_n\}$. We rank them by decreasing order, and introduce the convention:

$$x_{(1)} \geq x_{(2)} \geq \cdots \geq x_{(n)}$$

The largest of these realizations $x_{(1)}$ can be considered to be the realization of a new random variable $X_{(1)}$. The same idea prevails for the other observations $x_{(k)}$. Thus, we introduce n new random variables with the convention:

$$X_{(1)} = \max(X_1, \cdots, X_n)$$

which represents the random variable giving the largest value in a sample of size n. In the literature, this quantity is also sometimes denoted by $X_{(1:n)}$. Similarly:

$$X_{(n)} = \min(X_1, \cdots, X_n)$$

represents the random variable giving the smallest value observed in a sample of size n. More generally $X_{(k)}$ (or $X_{(k:n)}$) is the random variable attached to the k-th value $x_{(k)}$ obtained among n realizations. These n new random variables can be ordered in the following way:

$$X_{(1)} \geq X_{(2)} \geq \cdots \geq X_{(n)}$$

We want to determine the values that can be taken by maxima over a given observation period, i.e. we want to characterize the probability law of maxima. More generally, we aim to obtain the distribution of the k largest values. Two situations arise: either the c.d.f. $F(.)$ is known, or it is not.

When the c.d.f. $F(.)$ is known, it is possible to directly obtain the distribution maxima. Indeed, consider the c.d.f. of $X_{(k)}$, which we will denote by $F_{k:n}$ to indicate that the sample is of size n. We have:

$$F_{k:n}(x) = \Pr(X_{(k)} < x) = \sum_{p=0}^{k-1} C_n^p \left[F(x)\right]^{n-p} \left[1 - F(x)\right]^p$$

Using the fact that the random variables $X_1, \cdots, X_n$ are independent and have the same c.d.f. $F(x)$, we obtain the c.d.f. of the minimum and of the maximum for the values $k = 1$ (law of the maximum) and $k = n$ (law of the minimum):

$$\begin{cases} F_{1:n}(x) = \Pr(X_{(1)} < x) = \sum_{p=0}^{0} C_n^p \left[F(x)\right]^{n-p} \left[1 - F(x)\right]^p \\ F_{n:n}(x) = \Pr(X_{(n)} < x) = \sum_{p=0}^{n-1} C_n^p \left[F(x)\right]^{n-p} \left[1 - F(x)\right]^p \end{cases} \quad (8.1)$$

so that:

$$\begin{cases} F_{1:n}(x) = [F(x)]^n \\ F_{n:n}(x) = 1 - [1 - F(x)]^n \end{cases} \quad (8.2)$$

and their density functions are respectively:

$$\begin{cases} f_{1:n}(x) = n\left[F(x)\right]^{n-1} f(x) \\ f_{n:n}(x) = n[1 - F(x)]^{n-1} f(x) \end{cases} \quad (8.3)$$

We can also obtain these expressions directly. Consider for example the maximum. Then:

$$\{X_{(1)} < x\} \Leftrightarrow \{\max(X_1, \cdots, X_n) < x\} \Leftrightarrow \bigcap_{k=1}^{n} \{X_k < x\}$$

and, using the independence of the underlying random variables:

$$F_{1:n}(x) = \Pr\left(X_{(1)} < x\right) = \Pr\left(\bigcap_{k=1}^{n}(X_k < x)\right) = \prod_{k=1}^{n} \Pr\left(X_k < x\right)$$

so that indeed:

$$F_{1:n}(x) = [F(x)]^n$$

Therefore, we see that, as long as the c.d.f. $F(.)$ is known, the distribution of the largest values can be recovered readily.

However, if $F(.)$ is not known, what can we say about the behavior of $X_{(1)}$, about the shape of the distribution of the largest values? The purpose and interest of extreme value theory is that it allows us to answer such a question. This theory describes the asymptotic behavior of $F_{1:n}$ when n tends to infinity.

Because extreme values are by definition rare, one needs to be able to extrapolate from past observed values to future hypothetical large values. Extreme value theory proposes a framework adapted to this extrapolation: it describes the asymptotic behavior of $F_{1:n}$ when n tends to plus infinity. Due to technical reasons, this is not the maximum $X_{(1:n)}$ which is the object of study, but the centered and reduced maximum $(X_{(1:n)} - b_n)/a_n$. The Fisher–Tippett theorem provides the shape of all the possible asymptotic cumulative distribution functions of the centered reduced maximum of a sample of size n. Note that we remain in the context of a set of n random variables $\{X_1, \cdots, X_n\}$, independent, identically distributed, and having the c.d.f. $F_X(x)$. We stress the fact that this theory necessitates the validation of the i.i.d. hypothesis in order to be applied in an adequate way.

Theorem 8.1 (Laws of maxima: Fisher–Tippett (1928)).
Let a sequence of n i.i.d. random variables $\{X_1, \cdots, X_n\}$ having the c.d.f. $F_X(x)$. If there are two sequences $\{a_n\}_{n \in \mathbb{N}}$ and $\{b_n\}_{n \in \mathbb{N}}$ with $\forall n$, $a_n \in \mathbb{R}^{+*}$ and $b_n \in \mathbb{R}$, and a non-degenerated limit c.d.f. H such that:

$$\lim_{n \to \infty} \Pr\left(\frac{X_{(1:n)} - b_n}{a_n} \leq x\right) = H(x) \tag{8.4}$$

then H belongs to one of the three classes:

- *Fréchet.* $H(z) = \Phi_\alpha(z) = \exp\left(-z^{-\alpha}\right)$ if $z > 0$, 0 if $z \leq 0$
- *Weibull.* $H(z) = \Psi_\alpha(z) = \exp\left(-(-z)^\alpha\right)$ if $z \leq 0$, 1 if $z > 0$
- *Gumbel.* $H(z) = \Lambda(z) = \exp\left(-e^{-z}\right)$, $z \in \mathbb{R}$

with $z = \frac{x-b}{a}$ where $a > 0$, $b \in \mathbb{R}$, and $\alpha > 0$.

The three classes of distributions denoted by Φ_α, Ψ_α and Λ are called the laws of maxima. The Fréchet and Weibull laws are linked by the relationship:
$$\Phi_\alpha\left(-\frac{1}{z}\right) = \Psi_\alpha(z) \tag{8.5}$$
The parameters a and b are scale and localization parameters, whereas α is a shape parameter. We again encounter the double characterization size (scale)–shape.

These three laws are stable with respect to maxima. This means that for a sample of independent random variables $\{Z_1, \cdots, Z_n\}$ satisfying one of these laws, and for $Z_{(1:n)}$ the maximum of these variables, we have:
$$Z_{(1:n)} \stackrel{d}{=} a_n Z + b_n \tag{8.6}$$
where Z is a random variable having the same distribution as the $\{Z_1, \cdots, Z_n\}$ and being independent from these latter random variables.

In particular, we have:

- Fréchet law: $Z_{(1:n)} \stackrel{d}{=} n^{1/\alpha} Z$
- Weibull law: $Z_{(1:n)} \stackrel{d}{=} n^{-1/\alpha} Z$
- Gumbel law: $Z_{(1:n)} \stackrel{d}{=} Z + \ln n$

We illustrate the previous developments with a simple example. Suppose that $\mathbb{E}[Z] = 1$ and consider the Fréchet (with for instance $\alpha = \frac{1}{2}$) and Gumbel laws. We do not select the Weibull law because its support is bounded on the right (it is therefore of little use for financial problems). In the first case, we see that $\mathbb{E}[Z_{(1:n)}] = n^2$; whilst in the second case $\mathbb{E}[Z_{(1:n)}] \approx \ln n$. Written differently, with the Fréchet law the maximum increases much more quickly than with the Gumbel law, and consequently the same behavior is also observed for the contribution of the maximum within the sum. If we consider the liabilities of a non-life insurance company, it is critical to be able to predict the importance of the largest claim with respect to the aggregate claim. Thus, we first need to determine whether the data is more of the Fréchet or Gumbel type.

There is a formal analogy between this theorem for the maximum (expressed through the formula (8.4)) and the central limit theorem for the mean, that we recall here:
$$\lim_{n \to \infty} \Pr\left(\frac{\bar{X}_n - \mathbb{E}[X]}{\sigma(X)/\sqrt{n}} \leq x\right) = \Phi(x) \tag{8.7}$$

Table 8.1 Limit laws of means and maxima

Laws of large numbers	
Central limit theorems	Extreme values theorems
Convergence of empirical means $$\bar{X}_n = \frac{1}{n}\sum_{i=1}^{n} X_i$$ $$\Phi(x) = \lim_{n\to\infty} \Pr\left(\frac{\bar{X}_n - \mathbb{E}[X]}{\sigma(X)/\sqrt{n}} \leq x\right)$$	Convergence of empirical maxima $$X_{(n)} = \max(X_1, \cdots, X_n)$$ $$H(x) = \lim_{n\to\infty} \Pr\left(\frac{X_{(n)} - b_n}{a_n} \leq x\right)$$
Centered reduced Gaussian distribution if $\mathbb{E}[X]$ and $\sigma(X)$ exist	Standardized distribution of extreme values
Gaussian density: $$\Phi(x) = \frac{1}{\sqrt{2\pi}} \int_{-\infty}^{x} e^{-t^2/2} dt$$ Gaussian characteristic exponent: $$\Psi_X(u) = -\frac{u^2}{2}$$	Fréchet: $H(x) = \exp\left(-x^{-\alpha}\right)$ Weibull: $H(x) = \exp\left(-(-x)^{\alpha}\right)$ Gumbel: $H(x) = \exp\left(-e^{-x}\right)$
Moivre (1730)–Laplace (1812)	Fisher–Tippett (1928)
α-stable distributions if $\mathbb{E}[X]$ and/or $\sigma(X)$ do not exist	Generalized distribution of extreme values
$$\Psi_X(u) = i\mu u - \gamma^\alpha |u|^\alpha \left(1 - i\beta \frac{u}{|u|} \text{tg}\frac{\pi\alpha}{2}\right)$$	$$H_\xi(x) = \exp\left(-\left[1+\xi\left(\frac{x-\mu}{\sigma}\right)\right]^{-\frac{1}{\xi}}\right)$$
Lévy (1937)	Jenkinson–von Mises (1954–1955)
Fluctuations of means	Fluctuations of maxima

where Φ is the c.d.f. of the centered reduced Gaussian law.

This comparison between the two theorems, and the two formulas (8.4) and (8.7), allows us to interpret the parameters a_n and b_n. The parameter b_n corresponds to the mean $\mathbb{E}[X]$ in the central limit theorem and is therefore a localization parameter. The parameter a_n corresponds to $\sigma(X)/\sqrt{n}$ in the central limit theorem and is therefore a scale parameter. However, the parameter α has no equivalent in the central limit theorem. The elements of this analogy are displayed synthetically in Table 8.1.

8.1.1.2 The Generalized Jenkinson–von Mises Distribution

The three expressions of the Fréchet, Weibull, and Gumbel distributions can be united in a general expression using the condensed representation of Jenkinson–von Mises (1954–1955), which defines a generalized distribution for the maxima H_ξ satisfying:

$$H_\xi(x) = \exp\left(-\left[1+\xi\left(\frac{x-\mu}{\sigma}\right)\right]^{-\frac{1}{\xi}}\right) \tag{8.8}$$

for all x such that $1 + \xi(x-\mu)/\sigma > 0$.

The density function of this generalized law of maxima (derivative of the c.d.f. $H_\xi(x)$) is:

$$h_\xi(x) = \frac{1}{\sigma}\left[1+\xi\left(\frac{x-\mu}{\sigma}\right)\right]^{-\frac{1+\xi}{\xi}} \exp\left(-\left[1+\xi\left(\frac{x-\mu}{\sigma}\right)\right]^{-\frac{1}{\xi}}\right) \tag{8.9}$$

We can interpret the parameters μ and σ in an interesting way, comparing Eqs (8.4) and (8.8):

$$\lim_{n\to\infty} \Pr\left(\frac{X_{(n)} - b_n}{a_n} \leq x\right) = \exp\left(-\left[1+\xi\left(\frac{x-\mu}{\sigma}\right)\right]^{-\frac{1}{\xi}}\right)$$

This equality shows that the parameters μ and σ are the limits of b_n and a_n when n is large. These parameters therefore have the following meanings: $\mu \in \mathbb{R}$ is a localization parameter, $\sigma > 0$ is a scale parameter: in a way, this parameter indicates the size of the distribution tail.

Finally, the parameter $\xi \in \mathbb{R}$ models the shape of the distribution tail. More precisely, three situations can be encountered:

(1) $\xi > 0$, this is the Fréchet law. The distribution tail is thick and $\xi = \alpha^{-1} > 0$. The larger ξ, the thicker the distribution tail.
(2) $\xi < 0$, this is the Weibull law. The distribution tail is thin and $\xi = -\alpha^{-1} < 0$.
(3) $\xi = 0$, this is the Gumbel law. The distribution tail decreases exponentially. This is a limit case, between the two preceding situations, where $\xi \to 0$.

8.1.1.3 Maximum Domain of Attraction

Let us now reformulate the Fisher–Tippett theorem in order to introduce the notion of the maximum domain of attraction (MDA). Because:

$$\Pr\left(\frac{X_{(1:n)} - b_n}{a_n} \leq x\right) = \Pr\left(X_{(1:n)} < a_n x + b_n\right)$$

and, due to (8.2):

$$\Pr\left(X_{(1:n)} < a_n x + b_n\right) = [F(a_n x + b_n)]^n$$

the Fisher–Tippett theorem claims in fact that:

$$\lim_{n \to \infty} [F(a_n x + b_n)]^n = H(x) \qquad (8.10)$$

We say that the Fisher–Tippett theorem defines a maximum domain of attraction for the c.d.f. $F(.)$. We will sometimes write: $F \in \mathrm{MDA}(H)$.

This result is equivalent for the largest values to the central limit theorem for the mean values. A probability distribution of c.d.f. F belongs to the maximum domain of attraction of H if the distribution of the renormalized maximum converges to H.

Which types of probability distribution belong to which maximum domains of attraction and what are the renormalization constants? The answer to this question depends on the variation of the c.d.f. of the random variable. If a random variable is of regular variation, then it belongs to the MDA of Fréchet; if this random variable is of rapid variation, then it belongs to the MDA of Gumbel. We now examine these two cases.

In the first situation, $\bar{F}$ is a regular function of index $-\alpha$. Recall that f is a regular function of index α if for all $x > 0$ we have:

$$\lim_{t \to +\infty} \frac{f(tx)}{f(t)} = x^\alpha$$

In this case, $F \in \mathrm{MDA}(\Phi_\alpha)$: the maximum domain of attraction is that of the Fréchet law with $a_n = F^{-1}(1 - n^{-1})$ and $b_n = 0$. This result was obtained by Gnedenko in 1943.

In order to verify that $F \in \mathrm{MDA}(\Phi_\alpha)$, we will check that:

$$\lim_{t \to \infty} \frac{1 - F(tx)}{1 - F(t)} = x^{-\alpha}$$

Assume for instance that F is the c.d.f. of a Pareto distribution of type I, so that $F(x) = 1 - x^{-\alpha}$. Then:

$$\lim_{t \to \infty} \frac{1 - F(tx)}{1 - F(t)} = \lim_{t \to \infty} \frac{(tx)^{-\alpha}}{t^{-\alpha}} = x^{-\alpha}$$

A Pareto distribution of type I belongs to the maximum domain of attraction of the Fréchet law.

We can also check the Fisher–Tippett theorem by calculating F^n. We have here $a_n = F^{-1}(1 - n^{-1}) = n^{1/\alpha}$. This yields:

$$\lim_{n \to \infty} [F(a_n x + b_n)]^n = \lim_{n \to \infty} \left[F(n^{1/\alpha} x)\right]^n = \lim_{n \to \infty} \left(1 - \frac{x^{-\alpha}}{n}\right)^n$$

so that:

$$\lim_{n \to \infty} [F(a_n x + b_n)]^n = \exp\left(-x^{-\alpha}\right)$$

and we have found the Fréchet law.

The second situation is that in which the distributions belong to the MDA of the Gumbel law and have an infinite right endpoint. These distributions are of rapid variation. Recall that a function f is of rapid variation if it satisfies:
$$\begin{cases} \lim_{t \to +\infty} \frac{f(tx)}{f(t)} = 0 & x > 1 \\ \lim_{t \to +\infty} \frac{f(tx)}{f(t)} = +\infty & 0 < x < 1 \end{cases}$$

The infinitely divisible distributions whose underlying Lévy measure density is the product of an exponential function by a power function belong to the MDA of the Gumbel law. The asymptotic behavior of these distributions is studied in detail in Albin and Sundén (2009). Consider for instance the infinitely divisible distributions having the Lévy measure density:
$$\nu(x) = Cx^\rho e^{-\alpha x} \tag{8.11}$$

It appears that both the distribution densities and the cumulative distributions functions of these laws have, up to a constant, a mixed exponential-power form. It can also be noted that the distributions constructed from such Lévy measure densities (in particular the variance gamma, CGMY, or generalized hyperbolic distributions) possess cumulative distribution functions which can be written as:
$$F(x) = 1 - e^{-\int_{-\infty}^{x} c(t) dt}$$

where $\lim_{t \to +\infty} c(t) = \alpha$. This representation is that of a von Mises function. It readily implies that these distributions belong to the MDA of the Gumbel law.

Let us conclude on the behavior of distribution tails:

(1) The MDA of the Fréchet law contains all distributions having stretched or heavy tails, such as power functions.
(2) The MDA of the Gumbel law contains all thin and medium-tailed distributions. Examples of such distributions are the exponential distributions with semi-heavy tails thicker than the Gaussian, such as the distributions resulting from mixing a power function with an exponential function.
(3) The MDA of the Weibull law contains all distributions with a support bounded to the right. These distributions are said to be short-tailed.

These results are summarized in Table 8.2.

Table 8.2 Laws of maxima and maximum domain of attraction

Laws of maxima	Shape parameter	Basic distributions
Weibull	$\xi < 0$	Uniform Beta
Gumbel	$\xi = 0$	Gaussian Exponential Lognormal Gamma CGMY
Fréchet	$\xi > 0$	Cauchy Pareto α-stable Student Loggamma

8.1.2 The Maxima of Lévy Processes

To describe the behavior of the largest values of Lévy processes, it is sufficient to consider the Lévy measure. Indeed, this measure yields directly the asymptotic pattern of the density and c.d.f. of any marginal X_t.

8.1.2.1 CGMY and Variance Gamma Processes

Consider first the asymptotic behavior of the CGMY and Variance Gamma processes satisfying the relationship (8.11) with, respectively, $\rho < -1$ and $\rho = -1$. For a process X of the CGMY type having the Lévy measure density:

$$\nu(x) = C\frac{e^{Gx}}{(-x)^{1+Y}}\mathbf{1}_{x<0} + C\frac{e^{-Mx}}{x^{1+Y}}\mathbf{1}_{x>0}$$

where $Y > 0$, Albin and Sundén (2009) showed that, when u is large:

$$\Pr(X_t > u) \stackrel{+\infty}{\sim} \frac{Ct}{M}\,\mathbb{E}\left[e^{MX(t)}\right]\frac{e^{-Mu}}{u^{1+Y}}$$

From the computation of the moment generating function $\mathbb{E}\left(e^{MX(t)}\right)$, we deduce:

$$\Pr(X_t > u) \stackrel{+\infty}{\sim} \frac{Ct}{M}\, e^{Ct\Gamma(-Y)[(G+M)^Y - M^Y - G^Y]}\frac{e^{-Mu}}{u^{1+Y}}$$

so that:

$$\Pr(X_t > u) \stackrel{+\infty}{\sim} \tilde{C}(t)\,\frac{e^{-Mu}}{u^{1+Y}}$$

where:
$$\tilde{C}(t) = \frac{Ct}{M} e^{Ct\Gamma(-Y)[(G+M)^Y - M^Y - G^Y]}$$

This confirms that the distribution tail is a mix of power and exponential functions, similar to the Lévy measure density.

In the case of a Variance Gamma process $Z = Z_1 + Z_2$ having the Lévy measure density:
$$\nu(x) = C \frac{e^{Gx}}{(-x)} \mathbf{1}_{x<0} + C \frac{e^{-Mx}}{x} \mathbf{1}_{x>0}$$

where Z_1 is a gamma process describing the positive jumps and Z_2 is a gamma process describing the negative jumps, Albin and Sundén (2009) showed that, when u is large:
$$\Pr(X_t > u) \overset{+\infty}{\sim} \frac{M^{Ct-1}}{\Gamma(Ct)} \mathbb{E}\left[e^{MZ_2(t)}\right] \frac{e^{-Mu}}{u^{1-Ct}}$$

From the computation of the moment generating function $\mathbb{E}\left[e^{MZ_2(t)}\right]$ of the gamma process associated with negative jumps, we deduce:
$$\Pr(X_t > u) \overset{+\infty}{\sim} \frac{M^{Ct-1}}{\Gamma(Ct)} \left(\frac{G}{G+M}\right)^{Ct} \frac{e^{-Mu}}{u^{1-Ct}}$$

so that:
$$\Pr(X_t > u) \overset{+\infty}{\sim} \tilde{C}(t) \frac{e^{-Mu}}{u^{1-Ct}}$$

where:
$$\tilde{C}(t) = \frac{M^{Ct-1}}{\Gamma(Ct)} \left(\frac{G}{G+M}\right)^{Ct}$$

Recall that:
$$C = \frac{1}{v}, \quad G = \left(\sqrt{\frac{\theta^2 v^2}{4} + \frac{\sigma^2 v}{2}} - \frac{\theta v}{2}\right)^{-1} \quad M = \left(\sqrt{\frac{\theta^2 v^2}{4} + \frac{\sigma^2 v}{2}} + \frac{\theta v}{2}\right)^{-1}$$

Again, the distribution tail is a mix of power and exponential functions.

8.1.2.2 Alpha-Stable Motions

In the case of an alpha-stable motion, described by the Lévy measure (5.4), we have, extending proposition 1.2.15 in Samorodnitsky and Taqqu (1994) to an unconstrained time:
$$\Pr(X_t > u) \overset{+\infty}{\sim} \tilde{C}(t) \frac{1+\beta}{u^\alpha} \quad (8.12)$$

and:
$$\Pr(X_t < u) \overset{-\infty}{\sim} \tilde{C}(t) \frac{1-\beta}{(-u)^\alpha} \quad (8.13)$$

where $\tilde{C}(t) = \frac{(1-\alpha)}{2\,\Gamma(2-\alpha)} \frac{\sigma^\alpha \, t}{\cos\left(\frac{\pi\alpha}{2}\right)}$ in both situations, provided $\alpha \neq 1$.

8.1.2.3 Hyperbolic Motions

Recall that generalized hyperbolic distributions have the density:

$$f_{\text{HG}}(x) = a \, \frac{K_{\lambda-\frac{1}{2}}\left(\alpha\sqrt{\gamma^2 + (x-\mu)^2}\right) \, e^{\beta(x-\mu)}}{(\gamma^2 + (x-\mu)^2)^{(\frac{1}{2}-\lambda)/2}}$$

where a is defined by:

$$a = \frac{(\alpha^2 - \beta^2)^{\lambda/2}}{\sqrt{2\pi}\alpha^{\lambda-\frac{1}{2}} \gamma^\lambda K_\lambda\left(\gamma\sqrt{\alpha^2 - \beta^2}\right)}$$

From Abramowitz and Stegun (1965), we have the following relationship:

$$K_v(z) \overset{\pm\infty}{\sim} \sqrt{\frac{\pi}{2z}} \, e^{-z}$$

so that:

$$f_{\text{HG}}(x) \overset{\pm\infty}{\sim} a \, \frac{\sqrt{\frac{\pi}{2\alpha\sqrt{\gamma^2+(x-\mu)^2}}} \, e^{-\alpha\sqrt{\gamma^2+(x-\mu)^2}} \, e^{\beta(x-\mu)}}{(\gamma^2 + (x-\mu)^2)^{(\frac{1}{2}-\lambda)/2}}$$

Because:

$$\sqrt{\gamma^2 + (x-\mu)^2} \overset{\pm\infty}{\sim} |x|$$

we can write:

$$f_{\text{HG}}(x) \overset{\pm\infty}{\sim} a \, \frac{\sqrt{\frac{\pi}{2\alpha|x|}} \, e^{-\alpha|x|} \, e^{\beta(x-\mu)}}{|x|^{\frac{1}{2}-\lambda}}$$

so that, finally:

$$f_{\text{HG}}(x) \overset{\pm\infty}{\sim} \tilde{C} \, \frac{e^{-\alpha|x|+\beta x}}{|x|^{1-\lambda}} \tag{8.14}$$

with:

$$\tilde{C} = \frac{(\alpha^2 - \beta^2)^{\lambda/2} \, e^{-\beta\mu}}{2 \, \alpha^\lambda \, \gamma^\lambda \, K_\lambda\left(\gamma\sqrt{\alpha^2 - \beta^2}\right)}$$

This asymptotic expression corresponds to a mix of exponential and power functions. Consider the shape of the asymptotic density in the two particular cases below:

(1) $\lambda = 1$: hyperbolic distribution. The asymptotic behavior of the density is:

$$f_{\rm H}(x) \overset{\pm\infty}{\sim} \tilde{C}\, e^{-\alpha|x|+\beta x}$$

with:

$$\tilde{C} = \frac{\sqrt{\alpha^2 - \beta^2}\, e^{-\beta\mu}}{2\,\alpha\,\gamma\, K_1\left(\gamma\sqrt{\alpha^2 - \beta^2}\right)}$$

(2) $\lambda = -1/2$: normal inverse Gaussian (NIG) distribution. The asymptotic behavior of the density is:

$$f_{\rm NGI}(x) \overset{\pm\infty}{\sim} \tilde{C}\, \frac{e^{-\alpha|x|+\beta x}}{|x|^{\frac{3}{2}}}$$

with:

$$\tilde{C} = \frac{\sqrt{\alpha\gamma}\, e^{-\beta\mu}}{2\,(\alpha^2 - \beta^2)^{\frac{1}{4}}\, K_{-\frac{1}{2}}\left(\gamma\sqrt{\alpha^2 - \beta^2}\right)}$$

Compared to the alpha-stable distribution, whose tail is a power function, we observe here a faster asymptotic decrease, due to the presence of an exponential function in the tail. The distribution tail remains, however, thicker than in the Gaussian case. Generalized hyperbolic distributions, showing a medium asymptotic decrease, therefore represent a compromise between the Gaussian and alpha-stable distributions.

8.1.2.4 Completely Asymmetric Alpha-Stable Distributions

We examine the behavior of the distribution tail of X_t beyond the threshold x_0 in the Carr and Wu (2003) model. The tail function $\bar{F}(x)$ satisfies:

$$\bar{F}(x) = \frac{1}{\sqrt{2\pi\alpha(\alpha-1)}} \left(\frac{x}{\alpha\gamma^\alpha}\right)^{-\frac{1}{2}\frac{\alpha}{\alpha-1}} \exp\left(-(\alpha-1)\left(\frac{x}{\alpha\gamma^\alpha}\right)^{-\frac{\alpha}{\alpha-1}}\right) \tag{8.15}$$

The tail function (8.15) displays a faster decrease rate than $x^{-\alpha}$, due to the presence of an exponential function.

We exhibit in Table 8.3 several asymptotic behaviors of Lévy process marginals.

Table 8.3 Infinitely divisible distributions and asymptotic behavior

Basic distribution	Density function	Tail function
Gaussian	Exponential of parabola	Exponential of power
Hyperbolic	Exponential of hyperbola	Exponential-power mix
Truncated α-stable		Exponential-power mix
α-stable		Power

8.2 Threshold Approach

Large values are naturally considered "large" because they exceed a predefined threshold. The second approach of extreme values, which is studied in this section, consists of modeling threshold exceedances.

8.2.1 *The Law of Threshold Exceedances*

Whereas the laws of maxima can be represented by extreme value distributions, the law of threshold exceedances can be represented by the generalized Pareto distribution.

8.2.1.1 *Threshold Models*

We now examine the value of exceedances of the type $X - x_0$, that is, we introduce threshold models. This amounts to considering the conditioned random variable $\{X - x_0 \,|\, X > x_0\}$, whose maximum value is $x_{\max} - x_0$, and whose c.d.f. has the form:

$$F_{x_0}(x) = \Pr(X - x_0 \leq x \,|\, X > x_0) \qquad (8.16)$$

for all $0 \leq x < x_{\max} - x_0$.

Let us write this cumulative distribution function differently by using the formula of compounded probabilities. As:

$$\Pr(X - x_0 < x) = \Pr(X < x + x_0)$$

we readily obtain from (8.16), and by the definition of conditioning, that:

$$F_{x_0}(x) = \frac{\Pr(x_0 < X < x + x_0)}{\Pr(X > x_0)} = \frac{F(x + x_0) - F(x_0)}{1 - F(x_0)}$$

so that:

$$F_{x_0}(x) = 1 - \frac{1 - F(x + x_0)}{1 - F(x_0)}$$

and, under the conventions introduced above:

$$\bar{F}_{x_0}(x) = \frac{\bar{F}(x+x_0)}{\bar{F}(x_0)} \qquad (8.17)$$

We will show that if the distribution of the random variable X is a law of extremes, then the law of exceedances $X - x_0$ beyond the given threshold x_0 is a generalized Pareto distribution of parameters ξ and σ_{x_0}, which we will denote by $G_{\xi,\sigma_{x_0}}(x)$. The indexation of the dispersion parameter σ_{x_0} by the threshold x_0 indicates that the value of this parameter depends on the choice of threshold and highlights the importance of this choice.

The choice of threshold x_0 is difficult. This value should be as high as possible, but if it is too high, it leads to statistical issues in the estimation of the parameters necessary for the computation of risk measures.

8.2.1.2 *The Generalized Pareto Distribution*

If the values taken by X beyond the threshold x_0 can be modeled by extreme value theory, and for sufficiently high thresholds x_0, then the theorem developed by Balkema and De Haan (1974) and by Pickands (1975) shows that $F_{x_0}(x)$ can be approximated (with an increasing degree of precision as x_0 approaches $x_{\max}$) by $G_{\xi,\sigma_{x_0}}(x)$. More precisely, the generalized Pareto distribution is a good approximation of the limit distribution of the law of exceedances (8.17) when the threshold x_0 approaches the maximum attainable value $x_{\max}$. In mathematical terms:

$$\lim_{x_0 \to x_{\max}} \sup \left| \bar{F}_{x_0}(x) - G_{\xi,\sigma_{x_0}}(x) \right| = 0$$

Approximation Let us show how we can obtain a generalized Pareto distribution by means of an approximation. We look for an approximation of $H_\xi(x)$. We have seen that, using the Fisher–Tippett theorem, and by application of (8.10) and (8.8), we can write:

$$[F(a_n x + b_n)]^n \simeq \exp\left(-\left[1+\xi\left(\frac{x-\mu}{\sigma}\right)\right]^{-\frac{1}{\xi}}\right)$$

so that:

$$[F(x)]^n \simeq \exp\left(-\left[1+\xi\left(\frac{\frac{x-b_n}{a_n}-\mu}{\sigma}\right)\right]^{-\frac{1}{\xi}}\right)$$

or:

$$[F(x)]^n \simeq \exp\left(-\left[1+\xi\left(\frac{x-(a_n\mu+b_n)}{a_n\sigma}\right)\right]^{-\frac{1}{\xi}}\right)$$

We fix n and use the notation $\mu := \mu(n) = a_n\mu + b_n$ and $\sigma := \sigma(n) = a_n\sigma$. Taking the logarithm of the above expression, we get:

$$n \ln F(x) \simeq -\left[1 + \xi\left(\frac{x-\mu}{\sigma}\right)\right]^{-\frac{1}{\xi}}$$

When x is large, $F(x)$ is close to 1, and therefore $\bar{F}(x) = 1 - F(x)$ is close to 0. As $\ln(1+u) \simeq u$ when u is small, we have $\ln F(x) = \ln(1-\bar{F}(x)) \simeq -\bar{F}(x)$ when x is large. Consequently:

$$\bar{F}(x) \simeq \frac{1}{n}\left[1 + \xi\left(\frac{x-\mu}{\sigma}\right)\right]^{-\frac{1}{\xi}} \tag{8.18}$$

which is an approximation of the law of extreme values. We use this approximation in the computation of (8.17). This yields:

$$\bar{F}_{x_0}(x) \simeq \frac{\frac{1}{n}\left[1 + \xi\left(\frac{x+x_0-\mu}{\sigma}\right)\right]^{-\frac{1}{\xi}}}{\frac{1}{n}\left[1 + \xi\left(\frac{x_0-\mu}{\sigma}\right)\right]^{-\frac{1}{\xi}}} = \left(\frac{1 + \xi\left(\frac{x+x_0-\mu}{\sigma}\right)}{1 + \xi\left(\frac{x_0-\mu}{\sigma}\right)}\right)^{-\frac{1}{\xi}}$$

or:

$$\bar{F}_{x_0}(x) \simeq \left(\frac{1 + \xi\left(\frac{x_0-\mu}{\sigma}\right) + \frac{\xi x}{\sigma}}{1 + \xi\left(\frac{x_0-\mu}{\sigma}\right)}\right)^{-\frac{1}{\xi}} = \left(1 + \frac{\xi x}{\sigma + \xi(x_0-\mu)}\right)^{-\frac{1}{\xi}}$$

Setting $\sigma_{x_0} = \sigma + \xi(x_0 - \mu)$, we obtain:

$$\bar{F}_{x_0}(x) \simeq \left(1 + \frac{\xi x}{\sigma_{x_0}}\right)^{-1/\xi}$$

which is the tail function of the generalized Pareto distribution when $\xi \neq 0$. More precisely, the tail function of exceedances $X - x_0$ follows a generalized Pareto distribution if:

$$\bar{F}_{x_0}(x) = \begin{cases} (1 + \xi x/\sigma_{x_0})^{-1/\xi} & \xi \neq 0 \\ \exp(-x/\sigma_{x_0}) & \xi = 0 \end{cases} \tag{8.19}$$

or:

$$G_{\xi,\sigma}(x) = \begin{cases} 1 - (1 + \xi x/\sigma)^{-1/\xi} & \xi \neq 0 \\ 1 - \exp(-x/\sigma) & \xi = 0 \end{cases} \tag{8.20}$$

Its domain of definition is $x \geq 0$ when $\xi > 0$ and $0 \leq x \leq -\sigma/\xi$ when $\xi < 0$.

We can localize the generalized Pareto distribution using the parameter m: $G_{\xi,\sigma,m}(x) = G_{\xi,\sigma}(x-m)$. Assume that the law of exceedances is localized by the threshold x_0. Then:

$$\bar{F}(x|X > x_0) = \left(1 + \xi\left(\frac{x-x_0}{\sigma_{x_0}}\right)\right)^{-\frac{1}{\xi}} \tag{8.21}$$

A reduced form can be obtained simply by setting $y = (x - x_0)/\sigma_{x_0}$. In this case, the preceding expression becomes:

$$\bar{F}(y) = (1 + \xi y)^{-\frac{1}{\xi}}$$

The tail functions are:

$$\bar{F}(y) = \begin{cases} (1+\xi y)^{-1/\xi} & \xi \neq 0 \\ e^{-y} & \xi = 0 \end{cases}$$

or equivalently:

$$F(y) = \begin{cases} 1 - (1+\xi y)^{-1/\xi} & \xi \neq 0 \\ 1 - e^{-y} & \xi = 0 \end{cases}$$

with:

$$\begin{cases} y \geq 0 & \text{if } \xi \geq 0 \\ 0 \leq y \leq -1/\xi & \text{if } \xi < 0 \end{cases}$$

8.2.1.3 The Pareto Distribution in the Fréchet Case

When $\xi > 0$ (in the case of a Fréchet law), we can perform a change of parameters by setting $\xi = 1/\alpha$ and $\sigma_{x_0} = x_0 \xi = x_0/\alpha$, and by replacing, in (8.21), the parameters ξ and σ_{x_0} by their new value. We recognize the classic form of the Pareto distribution with the characteristic exponent α quantifying the weight of the tail:

$$\bar{F}(x|X > x_0) = \left(\frac{x_0}{x}\right)^\alpha \qquad (8.22)$$

The tail function (8.22) corresponds to a Pareto distribution of type I, or simple Pareto distribution. This parameterization can be generalized by including a translation term c such that $\sigma = (x_0 + c)\xi$ (we keep $\xi = 1/\alpha$). The tail function can then be expressed as:

$$\bar{F}(x|X > x_0) = \left(\frac{x_0 + c}{x + c}\right)^\alpha \qquad (8.23)$$

This latter tail function corresponds to a Pareto distribution of type II.

Let us illustrate this attraction in the case of the MDA of Fréchet. We know that alpha-stable distributions belong to this MDA and we therefore choose them for our illustration. For all $0 \leq x < x_{\max}$, denote by $f(x|X > x_0)$ the density function of such a distribution beyond the threshold x_0. The Lévy measure (8.12) associated with an alpha-stable distribution leads to:

$$f(x|X > x_0) = \frac{C_+}{x^{1+\alpha}} \mathbf{1}_{[x_0, +\infty[}(x)$$

then, because:

$$F(x|X > x_0) = \int_{-\infty}^{x} f(t|t > x_0)\, dt = \int_{x_0}^{x} f(t)\, dt$$

we have:

$$F(x|X > x_0) = \left[-\frac{C_+}{\alpha t^\alpha}\right]_{x_0}^{x} = \frac{C_+}{\alpha x_0^\alpha} - \frac{C_+}{\alpha x^\alpha}$$

We choose a value of the threshold x_0 that depends on the scale parameter C_+ of the distribution tail by setting $x_0 = (C_+/\alpha)^{1/\alpha}$. Indeed, for power functions, the order of magnitude of the threshold depends on the size of distribution tail which is given by C_+. We can write:

$$F(x|X > x_0) = 1 - \left(\frac{x_0}{x}\right)^\alpha$$

and we come back to Eq. (8.22):

$$\bar{F}(x|X > x_0) = \left(\frac{x_0}{x}\right)^\alpha$$

The behavior of distribution tails of X, beyond a threshold x_0 and for x_0 sufficiently large, is that of a Pareto law of type I. Written differently, from the point of view of their distribution tails, alpha-stable distributions present an asymptotic Paretian behavior.

We can summarize these results with the two following equivalent assertions:

(1) The values taken by X beyond the threshold x_0 can be modeled by Fréchet distribution.
(2) The exceedances $X - x_0$ follow a Pareto distribution of type I or II.

8.2.2 Linearity of Means beyond Thresholds

The generalized Pareto distribution has a very interesting property: its conditional means, defined as follows:

$$M(x_0) = \mathbb{E}\left[X\,|X > x_0\right] \qquad (8.24)$$

are linear.

Another approach (see Barbut (1989, 1998)), distinct from that relying on the generalized Jenkinson–von Mises distribution, leads to a generalized Pareto distribution. This approach looks for (when the threshold x_0 is given) a linear relationship of the type:

$$M(x_0) = a\,x_0 + b \qquad (8.25)$$

This functional relationship asserts that the conditional mean beyond a threshold x_0 is a multiple of that threshold up to a given constant, or put differently, a linear function of that threshold.

If a linear relationship of this type is satisfied by a random variable X, then we show that the probability distribution of this random variable is a generalized Pareto distribution. Conversely, if X is a random variable that follows a generalized Pareto distribution, then the preceding linear relationship (8.25) holds.

We can develop the conditional mean given in formula (8.24) as follows:

$$M(x_0) = \frac{\mathbb{E}[X \mathbf{1}_{X > x_0}]}{\Pr(X > x_0)} = \frac{1}{\Pr(X > x_0)} \int_{x_0}^{\infty} t \, dF(t) = \frac{1}{\bar{F}(x_0)} \int_{x_0}^{\infty} t \, dF(t) \tag{8.26}$$

We want to obtain the shape of the tail function $\bar{F}$, such that, for an arbitrary threshold x, the relationship (8.25) is verified. Combining (8.25) and (8.26) yields:

$$\frac{1}{\bar{F}(u)} \int_{u}^{\infty} t \, dF(t) = a\,u + b$$

or:

$$\int_{u}^{\infty} t \, dF(t) = (a\,u + b)\, \bar{F}(u)$$

Differentiating with respect to u, we find:

$$-u \frac{dF(u)}{du} = a\, \bar{F}(u) + (a\,u + b)\frac{d\bar{F}(u)}{du}$$

and (note that $dF = -d\bar{F}$):

$$a\, \bar{F}(u)\, du = -(u(a-1) + b)\, d\bar{F}(u) \tag{8.27}$$

which is a differential equation in $\bar{F}(u)$. Solving it will yield all the possible researched laws, depending on the values taken by the parameters a and b. Note that $a < 1$ produces solutions that cannot be survival functions, whereas the case $a = 1$ and $b = 0$ is degenerate. This therefore leads us to examine the following situations:

(1) First case: $a = 1$ and $b > 0$

The differential equation (8.27) can be written as:

$$\bar{F}(u)\, du = -b\, d\bar{F}(u)$$

so that:

$$\frac{d\bar{F}}{\bar{F}} = -\frac{du}{b}$$

Integrating this equation between the bounds x_0 and x, we obtain:
$$\bar{F}(x) = e^{-(x-x_0)/b}, \qquad x > x_0$$
The tail function has an exponential distribution whose standard deviation is equal to b. The conditional mean beyond the threshold x_0 can be obtained by computing (8.26), so by simple substitution of $a = 1$ in (8.25):
$$M(x_0) = x_0 + b$$
which shows that, in the exponential case, the conditional mean is equal to the sum of the threshold and of the standard deviation of the tail distribution.

(2) Second case: $a > 1$ and $b \neq 0$

The differential equation (8.27) can be written as:
$$a\,\bar{F}(u)\,du = -(u(a-1) + b)\,d\bar{F}(u)$$
so that:
$$\frac{d\bar{F}}{\bar{F}} = -\frac{a}{a-1}\frac{du}{u + \frac{b}{a-1}}$$
Set $\alpha = \frac{a}{a-1}$ and $c = \frac{b}{a-1}$. We then have:
$$\frac{d\bar{F}}{\bar{F}} = -\alpha\,\frac{du}{u+c}$$
Integrating between the bounds x_0 and x, we obtain:
$$\bar{F}(x) = \left(\frac{x_0 + c}{x + c}\right)^{\alpha} \qquad x > x_0$$
The tail function has a Pareto distribution of type II.

The relationship between α and a can be expressed as follows:
$$\frac{1}{\alpha} + \frac{1}{a} = 1$$
we have $a = \frac{\alpha}{\alpha-1}$ and $a - 1 = \frac{1}{\alpha-1}$. Therefore $b = c(a-1) = \frac{c}{\alpha-1}$. The conditional mean beyond the threshold x_0 is then:
$$M(x_0) = a\,x_0 + b = \frac{\alpha}{\alpha - 1}\,x_0 + \frac{c}{\alpha - 1}$$
For example, if $\alpha = 1.5$, then $a = 3$ and $M(x_0) = 3\,x_0 + 2c$.

Set $\xi = 1/\alpha$ and $\sigma_{x_0} = (x_0 + c)\,\xi$. With this change of parameterization, the tail function can be expressed as follows:
$$\bar{F}(x) = \left(\frac{x+c}{x_0+c}\right)^{-\alpha} = \left(\frac{x - x_0 + x_0 + c}{x_0 + c}\right)^{-\frac{1}{\xi}} = \left(1 + \frac{x - x_0}{x_0 + c}\right)^{-\frac{1}{\xi}}$$

so that:
$$\bar{F}(x) = \left(1 + \frac{\xi}{\sigma_{x_0}}(x - x_0)\right)^{-\frac{1}{\xi}}$$

Noting that $\frac{\alpha}{\alpha-1} = \frac{1}{1-\xi}$, we can express the conditional mean beyond x_0 as follows:
$$M(x_0) = \frac{\alpha}{\alpha - 1} x_0 + \frac{1}{\alpha}\frac{\alpha}{\alpha - 1} c$$

and then:
$$M(x_0) = \frac{x_0}{1-\xi} + \frac{\xi c}{1-\xi}$$

In addition, we deduce from $\sigma_{x_0} = (x_0 + c)\,\xi$ that $\xi c = \sigma_{x_0} - x_0\,\xi$ and that:
$$M(x_0) = \frac{x_0}{1-\xi} + \frac{\sigma_{x_0} - x_0 \xi}{1-\xi}$$

yielding another formulation of the conditional mean:
$$M(x_0) = x_0 + \frac{\sigma_{x_0}}{1-\xi}$$

(3) Third case: $a > 1$ and $b = 0$

The differential equation (8.27) can be written as:
$$a\,\bar{F}(u)\,du = -u(a-1)\,d\bar{F}(u)$$

and we obtain, by integration between the bounds x_0 and x:
$$\bar{F}(x) = \left(\frac{x_0}{x}\right)^{\alpha} \qquad x > x_0$$

The tail function has a Pareto distribution of type I (simple Pareto distribution). The conditional mean beyond the threshold can be expressed as:
$$M(x_0) = \frac{\alpha}{\alpha - 1} x_0$$

For example, if $\alpha = 1.5$, then $a = 3$ and $M(x_0) = 3\,x_0$.

Performing the same change of parameterization as before for $\xi = 1/\alpha$, but this time with $\sigma_{x_0} = x_0\,\xi$ (with $c = 0$), we obtain the tail function:
$$\bar{F}(x) = \left(1 + \frac{\xi}{\sigma_{x_0}}(x - x_0)\right)^{-\frac{1}{\xi}}$$

and:
$$M(x_0) = \frac{x_0}{1-\xi}$$

The different solutions of the differential equation (8.27) are displayed in Table 8.4.

Table 8.4 Cumulative distribution function $\bar{F}_{(x_0)}(x)$ parametrized in α or in ξ

	$M(x_0) = a\,x_0 + b$ $a > 1$	
	Pareto type I $b = 0$	Pareto type II $b \neq 0$
(α, c)	$\bar{F}_{x_0}(x) = \left(\frac{x_0}{x}\right)^{\alpha}$	$\bar{F}_{x_0}(x) = \left(\frac{x_0+c}{x+c}\right)^{\alpha}$
(ξ, σ)	$\bar{F}_{x_0}(x) = \left(1 + \frac{\xi}{\sigma_{x_0}}(x - x_0)\right)^{-\frac{1}{\xi}}$	
σ_{x_0}	$\sigma_{x_0} = x_0\,\xi$	$\sigma_{x_0} = (x_0 + c)\,\xi$

8.3 Statistical Phenomenon Approach

We examine here two main statistical characteristics of stock dynamics: a strong hierarchy of large values, and its corollary, the concentration of results over a very small number of values.

8.3.1 *Concentration of Results*

Fat-tailed distributions often display a phenomenon known as the law of "80/20": 20% of the values contribute to 80% of the total result, and *vice versa*. Extreme value theory gives a quantitative content to this intuition, by providing a relationship between the values of this ratio (80/20, 70/30, ...) and the values of the parameter ξ.

8.3.1.1 *The Law of 80/20*

We introduce the notion of the contribution of the k largest values to the total result. Let $\{x_1, \cdots, x_n\}$ be a sample of n realizations of the n random variables $\{X_1, \cdots, X_n\}$. Assume that the X_i are i.i.d. and have a finite mean μ. As before we denote by $X_{(k)}$ the random variable representing the k-th value of the sample ordered as follows $X_{(1)} \geq \cdots \geq X_{(n)}$.

The contribution of the $[np]$ largest values to the total result is defined by the random variable:

$$T_n(p) = \frac{X_{(1)} + \cdots + X_{([np])}}{X_{(1)} + \cdots + X_{(n)}} \quad (8.28)$$

where:

$$\frac{1}{n} < p \leq 1$$

and where $[np]$ is the integer part of np. The quantity $T_n(p)$ represents the ratio of the total contribution of the $[np]$ largest values to the sum of the n values sampled.

We want to examine whether the positive performance of a portfolio is concentrated over a small number of exchange days (this example is considered in greater detail on p. 208). In this case, $X_{(k)}$ represents a positive variation or return. If $p = 0.2$, and if $T_n(0,2) = 0.8$, then the total of the 20% largest positive variations contributes to 80% of the total positive performance of the portfolio. The ratio (8.28) expresses and generalizes the intuition of the law of 80/20.

8.3.1.2 The Concentration Index and the Lorenz Curve

In the previous paragraph, we did not specify the probability distribution of the random variables X_i. We now want to establish a relationship between this probability distribution and the structure of concentration: is this the law of 80/20 or another ratio? To do this, we need to show the c.d.f. of the probability distribution of the X_i.

Let F represent the c.d.f. of the positive returns of a financial asset, with a mean μ_F. For all $0 \leq p \leq 1$, we define the performance concentration index of this asset at the level p by:

$$D_F(p) = \frac{1}{\mu_F} \int_{1-p}^{1} F^{-1}(z) \, dz = \frac{1}{\mu_F} \int_{F^{-1}(1-p)}^{\infty} y \, dF(y) \qquad (8.29)$$

where $F(.)$ is the c.d.f. of a positive random variable Y, F^{-1} is the inverse of F and:

$$\mu_F = \mathbb{E}[Y] = \int_0^1 F^{-1}(z) \, dz = \int_0^{\infty} y \, dF(y)$$

with $\mu_F < \infty$.

The quantity D_F is an index measuring concentration. We will show that its value represents the way in which the 100 p% largest positive values contribute to the total positive performance.

Recall that the empirical c.d.f. (or cumulative frequency), denoted by $F_n(x)$, of a random variable X for a sample of size n is defined for all $x \in \mathbb{R}$ by:

$$F_n(x) = \frac{1}{n} \text{card}\{i : 1 \leq i \leq n, \, X_i \leq x\} = \frac{1}{n} \sum_{i=1}^{n} \mathbf{1}_{\{X_i \leq x\}} \qquad (8.30)$$

with, in particular:

$$F_n(X_{(k)}) = \frac{n-k+1}{n} = 1 - \frac{k-1}{n}$$

From this expression, we deduce the empirical inverse cumulative distribution function:
$$F_n^{-1}(t) = X_{(k)} \quad \forall\, t \in \left]1-\frac{k}{n}, 1-\frac{k-1}{n}\right] \tag{8.31}$$

We can show that:
$$T_n(p) = D_{F_n}\left(\frac{[np]}{n}\right)$$

where the right-hand side is the empirical index or performance concentration of the asset, defined as a function of the empirical cumulative distribution function. Indeed, we can write, from Eq. (8.29):
$$D_{F_n}\left(\frac{[np]}{n}\right) = \frac{1}{\mu_{F_n}} \int_{1-\frac{[np]}{n}}^{1} F_n^{-1}(z)\, dz$$

where μ_{F_n} is the empirical mean:
$$\mu_{F_n} = \frac{\sum_{k=1}^{n} X_k}{n}$$

We then have:
$$D_{F_n}\left(\frac{[np]}{n}\right) = \frac{n \sum_{k=[np]}^{1} \int_{1-\frac{k}{n}}^{1-\frac{k-1}{n}} F_n^{-1}(z)\, dz}{\sum_{k=1}^{n} X_k}$$

From the definition (8.31) of the empirical inverse c.d.f., we deduce:
$$D_{F_n}\left(\frac{[np]}{n}\right) = \frac{n \sum_{k=[np]}^{1} X(k)\left[\left(1-\frac{k-1}{n}\right) - \left(1-\frac{k}{n}\right)\right]}{\sum_{k=1}^{n} X_k}$$

so that:
$$D_{F_n}\left(\frac{[np]}{n}\right) = \frac{\sum_{k=[np]}^{1} X(k)}{\sum_{k=1}^{n} X_k} = \frac{\sum_{k=1}^{[np]} X(k)}{\sum_{k=1}^{n} X_k} = T_n(p)$$

We can show that $T_n(p) = D_{F_n}\left(\frac{[np]}{n}\right)$ is equivalent to $D_{F_n}(p)$, and then that:
$$\lim_{n \to \infty}\left(\sup_{p \in [0,1]} |T_n(p) - D_F(p)|\right) = 0 \quad p.\,s.$$

Table 8.5 Construction of the Lorenz curve

Ordered values	Cumulative population (rank)	Cumulative frequency $F_X(.)$	Cumulative mass (partial sums)	Cumulative frequency Fr(.)
$x_{(1)}$	1	$1/n$	$x_{(1)} = s_{(1)}$	$s_{(1)}/s_{(n)}$
$x_{(2)}$	2	$2/n$	$x_{(1)} + x_{(2)} = s_{(2)}$	$s_{(2)}/s_{(n)}$
$\vdots$	$\vdots$	$\vdots$	$\vdots$	$\vdots$
$x_{(k)}$	k	k/n	$x_{(1)} + \cdots + x_{(k)} = s_{(k)}$	$s_{(k)}/s_{(n)}$
$\vdots$	$\vdots$	$\vdots$	$\vdots$	$\vdots$
$x_{(n)}$	n	1	$x_{(1)} + \cdots + x_{(n)} = s_{(n)}$	1

for i.i.d. positive random variables of finite mean.

Setting $L_F(1-p) = 1 - D_F(p)$ implies that L_F is the Lorenz curve exhibiting the concentration of results (for instance, the concentration of the performance of a portfolio). We can construct this curve as shown in Table 8.5.

8.3.1.3 Application to a Generalized Pareto Distribution

We display in Table 8.6, p. 207, the values of $D_F(p)$ with respect to the values of ξ and p for a generalized Pareto distribution. We have in that case:

$$D_F(p) = \frac{1-\xi}{\xi}\left(\frac{p^{1-\xi}}{1-\xi} - p\right) \qquad (8.32)$$

To obtain this result, we start by computing the term μ_F in formula (8.29). We use the generalized Pareto c.d.f.:

$$F(y) = 1 - \left(1 + \frac{\xi y}{\sigma}\right)^{-\frac{1}{\xi}}$$

defined for all positive y. This amounts to using the density function:

$$f(y) = \frac{1}{\sigma}\left(1 + \frac{\xi y}{\sigma}\right)^{-\frac{1}{\xi}-1}$$

and we have:

$$\mu_F = \mathbb{E}[Y] = \int_0^{+\infty} yf(y)dy = \int_0^{+\infty} y\frac{1}{\sigma}\left(1 + \frac{\xi y}{\sigma}\right)^{-\frac{1}{\xi}-1} dy$$

The change of variable:

$$x = 1 + \frac{\xi y}{\sigma} \qquad (8.33)$$

for x going from 1 to $+\infty$ yields:
$$\mu_F = \int_1^{+\infty} \frac{\sigma}{\xi}(x-1)\frac{1}{\sigma}x^{-\frac{1}{\xi}-1}\frac{\sigma}{\xi}dx$$
so that:
$$\mu_F = \frac{\sigma}{\xi^2}\int_1^{+\infty}(x-1)x^{-\frac{1}{\xi}-1}dx \tag{8.34}$$
which gives:
$$\mu_F = \frac{\sigma}{\xi^2}\left(\left[\frac{x^{1-\frac{1}{\xi}}}{1-\frac{1}{\xi}}\right]_1^{+\infty} - \left[\frac{x^{-\frac{1}{\xi}}}{-\frac{1}{\xi}}\right]_1^{+\infty}\right) = \frac{\sigma}{\xi^2}\left(\frac{1}{\frac{1}{\xi}-1}-\xi\right)$$
and finally:
$$\mu_F = \frac{\sigma}{\xi^2}\left(\frac{\xi}{1-\xi} - \frac{(1-\xi)\xi}{1-\xi}\right) = \frac{\sigma}{1-\xi} \tag{8.35}$$

For the computation of formula (8.29), we then need to compute $F^{-1}(1-p)$. After the change of variable (8.33), we obtain:
$$F(x) = 1 - x^{-\frac{1}{\xi}}$$
so that:
$$F(x) = 1-p \Leftrightarrow x = p^{-\xi}$$
We can then write, by analogy with formula (8.34):
$$\int_{F^{-1}(1-p)}^{\infty} y\, dF(y) = \frac{\sigma}{\xi^2}\int_{p^{-\xi}}^{+\infty}(x-1)x^{-\frac{1}{\xi}-1}dx$$
and:
$$\int_{F^{-1}(1-p)}^{\infty} y\, dF(y) = \frac{\sigma}{\xi^2}\left(\left[\frac{x^{1-\frac{1}{\xi}}}{1-\frac{1}{\xi}}\right]_{p^{-\xi}}^{+\infty} - \left[\frac{x^{-\frac{1}{\xi}}}{-\frac{1}{\xi}}\right]_{p^{-\xi}}^{+\infty}\right) = \frac{\sigma}{\xi^2}\left(\frac{p^{1-\xi}}{\frac{1}{\xi}-1}-p\xi\right)$$
so, finally:
$$\int_{F^{-1}(1-p)}^{\infty} y\, dF(y) = \frac{\sigma}{\xi}\left(\frac{p^{1-\xi}}{1-\xi}-p\right) \tag{8.36}$$

Dividing formula (8.36) by formula (8.35), we directly obtain formula (8.32).

The law of 80/20, according to which 20% of individuals contribute to 80% of the total result, corresponds to a c.d.f. F such that $D_F(0.2) = 0.8$. If we now look at Table 8.6, we come back to this law with a generalized Pareto distribution of index $\xi \approx 0.7$ so that $\alpha \approx 1.4$. These values are

Table 8.6 Concentration level w.r.t. parameter value of ξ

p \ ξ	0.1	0.2	0.3	0.4	0.5	0.6	**0.7**	0.8
0.00	0.0000	0.0000	0.0000	0.0000	0.0000	0.0000	0.0000	0.0000
0.01	0.0685	0.0856	0.1094	0.1427	0.1900	0.2575	0.3546	0.4951
0.02	0.1158	0.1387	0.1689	0.2091	0.2628	0.3352	0.4332	0.5666
0.03	0.1560	0.1825	0.2163	0.2599	0.3164	0.3899	0.4861	0.6124
0.04	0.1919	0.2207	0.2569	0.3024	0.3600	0.4332	0.5268	0.6466
0.05	0.2246	0.2551	0.2927	0.3393	0.3972	0.4695	0.5601	0.6741
0.06	0.2549	0.2866	0.3251	0.3722	0.4299	0.5009	0.5885	0.6971
0.07	0.2832	0.3157	0.3548	0.4020	0.4592	0.5286	0.6133	0.7169
0.08	0.3099	0.3429	0.3822	0.4293	0.4857	0.5535	0.6353	0.7343
0.09	0.3350	0.3684	0.4078	0.4545	0.5100	0.5761	0.6551	0.7498
0.10	0.3589	0.3924	0.4318	0.4780	0.5325	0.5968	0.6731	0.7637
0.15	0.4634	0.4961	0.5334	0.5759	0.6246	0.6803	0.7443	0.8178
0.20	0.5492	0.5797	0.6138	0.6518	0.6944	0.7422	**0.7958**	0.8560
0.25	0.6217	0.6494	0.6798	0.7132	0.7500	0.7906	0.8354	0.8848
0.30	0.6838	0.7084	0.7350	0.7640	0.7954	0.8297	0.8669	0.9075
0.35	0.7374	0.7589	0.7819	0.8066	0.8332	0.8618	0.8926	0.9258
0.40	0.7838	0.8022	0.8218	0.8427	0.8649	0.8886	0.9138	0.9407
0.45	0.8241	0.8396	0.8560	0.8733	0.8916	0.9110	0.9314	0.9530
0.50	0.8589	0.8717	0.8852	0.8994	0.9142	0.9298	0.9461	0.9632
0.55	0.8888	0.8993	0.9101	0.9215	0.9332	0.9455	0.9583	0.9716
0.60	0.9145	0.9227	0.9312	0.9401	0.9492	0.9587	0.9685	0.9786
0.65	0.9361	0.9424	0.9489	0.9556	0.9625	0.9695	0.9768	0.9843
0.70	0.9542	0.9588	0.9635	0.9684	0.9733	0.9784	0.9836	0.9889
0.75	0.9689	0.9721	0.9753	0.9787	0.9821	0.9855	0.9890	0.9926
0.80	0.9805	0.9826	0.9846	0.9867	0.9889	0.9910	0.9932	0.9954
0.85	0.9893	0.9904	0.9916	0.9927	0.9939	0.9951	0.9963	0.9975
0.90	0.9953	0.9958	0.9963	0.9969	0.9974	0.9979	0.9984	0.9989
0.95	0.9989	0.9990	0.9991	0.9992	0.9994	0.9995	0.9996	0.9997
1.00	1.0000	1.0000	1.0000	1.0000	1.0000	1.0000	1.0000	1.0000

indicated in bold in the table. Other situations are possible: for instance, setting p to 20%, for $\xi = 0.3$, the law of 80/20 instead resembles a law of 60/20, and so on.

Note that if the condition $\mu_F < \infty$ is not verified, then $T_n(p) \to 1$, where the convergence holds in probability.

Furthermore, we can observe that, in the standard Paretian case, we have the following simplified formula:

$$D_F(p) = p^{1-\xi}$$

8.3.1.4 Illustration

We present an empirical illustration of the performance concentration phenomenon. The analysis is conducted based on the performance structure of the stock portfolio of the insurance company, SMABTP, between December 31, 2001 and September 23, 2004. We follow Walter (2004b).

This portfolio was chosen because of SMABTP's management style. This company has enforced a management process that is homogeneous and stable, that is not constructed in reference to benchmarks, is not oriented towards a *top down* approach, and that does not make use of the diversification theory as it is usually applied by the financial industry (see SMABTP (2004a, 2004b)). On the contrary, this company manages its portfolio in a precise and concentrated way.

The results of this analysis confirm the concentration of performance, for a sample of individual days and assets. More precisely:

(1) The concentration of the portfolio's performance over a small number of days reveals concentration structures of the type 70/30.
(2) The concentration of performance over a small number of days reveals a disproportionate contribution of a very small number of stocks to the total performance. Removing the 7 worst-performing stocks from the portfolio of 114 stocks leads to an increase in performance of 5%.

With this level of concentration, the performance of the portfolio between December 31, 2001 and September 23, 2004 was −3.76%, compared to the CAC40 index (−21.00%) or the Eurostoxx index (−28.13%). The weak diversification has led to a strong performance with respect to these indices. From these results, it appears that a decision by management to diversify the portfolio would have yielded bad performances. We now set out in detail the calculation of performance concentration.

Performance concentration over a few days

The market value of the stock portfolio of the insurance company SMABTP is denoted by $V(t)$ at the valuation date t. Recall from (2.13) that the rate of return of the portfolio between the dates 0 and t is:

$$X(t) = \ln V(t) - \ln V(0) \qquad (8.37)$$

The process $\{X_t,\ t \geq 0\}$ represents the evolution of the return in base 100. The periodic return over a period τ is:

$$\Delta X(t, \tau) = X(t) - X(t - \tau) \qquad (8.38)$$

The behavior of the portfolio is examined over the total period $[0, T]$ where τ is equal to one day: $\Delta X(t, 1) = \Delta X_k = X_k - X_{k-1}$. As already noted in Chapter 3, p. 35, to simplify the notation when it is not necessary to recall the characteristics (τ, T, Δ) of the difference process, we denote by $\Delta X_k = x_k$ the k-th periodic return obtained over a chosen scale τ.

We constitute two series: one for the positive daily returns, and one for the negative daily returns. The samples are then ranked by decreasing order. There are n positive variations and n' negative or zero variations, with $n + n' = N$. Because the same method is used in both cases, we only describe the method applied to the n positive variations.

We operate over a sample of size n:
$$\{x_1, x_2, \cdots, x_n\}$$
that is ranked in decreasing order:
$$x_{(1)} \geq x_{(2)} \geq \cdots \geq x_{(n)}$$
As observed before, the largest of these realizations, $x_{(1)}$, can be considered the realization of a new random variable $\Delta X_{(1)}$, and similarly for the other realizations. For $x_{(k)}$, $\Delta X_{(k)}$ represents the k-th largest increase over the period considered. In particular, $\Delta X_{(1)} = \max(\Delta X_1, \cdots, \Delta X_n)$, and $\Delta X_{(n)} = \min(\Delta X_1, \cdots, \Delta X_n)$. The n random variables are then ranked as follows:
$$\Delta X_{(1)} > \Delta X_{(2)} > \Delta X_{(3)} > \cdots > \Delta X_{(n)}$$

The cumulative performance of the k largest positive variations over the period $[0, T]$ is denoted by $X_{(k)}$, if it is a random variable, or $s_{(k)}$, if it is in the form of empirical data computed from the sample. This cumulative performance satisfies:
$$X_{(k)} = \sum_{i=1}^{k} \Delta X_{(i)} \qquad s_{(k)} = \sum_{i=1}^{k} x_{(i)}$$
The total cumulative positive performance of the portfolio is then:
$$X_{(n)} = \sum_{i=1}^{n} \Delta X_{(i)} \qquad s_{(n)} = \sum_{i=1}^{n} x_{(i)}$$
We can now examine the way in which performance is concentrated over the largest variations.

Applying (8.28) directly, the contribution of the $100p$ (p is defined as a percentage) largest positive daily variations to total performance is the ratio:
$$J_+(p) = \frac{\Delta X_{(1)} + \cdots + \Delta X_{([np])}}{X_{(n)}} \qquad \frac{1}{n} < p < 1 \qquad (8.39)$$

Table 8.7 Performance concentration over a sample of days (values corresponding to the Lorenz curves displayed in Fig. 8.1), with p equal to 10%, 30%, and 50%

	Total period		Period 1		Period 2		Period 3	
	−	+	−	+	−	+	−	+
$J(0.10)$	0.30	0.35	0.30	0.35	0.30	0.30	0.30	0.25
$J(0.30)$	0.65	0.68	0.63	0.70	0.62	0.65	0.60	0.58
$J(0.50)$	0.85	0.85	0.83	0.87	0.82	0.85	0.80	0.80

where $[np]$ is the integer part of np.

The Lorenz curve of performance concentration is constructed as shown in Table 8.5 using the empirical contribution of the k largest positive variations to total performance, given by the ratio:

$$J_n(k) = \frac{x_{(1)} + \cdots + x_{(k)}}{x_{(1)} + \cdots + x_{(n)}} = \frac{s_{(k)}}{s_{(n)}}$$

The same methodology holds for negative performances.

Recall that the period of study of the portfolio is from December 31, 2001 to September 23, 2004, so is made of 719 work days. We examine the following sub-periods: the year 2002 (266 observations), the year 2003 (262 observations), and the year 2004 (191 observations). Figure 8.1 displays the Lorenz curves corresponding to these sub-periods. Particular values of these curves are given in Table 8.7, which presents the results on the temporal concentration of performance.

The columns of this table represent: the total period (12/31/2001–09/23/2004), the first sub-period (the year 2002), the second sub-period (the year 2003), and the third sub-period (12/31/2003–09/23/2004). For each period, we give the value of the negative concentration $J_-(p)$ and of the positive concentration $J_+(p)$, for three values of p: 10%, 30% and 50%. A relative stability of results can be observed betwen the sub-periods 1 and 2 and the total period. The third sub-period is slightly different, perhaps due to the shorter time length of the observation period.

It appears that 10% of the positive (resp. negative) variations contribute to approximately 30% of the positive (resp. negative) performance, that 30% of the positive (resp. negative) variations contribute to approximately 65% of the positive (resp. negative) performance, and that 50% of the positive (resp. negative) variations contribute to approximately 85% of the positive (resp. negative) performance.

We now seek to assess the effect on the total performance of removing the k largest variations. The principle of this approach consists in recalcu-

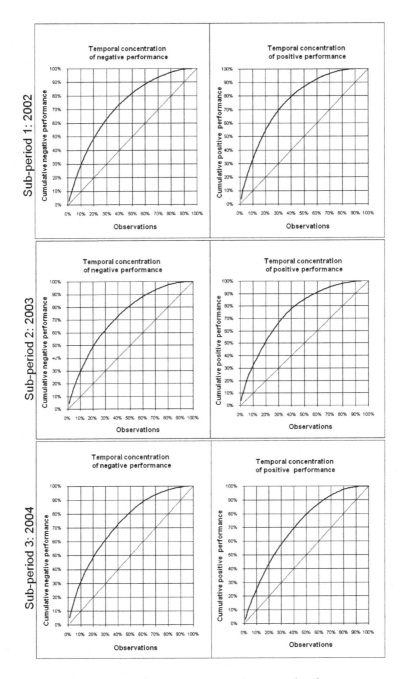

Figure 8.1 Performance concentration over a few days

lating the total performance if the worst day can be avoided, if the worst two days can be avoided, etc. A similar approach can be developed with the best days. This computation allows us to assess the effect of these days on the total performance.

It is not possible in practice to completely avoid large market downturns, but it is possible to limit their effect, for instance with a strategy making use of options. As shown below, the performance discrepancy obtained by reallocations is so large that, even by using options to limit the effect of the worst days at 20% of their level, it is possible to greatly increase the final performance, up to the cost of the derivatives. The description of these methods goes beyond the aims of this book.

The concentration of total performance can be measured by progressively removing the best (or worst) days, in the following manner. We start from the total performance:

$$s_N = \underbrace{x_{(1)} + \cdots + x_{(n)}}_{\text{ranked positive variations}} + \underbrace{x_{(n+1)} + \cdots + x_{(N)}}_{\text{ranked negative variations}}$$

Then, we subtract the largest increase, in order to obtain the new total performance, reduced by this largest increase, as:

$$s_N^{(-1)} = s_N - x_{(1)} = x_{(2)} + \cdots + x_{(n)} = s_N - s_{(1)}$$

Similarly with the two largest increases:

$$s_N^{(-2)} = s_N - \left(x_{(1)} + x_{(2)}\right) = x_{(3)} + \cdots + x_{(n)} = s_N - s_{(2)}$$

and so on. Set $s_N^{(0)} = s_N$. We obtain a sequence of new corrected performances which decreases with the number of subtracted days of stock increase:

$$s_N^{(0)} > s_N^{(-1)} > s_N^{(-2)} > \cdots > s_N^{(-n)}$$

Removing days of stock decrease can be performed in a similar fashion.

Particular values of the effect of such removals are given in Table 8.8. The three lines $s(-.)$ indicate the value of the annualized performance recomputed after removing the 5, 10, and 15 largest decreases (resp. increases) of the period analyzed.

Performance concentration for a sample of assets

The analysis of performance concentration for a sample of assets employs the same method as that described in the preceding paragraphs. Over a given period of study $[0, t_1]$, the performance of the portfolio is decomposed into the contributions of each line j. From a notation point of view, this

Table 8.8 Performance concentration over a sample of days: effect of removing the 5, 10, and 15 largest decreases (resp. increases)

	Total period	
	−	+
$s(-5)$	+8%	−8%
$s(-10)$	+15%	−14%
$s(-15)$	+20%	−20%

analysis is conducted over a unique period, which allows us to suppress the time indexation. However, an indexation for assets does need to be introduced.

Denoting by X_P the total performance of the portfolio over the period (one day, one year), the contribution of each line j to this performance is denoted by C_j. We have:

$$X_P = \sum_{j=1}^{N} C_j$$

As before, we distinguish between two series of contributions: the positive and the negative ones, ranked by decreasing order. There are n positive contributions and n' negative or zero contributions, with $n + n' = N$. The k-th largest contribution over the period of study is denoted by $C_{(k)}$:

$$C_{(1)} > C_{(2)} > C_{(3)} > \cdots > C_{(n)}$$

The core variable is now the contribution of one asset.

To show the concentration of the positive performance for a sample of assets, we consider the aggregation of the $100\,p$ (p is defined as a percentage) largest positive contributions, and we divide this quantity by the total positive performance. We therefore consider the ratio:

$$T_+(p) = \frac{C_{(1)} + \cdots + C_{([np])}}{C_{(1)} + \cdots + C_{(n)}} \qquad \frac{1}{n} < p < 1 \qquad (8.40)$$

where $[np]$ is the integer part of np.

As with temporal concentration, we construct a Lorenz concentration curve of the positive contributions of asset choices using the empirical computation of the k largest positive contributions to the total performance. This construction is presented in Table 8.9.

We now assess the effect of removing the p strongest positive contributions to the total performance. This latter quantity is equal to:

$$X_P = \underbrace{C_{(1)} + C_{(2)} + \cdots + C_{(n)}}_{\text{ranked positive contributions}} + \underbrace{C_{(n+1)} + C_{(n+2)} + \cdots + C_{(N)}}_{\text{ranked negative contributions}}$$

Table 8.9 Construction of the Lorenz curve of positive contributions of individual assets within the portfolio

Frequency	Positive performance mass	Contribution
$1/n$	$C_{(1)}$	$T_n(1)$
$2/n$	$C_{(1)} + C_{(2)}$	$T_n(2)$
$\vdots$	$\vdots$	$\vdots$
k/n	$C_{(1)} + C_{(2)} + \cdots + C_{(k)}$	$T_n(k)$
$\vdots$	$\vdots$	$\vdots$
1	$C_{(1)} + C_{(2)} + \cdots + C_{(k)} + \cdots + C_{(n)}$	1

The total performance diminished by the strongest variation is:

$$X_P - C_{(1)} = C_{(2)} + \cdots + C_{(N)}$$

Subtracting the two strongest variations:

$$X_P - \left(C_{(1)} + C_{(2)}\right) = C_{(3)} + \cdots + C_{(N)}$$

and so on. The same construction holds for negative contributions.

To analyze the performance concentration that is due to the choice of assets in the portfolio, we examine the total period from December 31, 2001 to September 23, 2004. Over this period, the total performance (without reinvesting dividends) is -3.76%, and stems from the aggregation of 114 lines:

$$X_P = \sum_{j=1}^{114} C_j = -3.76\%$$

This performance can be compared to those of the CAC40 index (-21.00%) or the Eurostoxx index (-28.13%). To analyze the sources of performance, we rank the contributions by decreasing order:

$$C_{(1)} > C_{(2)} > C_{(3)} > \cdots > C_{(n)}$$

so here:

$$C_{\text{Areva}} > C_{\text{PernodRicard}} > C_{\text{Essilor}} > C_{\text{CréditLyonnais}} > \cdots > C_{\text{Sanofi}}$$

We compute as before the concentrations of positive, and negative or zero, contributions. The results are given in Tables 8.10 and 8.11.

In Table 8.10, we present the performance concentration that is observed during the the total period. We compute both negative ($T_-(p)$) and positive ($T_-(p)$) concentrations for the following three values of p: 10%, 30%,

Table 8.10 Performance concentration over a few assets

	−	+
$T(0.10)$	0.40	0.45
$T(0.30)$	0.75	0.75
$T(0.50)$	0.88	0.90
$X(-5)$	+3 %	−9 %
$X(-10)$	+7 %	−11 %
$X(-15)$	+9 %	−13 %

and 50%. It appears that 10% of the positive (resp. negative) contributions represent between 40% and 45% of the total positive (resp. negative) performance, that 30% of the positive (resp. negative) contributions represent approximately 75% of the total positive (resp. negative) performance, and that 50% of the positive (resp. negative) contributions represent approximately 90% of the total positive (resp. negative) performance. The three lines $X(-.)$ give the annualized total performance obtained after withdrawing the 5, 10, and 15 best performing stocks.

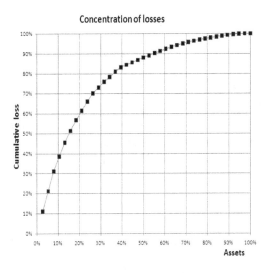

Figure 8.2 Concentration of losses

We complement this analysis in Table 8.11, which displays the seven largest positive and negative contributions within the SMABTP portfolio over the total period, and their effect on the final financial result. Recall that the total performance of the portfolio (without reinvesting dividends) is −3.76%, by aggregation of the contributions of the 114 underlying assets. The seven largest positive contributions represent a performance of 6.48%, whereas the seven largest negative contributions represent a performance of −8.89%. In Fig. 8.2, we plot the Lorenz concentration curve for the losses incurred on this portfolio.

These results illustrate the effect of a very small number of assets on performance. Avoiding investing in the seven worst-performing assets would see performance rise from −3.76% to +5.13%, without considering the reinvestment of dividends. Failing to invest in the seven best-performing assets would lead the performance to fall from −3.76% to −10.24%. Note that the seven assets, out of a total of 114, represent 6% of the portfolio lines. We therefore observe that 6% of the portfolio could produce a result that is twice, in absolute value, the total performance.

Such a performance concentration weakens the argument in favor of strategic allocation with respect to the specific choice of assets. This also highlights the importance of the largest values, and their strong hierarchy. We will now examine this aspect.

8.3.2 Hierarchy of Large Values

The hierarchy of large values is another aspect of the concentration of results: it is precisely because these values are the largest that results are concentrated on the partial sums T_n. There are important discrepancies between large values. One of the ways to measure the effect of the spacings between large values is to introduce an interesting quantity called the hierarchy ratio between two successively ranked values. We now define this ratio and we show what can be deduced from it in the case of Pareto distributions.

8.3.2.1 The Hierarchy Ratio

For a sample of size n, $\{X_1, \cdots, X_n\}$, the hierarchy ratio at rank k is defined by:

$$\frac{X_{(k)}}{X_{(k+1)}} \qquad k = 1, \cdots, n-1$$

where $X_{(k)}$ is the k-th value of the ordered sample $X_{(1)} \geq \cdots \geq X_{(n)}$.

Table 8.11 The strongest contributions in the portfolio

Rank	Stock	Contribution	Total performance obtained after removing asset
1	Areva	+1.94%	−5.70%
2	Pernod Ricard	+1.08%	−6.78%
3	Essilor	+0.92%	−7.70%
4	Crédit Lyonnais	+0.70%	−8.40%
5	Technip	+0.68%	−9.08%
6	Spir comm.	+0.59%	−9.67%
7	Vinci	+0.57%	−10.24%
⋮	⋮	⋮	⋮
108	Lafarge	−0.79%	+5.13%
109	France Telecom	−0.92%	+4.34%
110	Suez	−1.11%	+3.42%
111	Aventis	−1.32%	+2.31%
112	Scor	−1.37%	+0.99%
113	Vivendi	−1.48%	−0.38%
114	Sanofi	−1.90%	−1.86%
	Total performance	−3.76%	

An important question in risk management and portfolio management is that of the sensitivity of the risk measure to the largest value that is observed during a given period. The hierarchy ratio brings an interesting answer to this question. We will show this for the case of power functions (Pareto distributions), whose particularity is precisely that they offer a strong hierarchy of large values.

Let us construct the deterministic quantities $\{Y_k \quad k = 1, \cdots, n\}$:

$$\Pr(X \geq Y_k) = \frac{k}{n}$$

When X follows an exponential distribution, we show that the Y_k are logarithmically distributed. So, under the hypothesis:

$$\Pr(X \geq x) = e^{-\alpha x}$$

we can write:

$$\Pr(X \geq Y_k) = e^{-\alpha Y_k} = \frac{k}{n}$$

yielding directly:

$$Y_k = \frac{1}{\alpha} \ln\left(\frac{n}{k}\right)$$

However, when X follows a power distribution, we show that the Y_k have a distribution of the same type. Indeed, with:

$$\Pr(X \geq x) = \left(\frac{a}{x}\right)^\alpha$$

we have:

$$\Pr(X \geq Y_k) = \left(\frac{a}{Y_k}\right)^\alpha = \frac{k}{n}$$

so that:

$$Y_k = a \left(\frac{n}{k}\right)^{1/\alpha} \qquad (8.41)$$

which gives the order of magnitude of the k-th encountered value. We can now compute:

$$\frac{Y_k}{Y_{k+1}} = \left(\frac{k+1}{k}\right)^{\frac{1}{\alpha}} \qquad (8.42)$$

which gives an approximation of the hierarchy ratio between a large value and the value that follows it. From (8.42), we have:

$$\frac{Y_1}{Y_2} = 2^{\frac{1}{\alpha}}$$

The ratio between the first two largest values is of the order of magnitude of $2^{1/\alpha}$. If $\alpha = 1.5$, we obtain $2^{2/3} \simeq 1.6$: the largest value should be 1.6 times more important than the second largest one. If $\alpha = 0.5$, we obtain $2^2 = 4$: the largest value should now be four times more important than the second largest one. The more α diminishes, the more this ratio (and therefore the ranking of values) increases. Conversely, this ratio tends to 1 when α tends to infinity (exponential tails).

This ranking of the largest values has an important consequence in risk management: the weakening of the relevance of classic statistical values, such as the mean and the variance, due to the strong contribution of the largest values. In the case of power functions, the behavior of the sum is mainly determined by the behavior of the first largest values $X_{(k)}$, which invalidates the direct use of the mean. As far as standard deviation is concerned, a similar rationale can be conducted when $\alpha < 2$. Indeed, in this case, $n^{1/\alpha} > \sqrt{n}$ and $X_{(1)}$ increase faster than $\sqrt{n}$: the standard deviation is also not a valid indicator.

8.3.2.2 The Cumulative Frequency Curve

It is very convenient to display the largest values of a sample graphically. This can be performed by means of a graphical representation that uses an order statistic: the amplitude-rank diagram, or cumulative frequency curve. We plot the values taken by X according to their rank, from the largest to the smallest (the largest has rank 1, etc). The first point will therefore have the abscissa $X_{(1)}$ and the coordinate 1, and so on. We obtain the following graph, or set of points:

$$\{(X_{(k)}, k) : k = 1, \cdots, n\} \qquad (8.43)$$

The graph that can be constructed from the relationship (8.43) is called the cumulative frequency curve. This is the empirical c.d.f. of X (in fact, its distance from 1).

Let us now make this idea more precise. We ask the following question, "How many observed values are superior or equal to a given value x?" and accordingly we define $\mathrm{Nb}(X_i \geq x)$. If $x = X_{(1)}$ is the maximum of the ordered sample $X_{(1)} \geq X_{(2)} \geq \cdots \geq X_{(n)}$, then it is clear that there is only one value above or equal to this level: $\mathrm{Nb}(X_i \geq X_{(1)}) = 1$. The point $(X_{(1)}, 1)$ therefore belongs to the amplitude-rank plot. Similarly, if $x = X_{(n)}$ is the last (smallest) value of the ordered sample, then we have $\mathrm{Nb}(X_i \geq X_{(n)}) = n$, so that we have reached the last rank. There is therefore a point $(X_{(n)}, n)$ in the amplitude-rank graph. More generally, for $x = X_{(k)}$, the k-th value of the ordered sample, we have $\mathrm{Nb}(X_i \geq X_{(k)}) = k$, yielding the point $(X_{(k)}, k)$ in the amplitude-rank graph.

We now introduce the empirical frequency of a sample of size n denoted by $\mathrm{Fr}(X_i \geq x) = \mathrm{Nb}(X_i \geq x)/n$. We therefore have:

$$\mathrm{Fr}(X_i \geq X_{(k)}) = \frac{k}{n}$$

The graph:

$$\left\{\left(X_{(k)}, \frac{k}{n}\right) : k = 1, \cdots, n\right\} \qquad (8.44)$$

is the same as that displayed in Fig. (8.43), the coordinate being normalized to 1.

Knowing that there are $n - k$ values strictly inferior to $X_{(k)}$, we have:

$$F_n(X_{(k)}) = \frac{1}{n}\sum_{i=1}^{n} \mathbf{1}_{\{X_i < X_{(k)}\}} = \frac{n-k}{n} = 1 - \frac{k}{n}$$

We see that:

$$\mathrm{Fr}(X_{(k)}) = 1 - F_n(X_{(k)}) \qquad (8.45)$$

In other words, the rank k, or the cumulative frequency k/n, are two expressions of the tail function $\bar{F}(.) = 1 - F(.)$, up to a transformation. Figures (8.43) and (8.44) are constructed in bilogarithmic scale, with the values of X ranked in decreasing order (so the ordered sample $X_{(1)} \geq X_{(2)} \geq \cdots \geq X_{(n)}$) in the abscissa, and the corresponding value of k or k/n in the ordinate. These graphs allow us to visualize the shape of the c.d.f., zooming in on the distribution tails.

The relationship (8.43) is represented in Fig. 8.3 for the daily negative and positive returns of the CAC40 index between January 1, 1988 and June 30, 2001 (on the left, $n = 1,604$ work days is the maximum value of the coordinate; and on the right, $n = 1,856$ work days is the maximum value of the coordinate). We plotted the values k (ranks), but not the values k/n (frequencies). However, the interpretation is identical. This graph is a reversed empirical c.d.f. curve (of the cumulative frequency), or $1 - F_n(x)$.

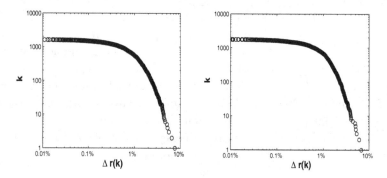

Figure 8.3 Daily negative (left) and positive (right) returns of the CAC 40 index over a period of 13 years in an amplitude-rank representation

8.4 Estimation of the Shape Parameter

The estimation of the parameter ξ is essential for the determination of the shape of the distribution tail and of the maximum domain of attraction to which the studied phenomenon belongs. In general, the Hill (1975) method is the most widely used. This method was initially devised to estimate the parameter of the distribution tail of a law belonging to the domain of attraction of extreme laws (see Embrechts, Klüppelberg, and Mikosch (1997)). However, it appears that the results of estimations are very sensitive to the position of the threshold x_0 beyond which the behavior of the portfolio can be modeled by a generalized Pareto distribution.

We present a simple algorithm that allows us to estimate the threshold x_0. This algorithm relies on the work undertaken by the actuarial team of the insurance company SMABTP (see Majri (2005)). It appears (see Gilli and Këllezi (2003)) that there does not yet exist a satisfactory and automatic algorithm allowing us to choose the threshold x_0. In this context, we now propose and describe a simple algorithm.

8.4.1 A New Algorithm

We first present the Hill method and then the way in which it can be used to automatically estimate the threshold x_0.

8.4.1.1 The Hill Method

The Hill method can be used for the estimation of a strictly positive shape parameter ($\xi > 0$). It is presented in Hill (1975), in Drees, De Haan and Resnick (2000), and in Embrechts, Klüppelberg and Mikosch (1997).

The starting point is formula (8.21), giving the generalized Pareto distribution, which we recall:

$$\bar{F}(x|X > x_0) = \left(1 + \xi \left(\frac{x - x_0}{\sigma_{x_0}}\right)\right)^{-\frac{1}{\xi}} \quad (8.46)$$

Working on a sample $x_i, i = 1, \cdots, n$ of i.i.d. observations having the same sign, the Hill method consists in representing graphically the values of the quantity:

$$\hat{\alpha}_{k,n} = \left(\frac{1}{k} \sum_{j=1}^{k} \ln x_{j,n} - \ln x_{k,n}\right)^{-1}$$

with respect to k where $k = 1, \cdots, n$. Following the convention introduced at the beginning of the chapter, $x_{k,n}$ is the k-th observation in decreasing order (written differently $x_{1,n}$ is the largest observed value). The coefficients $\hat{\alpha}_{k,n}$ describe the mean behavior of exceedances beyond the thresholds $x_{k,n}$.

If the random variable X follows a generalized Pareto distribution beyond a threshold x_0, then, when $n \to +\infty$, we have $\hat{\alpha}_{k,n} \to 1/\xi$ under the asymptotic conditions $k = k(n) \to +\infty$ and $k/n \to 0$. Thus, the Hill method could allow us to estimate the threshold ξ by analyzing the stability zone of the obtained graph, where the asymptotic conditions are more likely to hold.

However, the practical use of this method is problematic when asymptotic conditions from the sample are difficult to enforce.

The Hill plot is most efficient when the underlying distribution is a Pareto law of type I, or very close to it (see Drees, De Haan and Resnick (2000)). The reason is that (see Embrechts, Klüppelberg and Mikosch (1997), p. 331), when the random variable follows a simple Pareto distribution of exponent α, for a given value of k (where $(0 < k < n)$), the quantity $\hat{\alpha}_{k,n}$ corresponds to an estimation of α with the conditional maximum likelihood method applied to the observation of the k largest observations $x_{k,n}$. In this particular case, the values $\hat{\alpha}_{k,n}$ happen to be very stable with respect to k and allow a very good approximation of the coefficient ξ to be obtained.

However, as already indicated, this situation is not often encountered in practice, which means that we need to find an automatic method for estimating the threshold x_0.

8.4.1.2 *An Algorithm: Automatic Estimation of the Threshold*

The proposed algorithm relies on the facts that 1) the Hill plot is very stable when the underlying distribution is a simple Pareto one, and 2) a relationship exists that allows us to move from a simple Pareto distribution to a generalized one by means of a simple translation. We describe below the four steps of the algorithm.

(1) First step, construction of the candidate thresholds that depend on the level of precision h chosen by the user:

$$\hat{u}(i) = x_{n,n} + i\,h$$

with:

$$0 \leq i \leq \frac{x_{1,n} - x_{n,n}}{h}$$

(2) Second step, computation of the estimators $\hat{\sigma}(\hat{u}(i))$ and $\hat{\xi}(\hat{u}(i))$ of the parameters σ and ξ of the potential generalized Pareto distribution using the method of Hosking and Wallis (1987). This computation is performed using the values of the sample that exceed $\hat{u}(i)$ for each value of i.

(3) Third step, computation of the Hill coefficients obtained from the sample of translated ordered observations:

$$\tilde{x}_{k,n} = x_{k,n} - \hat{u}(i) + \frac{\hat{\sigma}(\hat{u}(i))}{\hat{\xi}(\hat{u}(i))}$$

for values of k such that $x_{k,n} > \hat{u}(i)$.

So, for i given, we compute the following Hill coefficients:

$$\hat{\alpha}_{i,k,n} = \left(\frac{1}{k} \sum_{j=1}^{k} \ln \tilde{x}_{j,n} - \ln \tilde{x}_{k,n} \right)^{-1}$$

for all k ($1 \leq k \leq m(i)$) with $m(i)$ defined such that $x_{m(i)+1,n} \leq \hat{u}(i)$. Then, we compute a coefficient of dispersion S_i taking account of the stability of the Hill plot comprised of the points $(k, \hat{\alpha}_{i,k,n})$. This coefficient is defined as follows:

$$S_i = \frac{1}{m(i)\,\hat{\xi}(\hat{u}(i))} \sum_{j=1}^{m(i)} \left| \hat{\alpha}_{i,j,n} - \frac{1}{m(i)} \sum_{k=1}^{m(i)} \hat{\alpha}_{i,k,n} \right|$$

(4) Fourth step, computation of the threshold. The optimal threshold provided by the algorithm is $\hat{u}(i_{\text{opt}}) = r_{n,n} + i_{\text{opt}}\, h$ where i_{opt} is such that:

$$S_{i_{\text{opt}}} = \inf\left(S_i \middle|\ 0 \leq i \leq \frac{x_{1,n} - x_{n,n}}{h} \right)$$

This threshold is optimal because the Hill plot obtained is the most stable one. In this case, the distribution associated with the sample of the translated observations is a simple Pareto distribution. The closer we get to the optimal threshold, the more the Hill coefficients tend to the underlying coefficients of a simple Pareto distribution, and the more the dispersion coefficients S_i diminish.

What happens when we move away from this threshold?

- For thresholds inferior to the optimal threshold, the computed Hill coefficients do not correspond to the coefficients associated with a simple Pareto distribution. Consequently, the Hill plots are not very stable and the coefficients of dispersion S_i are quite high.
- For thresholds exceeding the optimal threshold, although the computed Hill coefficients correspond to a simple Pareto distribution, the algorithm will yield coefficients of dispersion superior to that obtained with the optimal threshold, because the observations considered become less and less numerous and increasingly dispersed due to their extremal nature.

8.4.2 Examples of Results

We start by considering theoretical cases, before providing a practical application using the S&P 500 index.

8.4.2.1 Theoretical Illustration

We produce by Monte Carlo simulation two samples of $5,000$ i.i.d. values of a Pareto distribution.

First Case

We consider a generalized Pareto distribution endowed with the threshold $x_0 = 1.50\%$, the scale parameter $\sigma = 0.10\%$, and the shape parameter $\xi = 0.35$.

We apply the algorithm that allows us to compute the optimal threshold for this sample. x_0 is then equal to 1.50% and corresponds perfectly to the theoretical threshold. The coefficient of dispersion is equal to 1.46%.

The coefficient of dispersion increases if we consider a larger or smaller threshold. For instance, with a threshold of 1.49%, the coefficient of dispersion is equal to 7.95%, so to five times that of the theoretical threshold. For a threshold of 1.47%, the coefficient of dispersion is equal to 28.42%, so to 19 times the coefficient of dispersion of the "true" threshold. These cases are illustrated in Fig. 8.4.

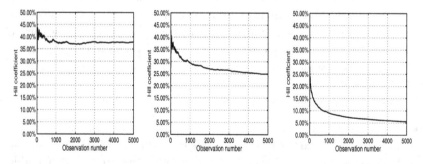

Figure 8.4 Hill plot for translated observations with $x_0 = 1.50\%$ (left), $x_0 = 1.49\%$ (center) and $x_0 = 1.47\%$ (right)

Second Case

We consider a generalized Pareto distribution having the threshold $x_0 = 1.50\%$, the scale parameter $\sigma = 0.10\%$, and the shape parameter $\xi = 0.35$. We replace the values inferior to the threshold 1.65% by values drawn randomly between 0 and 1.65%. The sample is therefore produced based on a generalized Pareto distribution of shape parameter $\xi = 0.35$, only for values superior or equal to the threshold $x_0 = 1.65\%$.

In Fig. 8.5, we present a POT (for Peaks Over Threshold, see Embrechts et al. (1997)) plot showing, for a given threshold, the mean of exceedances beyond this threshold.

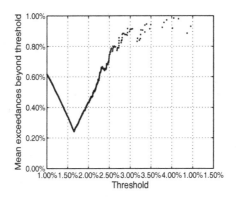

Figure 8.5   POT Plot

We observe from this POT plot that the optimal threshold ranges between 1.60% and 1.70% (consistently with the real result). The POT plot also allows us to verify that below this threshold, the observations were obtained from a completely different distribution (again, consistently with our construction).

We apply the algorithm for calculating x_0 to this sample. The optimal threshold is then 1.65% and corresponds perfectly to the theoretical threshold. The coefficient of dispersion is equal to 0.95%. It increases if we consider a smaller or larger threshold.

For example, for a threshold of 1.60%, the coefficient of dispersion is equal to 15.17%, so to 15 times that corresponding to the optimal threshold. For a threshold of 1.69%, the coefficient of dispersion is equal to 1.58%.

8.4.2.2   Application to the S&P 500 Index

We now conduct a test using market data. We consider a sample comprised of daily quotes from the S&P 500 index observed between January 3, 1928 and August 19, 2005. This sample consists of 20,098 observations and was obtained from the JCF database.

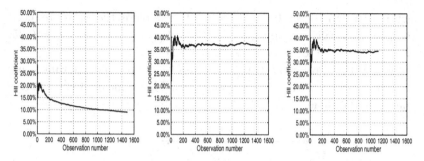

Figure 8.6 Hill plot for translated observations with $x_0 = 1.60\%$ (left), $x_0 = 1.65\%$ (center) and $x_0 = 1.69\%$ (right)

The application of the algorithm on negative returns yields the following results. The optimal threshold for the large negative values is -1.337%. There are $1,394$ observations below this threshold. The resulting tail coefficient is 0.27. This coefficient is coherent with the results usually obtained on this index during the period of study. The coefficient of dispersion accounts for the stability of the curve. It is very weak, meaning that the exceedances of negative returns beyond the optimal threshold x_0 can be correctly modeled by a generalized Pareto distribution. However, one should keep in mind that, although this method seems to function well for this sample from the S&P 500 index data, it relies on the hypothesis of i.i.d. observations in order to make short-term predictions reliable.

Let us conclude on the results obtained. The proposed algorithm is simple, intuitive, and can be computed very quickly. It relies on two principles: the efficiency of Hill plots for a simple Pareto distribution; and the passage, by a simple operation of linear translation, from a simple Pareto distribution to a generalized one, and *vice versa*. The results are conclusive for the tests performed. Other tests were conducted on actual financial portfolios, giving rise to equally convincing results.

Chapter 9

Risk Budgets

This chapter presents the main risk measures used for the assessment of risk budgets, and their practical computation. We first recall the conceptual framework within which risk measures are defined. We then highlight the importance of probabilistic hypotheses for the construction of risk measures. Finally, we examine these measures, and the numerical methods that allow us to compute them. This chapter concludes with examples of computations using Laplace processes, in calendar or in modified time, and with alpha-stable motions.

The developments that follow are limited to the sole case, as employed by practitioners, of risk measures computed in a static world. We do not present here more recent works displaying dynamic or multiperiodic risk measures (see for instance the first chapter of the book of Di Nunno and Oksendal (2011) written by Acciaio and Penner). The goal here is instead to show the effect of modeling with Laplace processes on usual risk measures.

9.1 Risk Measures

We begin this chapter with a general presentation of risk measures.

9.1.1 *Main Issues*

Computing a risk budget amounts to assessing the potential losses that can be related to a specific random variable X. Two main kinds of measure allow us to achieve such a goal. The first family of risk measures is known as "coherent" in finance literature and was introduced by Artzner, Delbaen, Eber, and Heath (1999). The second family stems from an actuarial computation of insurance premia developed by Wang, Young, and Panjer (1997). Indeed, an insurance premium is simply a computation that transforms random future losses into a unique number. This is also a way to measure risk. The two approaches rely on the choice of axioms.

We now present these two families of risk measures. Ω is the collection of all the possible states of the world and $\mathcal{X}$ is the collection of all the potential financial losses that can occur on Ω. A risk measure is a mapping from $\mathcal{X}$ to $\mathbb{R}$. We denote by $\rho(X)$ a measure of the risk X.

9.1.1.1 Coherent Risk Measures in Finance

We follow the presentation of Artzner *et al.* (1999). The three axioms that define the family of coherent risk measures are:

Axiom A1 Invariance by translation and positive homogeneity.

$$\rho(aX + b) = a\rho(X) + b \qquad \forall a \geq 0,\, b \in \mathbb{R}$$

It combines two distinct features. On the one hand, $\rho(aX) = a\rho(X)$ means that the risk of a position X is linearly proportional to its size. On the other hand, $\rho(b) = b$ means that the risk of a certain loss b is simply equal to this loss.

Axiom A2 Monotonicity.

$$\rho(X) \leq \rho(Y) \qquad \text{if} \quad X \leq Y \quad \text{a.s.}$$

It is intuitive and expresses a minimum requirement: if the risk Y is larger than the risk X, then the corresponding measures should satisfy the same inequality.

Axiom A3 Sub-additivity.

$$\rho(X+Y) \leq \rho(X) + \rho(Y), \quad \text{for all} \quad X, Y \in \mathcal{X}$$

This axiom has been the subject of an abundant literature since its definition. Intuitively, it expresses the idea that, along the lines of Artzner et al. (1999) p. 3, grouping together assets does not create additional risk. More precisely, adding an asset Y to a portfolio X does not increase the global risk: the risk $\rho(X+Y)$ of the new portfolio composed of the old portfolio and of the new asset is inferior to the sum of the two individual risks $\rho(X) + \rho(Y)$. We find here the idea of the potential benefits of diversification, presented in a different manner.

If Ω is finite, then a risk measure ρ will be coherent in the sense of Artzner et al. (1999) if a family $\mathcal{P}$ of probability measures exists on Ω such that:

$$\rho(X) = \sup_{P \in \mathcal{P}} \{\mathbb{E}_P[X]\}, \quad \forall X \in \mathcal{X}$$

where the convention $\mathbb{E}_P[X]$ denotes the expectation of X computed with the probability P. It appears that the coherence criterion of Artzner et al. (1999) forbids all risk measures relying on quantiles, as for instance VaR. See also Föllmer and Schied (2011).

9.1.1.2 Coherent Evaluation of Risk in Insurance

We present now the conceptual framework developed by Wang et al. (1997). The four axioms that define the coherent evaluation of risk in insurance are:

Axiom B1 Invariance in law.

$$\rho(X) = \rho(Y) \quad \text{if} \quad X \text{ and } Y \text{ have the same distribution}$$

This axiom simply means that the risk of a position is completely determined by the structure of its loss distribution.

Axiom B2 Monotonicity.

$$\rho(X) \leq \rho(Y) \quad \text{if} \quad X \leq Y \quad \text{a.s.}$$

This axiom is identical to axiom A2.

Axiom B3 Comonotonic additivity.

$$\rho(X+Y) = \rho(X) + \rho(Y), \quad \text{if} \quad X \text{ and } Y \text{ are comonotonic}$$

The comonotonicity of two random variables X and Y corresponds to the fact that, when the state of the world ω changes, then $X(\omega)$ and $Y(\omega)$ evolve in the same manner (formally, X and Y are comonotonic if there exists a random variable Z and two non-decreasing functions f_X and f_Y such that $X = f_X(Z)$ and $Y = f_Y(Z)$). For example, a call option and its underlying evolve similarly; they are comonotonic. The risk $\rho(X+Y)$ is merely the sum of the individual risks $\rho(X) + \rho(Y)$.

Axiom B4 Continuity.

$$\begin{cases} \lim_{d \to 0} \rho((X-d)^+) = \rho(X^+) \\ \lim_{\rho \to +\infty} (\min(X,d)) = \rho(X) \quad \lim_{\rho \to -\infty} (\max(X,d)) = \rho(X) \end{cases}$$

where $(X-d)^+ = \max(X-d, 0)$.

While this was not the case with the coherence criterion of Artzner et al. (1999), VaR satisfies the coherence criterion of Wang et al. (1997).

9.1.1.3 The Debate on Sub-Additivity

This book does not aim to discuss the pros and cons of sub-additivity. Instead, it highlights the limits of diversification when markets become violent. In particular, in markets where dynamics display many large positive and negative jumps, i.e. when it is reasonable to consider models based on alpha-stable motions with $\alpha < 1$ (see the Appendix), diversification increases risks. On such a market, adding an asset Y to a portfolio X leads to an increase in the total risk, such that:

$$\rho(X+Y) \geq \rho(X) + \rho(Y), \quad \text{for all} \quad X, Y \in \mathcal{X}$$

and axiom A3 is not valid. This shows that axiom A3 is valid only on tempered markets. Written differently, axiom A3 is conditioned by a probabilistic hypothesis.

9.1.2 Definition of the Main Risk Measures

We now introduce the two principal risk measures that can be applied to distribution tails, using the developments of the previous chapter. We start by defining a notion of capital requirement based on the probable loss at a given probability threshold: Value-at-Risk, or VaR.

9.1.2.1 VaR

The notion of Value-at-Risk (VaR) being well known, we only briefly recall its definition. VaR is simply the level x_0 such that the probability of exceeding this level is equal to $1-c$ where c is a given confidence threshold. We use the convention VaR_c. This is the value such that:

$$\Pr(X \leq -\text{VaR}_c) = F_X(-\text{VaR}_c) = 1 - c \tag{9.1}$$

If for instance $c = 99\%$ and VaR is equal to 600 euros, then there is a probability of 1% that the incurred loss will be larger in absolute value than 600 euros:

$$\Pr(X \leq -600) = 0.01$$

In practice, we want to obtain the value VaR_c as a function of c. From (9.1), we deduce that:

$$\text{VaR}_c(X) = -F_X^{-1}(1-c)$$

where F^{-1} is the reciprocal function of the cumulative distribution function F. If the distribution of X is symmetrical, then $-F^{-1}(1-c) = F^{-1}(c)$ and we have:

$$\text{VaR}_c(X) = F_X^{-1}(c) = q_c \tag{9.2}$$

where q_c is the quantile of order c of the c.d.f. F_X.

We now present two risk measures constructed based on VaR. With reference to the previous chapter, the first of these measures uses the approach of the largest values, whilst the second of these measures is based on threshold exceedances. In both cases, we use the property of the linearity of conditional means to define these measures.

9.1.2.2 Mean Loss if VaR is Exceeded

The c.d.f. of the largest values is:

$$F(x|X > x_0) = \Pr(X < x \mid X > x_0) \tag{9.3}$$

for all $0 \leq x < x_{\max}$.

Recall that the conditional mean beyond the threshold x_0 satisfies:

$$M(x_0) = \mathbb{E}[X \mid X > x_0] \tag{9.4}$$

The quantity $M(x_0)$ provides a convenient way of presenting unequal distributions. For example, if X is an income, $M(x_0)$ represents the mean income of individuals having an income superior to x_0. If X is an insurance claim, then $M(x_0)$ represents the mean value of claims larger than x_0.

Also, when X is the return of a financial asset and $x_0 = \text{VaR}_c(X)$, where $\text{VaR}_c(X)$ is the Value-at-Risk at the confidence threshold c (as defined in Eq. (9.1)), then $M(x_0)$ represents the mean loss if VaR_c is exceeded. It is called the Tail Conditional Expectation (TCE) or Conditional Tail Expectation (CTE), and satisfies:

$$\text{TCE}_c(X) = \mathbb{E}[X \mid X > \text{VaR}_c(X)] \tag{9.5}$$

9.1.2.3 Mean Excess Loss if VaR is Exceeded

We first recall the c.d.f. of exceedances:

$$F_{x_0}(x) = \Pr(X - x_0 < x \mid X > x_0) \tag{9.6}$$

for all $0 \leq x < x_{\max} - x_0$.

We obtain an equivalent to the conditional mean, computed on the random variable $X - x_0$. This is the mean value of exceedances beyond the threshold x_0, denoted by:

$$e(x_0) = \mathbb{E}[X - x_0 \mid X > x_0] \tag{9.7}$$

In insurance theory, $e(x_0)$, called the excess of loss, represents the mean value of loss exceedances beyond a threshold x_0.

In finance, if $x_0 = \text{VaR}_c(X)$, $e(x_0)$ represents the mean value of losses beyond $\text{VaR}_c(X)$, provided this level is crossed at the confidence threshold c. This quantity, called Conditional VaR or CVaR, verifies:

$$\text{CV}_c(X) = e(\text{VaR}_c(X)) \tag{9.8}$$

so that:

$$\text{CV}_c(X) = \mathbb{E}[X - \text{VaR}_c(X) \mid X > \text{VaR}_c(X)] \tag{9.9}$$

The expected shortfall is a quantity that can be expressed, depending on the authors, either as in formula (9.9), or as follows:

$$\text{ES}_c(X) = \mathbb{E}[(X - \text{VaR}_c(X)) \mathbf{1}_{X > \text{VaR}_c}] \tag{9.10}$$

i.e. as:

$$\begin{aligned}\text{ES}_c &= \mathbb{E}[X - \text{VaR}_c \mid X > \text{VaR}_c] \Pr(X > \text{VaR}_c) \\ &= \text{CV}_c(X) \Pr(X > \text{VaR}_c)\end{aligned}$$

9.1.2.4 Connection Between the Two Approaches

A direct link exists between the two cumulative distribution functions. Because it is clear that:

$$\Pr(X \leq x | X > x_0) = \Pr(X - x_0 \leq x - x_0 | X > x_0)$$

we have:

$$F(x|X > x_0) = F_{x_0}(x - x_0) \qquad (9.11)$$

Using the notions of VaR, TCE, and CV exposed previously, and combining (9.5) and (9.9), we obtain a very useful relationship between the various measures of large risks:

$$\text{TCE}_c(X) = \text{VaR}_c(X) + \text{CV}_c(X) \qquad (9.12)$$

The measure of VaR and its related quantities depends directly on the probability distribution of X. We introduce in Section 9.1.4 the notion of model risk related to the choice of probability distribution.

9.1.3 VaR, TCE, and the Laws of Maximum

We now aim to obtain general expressions for VaR and TCE.

9.1.3.1 TCE as a Simple Multiple of VaR

Following the analysis of Marc Barbut, we look for a linear relationship between VaR and TCE of the type:

$$\text{TCE}_c = a \times \text{VaR}_c + b \qquad (9.13)$$

This functional relationship shows that the conditional mean beyond a threshold x_0 (TCE beyond VaR) is a multiple of this threshold, up to an additive constant, i.e. a linear function of this threshold. We have the relationship:

$$\text{TCE}_c = \frac{1}{\Pr(X > \text{VaR}_c)} \int_{\text{VaR}_c}^{\infty} t \, dF(t) = \frac{1}{\overline{F}(\text{VaR}_c)} \int_{\text{VaR}_c}^{\infty} t \, dF(t)$$

Its solutions are:

(1) First case: $a = 1$ and $b > 0$
We obtain:

$$\overline{F}(x) = e^{-(x-\text{VaR}_c)/b}, \qquad x > x_0$$

The tail function possesses an exponential distribution whose standard deviation is b. The conditional mean beyond the threshold VaR_c can be written as:

$$\text{TCE}_c = \text{VaR}_c + b$$

which shows that, in the exponential case, TCE is equal to the sum of VaR and of the standard deviation of the distribution.

(2) Second case: $a > 1$ and $b \neq 0$
We obtain:

$$\overline{F}(x) = \left(\frac{\text{VaR}_c + d}{x + d}\right)^\alpha \qquad x > \text{VaR}_c$$

The tail function possesses a Pareto distribution of type II. The conditional mean beyond VaR can be written as:

$$\text{TCE}_c = \frac{\alpha}{\alpha - 1}\text{VaR}_c + \frac{d}{\alpha - 1} = \frac{1}{\alpha - 1}(\alpha \text{VaR}_c + d)$$

For example, if $\alpha = 1.5$, then $a = 3$ and $\text{TCE}_c = 3\,\text{VaR}_c + 2d$. With the Jenkinson–von Mises parameterization, we have:

$$\overline{F}(x) = \left(1 + \frac{\xi}{\sigma}(x - \text{VaR}_c)\right)^{-\frac{1}{\xi}}$$

We can now express the conditional mean beyond VaR as follows:

$$\text{TCE}_c = \frac{1}{1 - \xi}\text{VaR}_c + \frac{\sigma}{1 - \xi} = \frac{\text{VaR}_c + \sigma}{1 - \xi}$$

(3) Third case: $a > 1$ and $b = 0$
We obtain:

$$\overline{F}(x) = \left(\frac{\text{VaR}_c}{x}\right)^\alpha \qquad x > \text{VaR}_c$$

The tail function possesses a Pareto distribution of type I, or simple Pareto distribution. The conditional mean beyond VaR can be written as:

$$\text{TCE}_c = \frac{\alpha}{\alpha - 1}\text{VaR}_c$$

For example, if $\alpha = 1.5$, then $a = 3$ and $\text{TCE}_c = 3\,\text{VaR}_c$.
Performing the same change of parameterization as before for $\xi = 1/\alpha$, but this time with $\sigma = \text{VaR}_c\,\xi$ (so with $d = 0$), we obtain the tail function:

$$\overline{F}(x) = \left(1 + \frac{\xi}{\sigma}(x - \text{VaR}_c)\right)^{-\frac{1}{\xi}}$$

9.1.3.2 *Direct Computation with Pareto Distributions*

Recall that VaR_c is the quantile of order c of the distribution function. We thus have for the Jenkinson–von Mises (8.8) distribution:

$$H_\xi(q_c) = c$$

yielding:

$$\text{VaR}_c = \mu - \frac{\sigma}{\xi}\left[1 - (-\ln c)^{-\xi}\right] \qquad (9.14)$$

The sensitivity of VaR to the parameters σ and ξ is strong. Indeed, we have:

$$\frac{\partial \text{VaR}_c}{\partial c} = \frac{\sigma}{c(-\ln c)^{1+\xi}}$$

We see when ξ and/or σ increase in value, then VaR also increases in value.

Recall that the quantity TCE defined in (9.5) is:

$$\text{TCE}_c = \mathbb{E}[X \mid X > \text{VaR}_c]$$

so here:

$$\text{TCE}_c = \frac{\xi}{1-\xi}\text{VaR}_c + \frac{\sigma}{1-\xi} \qquad (9.15)$$

The quantity CV defined in (9.9) is:

$$\text{CV}_c = \mathbb{E}[X - \text{VaR}_c \mid X > \text{VaR}_c]$$

with:

$$\text{CV}_c = \text{TCE}_c - \text{VaR}_c$$

so here:

$$\text{CV}_c = \left(\frac{\xi}{1-\xi} - 1\right)\text{VaR}_c + \frac{\sigma}{1-\xi} \qquad (9.16)$$

Note that, when $\xi = 0$, we have $\text{TCE}_c = \sigma$ and $\text{CV}_c = \sigma - \text{VaR}_c$.

9.1.4 *Notion of Model Risk*

Let us now come to the study of model risk.

9.1.4.1 Main Issues and Ideas

Model risk is traditionally understood as the existence of a discrepancy between a given probabilistic model and the true nature of the financial phenomenon. This approach will not be discussed here. We shall only highlight that it neglects an important component of model risk: the way models influence professional practices, which is called "performativity" or "reflexitivy" in the sociology of science. This reflexive effect of models is crucial and must be taken into account in the management of financial risks (see Walter (2011)).

In this book, we limit ourselves to the classic conception of model risk. From this viewpoint, the choice of a stochastic process, and therefore the shape of the distribution chosen for the computation of risk budgets, is not neutral. Indeed, this distribution defines and quantifies the nature of the risk taken when investing in a given asset. It influences the definition of the relevant statistical tools. Finally, the choice of stochastic process is a core determinant of the link between short-term and long-term risks.

Implicit since the 1987 market crash, the notion of model risk has really only become popular since the year 2000. For example, a workshop entitled "Model Risk" was organized by the Finance and Insurance group of the French Statistical Association in collaboration with the French Federation of Actuaries on March 2, 2000. Its proceedings were published in the *Journal de la Société Française de Statistiques* (141, issue 1–2). At the same time, the professional literature was largely open to the notion of model risk: "Models imperiled by volatility turmoil" (*International Financing Review*, 1300, September 11, 1999, p. 85), "Faulty pricing models leave financial institutions at a loss" (*International Financing Review*, 1316, January 15, 2000, p. 100), "Report allays some model risk fears" (*International Financing Review*, 1319, February 5, 2000, p. 84).

Also in the year 2000, a study on capital allocation models conducted by an audit firm showed that 49% of banks were not satisfied with their model, and that the behavior of models during a crisis was imperfect for 58% of banks. The first surveys also appeared in the academic literature (see for example Gourieroux and Le Fol (1997)) and the academic community started giving up the volatility parameter as sole risk criterion, expanding the measure of risk beyond this parameter, and examining the model risk associated with volatility (Alami and Renault (2000)).

Although its formulation is new, the notion of model risk presented in Proposition 9.1 (see below) is consistent with the recommendation of

Gourieroux and Le Fol (1997), p. 9, warning market players against the limitations of volatility. It is also consistent with the notion of volatility risk defined by Alami and Renault (2000), p. 104. Indeed, as shown in the preceding chapters, in the presence of fat tails, volatility is affected by the weight of the distribution tails. It is therefore not certain that "all the relevant dynamics (of returns) are related to volatility" (Alami and Renault (2000), p. 132), and so other dynamics (intrinsic to innovations) remain to be modeled.

We suggest completing volatility by introducing a distribution tail parameter such as ξ. Risk then becomes bidimensional, along the pair (σ, ξ), and model risk stems from the unidimensional reduction of the pair (σ, ξ) to σ. This method of completing volatility is consistent with the recommendation of Gourieroux and Le Fol (1997), which was to introduce measures of transactions and volume. Indeed, the existence of distribution tails ($\xi > 0$) is related to the microstructure of markets, i.e. to volume effects or other characteristics of transactions.

Let us summarize our position. In a perspective that distinguishes between regular and violent markets, we consider that model risk stems from the reduction of risk to the sole dispersion parameter. This reduction is acceptable in regular markets but dangerous in markets displaying numerous jumps.

Proposition 9.1 (Model risk and violent markets). *Model risk appears in violent markets if we try to compensate for the absence of a shape parameter by the increase in a dispersion parameter.*

We give below three examples illustrating the importance of the shape of distributions, addressing the recurrence time, a computation of VaR, and a computation of TCE. In the case of the recurrence time, we compare two models that are very different in terms of practical consequences: the Gaussian and the alpha-stable models. The goal is to show that what is unrealistic using the Gaussian distribution (a time of 83 million years between two market crashes) can become realistic using an alpha-stable distribution (a time of 42 years between two market crashes). In terms of the computation of VaR, we choose a Brownian VaR and count exceedances in order to show a significant discrepancy (59% of values are too large) at the 99% threshold. In the computation of the TCE, we want to show how this quantity varies with the domain of attraction, which is either of the Fréchet or Gumbel type.

We first illustrate the importance of the choice of stochastic process.

9.1.4.2 A First Illustration

Consider a ten-day, 99% VaR, consistent with Basle banking regulations. We assume that the Brownian representation is applicable for the returns of the asset considered.

We therefore assume that $\Delta S_{t+1} \stackrel{d}{=} \mathcal{N}(0, S_t \sigma_1)$, where $\mathcal{N}$ is the Gaussian distribution. In this case, the one-day, 99% VaR of the financial asset having σ_1 as daily volatility, and being worth S_0 at time $t = 0$, is given by:

$$\Pr\left(\Delta S_1 \leq -\mathrm{VaR}_1^{0.99}\right) = 0.01$$

where $\Delta S_1 = S_1 - S_0 = S_0 \times \sigma_1 \times \Delta W_1$ with $\Delta W_1 \stackrel{d}{=} \mathcal{N}(0,1)$. Therefore:

$$\begin{aligned}\mathrm{VaR}_1^{0.99} &= F^{-1}_{\Delta S_1}(0.99) \\ &= S_0 \times \sigma_1 \times F^{-1}_{\Delta W_1}(0.99) \\ &= S_0 \times \sigma_1 \times 2.3263\end{aligned}$$

and more generally:

$$\mathrm{VaR}_1^c = S_0 \times \sigma_1 \times q_c$$

where q_c is the quantile of order c of the centered reduced Gaussian distribution.

For a ten-day VaR, it is necessary to use a passage rule. Here, we use the square root of time rule, which is the scaling law of Brownian motion. According to this rule: $\sigma_{10} = \sigma_1 \sqrt{10}$ and we obtain:

$$\mathrm{VaR}_{10}^c = S_0 \times q_c \times \sigma_1 \sqrt{10} = \mathrm{VaR}_1^c \times \sqrt{10} \qquad (9.17)$$

9.1.4.3 Recurrence Time of a Large Fall of the S&P 500 Index

We consider the monthly returns of the S&P 500 index between 1926 and 2004. If we model the evolution of the index by an exponential of Brownian motion, we need to estimate the parameters of a marginal Gaussian distribution. This estimation yields $\mu = 0.812\%$ and $\sigma = 5.82\%$ (where we observe that $\Delta X_{\max} = 35.46\%$ and $\Delta X_{\min} = -35.28\%$). The minimal value found during the observation period is at a distance of six standard deviations from the mean value. Let us compute the probability of a monthly decrease *beyond the mean* superior to six standard deviations under the Gaussian hypothesis. We find:

$$\Pr(\Delta X < -6\,\sigma) = F(-6\,\sigma) = 0.0000000009866 \simeq 10^{-9}$$

We call "recurrence time of an event" the inverse of its probability. Let the event "decrease ΔX superior or equal in absolute value to x". The recurrence time of this event is defined as:

$$d(x) = \frac{1}{\Pr(\Delta X < -x)}$$

So here: $d(6\,\sigma) = 1/10^{-9} = 10^9$. Such an event will occur on average every 10^9 month, so every 83 million years. It is clear that this result is not realistic.

If we now assume that the generating process of the variations of the index is an alpha-stable motion (a symmetrical alpha-stable Lévy process), then we need to estimate the parameters of a symmetrical alpha-stable distribution. This estimation yields: $\alpha = 1.661$ and $\gamma = 2.945$. If X follows an alpha-stable distribution parametrized as $S(1.661; 2.945; 0.00812)$, then $u = (X - 0.00812)/2.945$ is described by a standardized reduced alpha-stable distribution $S(1.661; 1; 0)$.

Under the alpha-stable hypothesis, we have:

$$\Pr(\text{decrease} > 35\%) = \Pr(u > 11.61\%) = 1 - F_\alpha(11.61)$$

where $F_\alpha(11;61)$ represents the value of the alpha-stable cumulative distribution function for the quantile 11.61.

See Table 9.2 (p. 252) for the values of $1 - F_\alpha$, i.e. for the loss probability with respect to α and x. To simplify the computation, we round the value of α to 1.7 (this increase in α slightly underweights large return decreases) and we round the threshold 11.61% to 12%. We obtain: $F_{1.7}(12) = 0.9980$, and therefore:

$$\Pr(\text{decrease} > 35\%) = 1 - 0.9980 = 0.0020 = 0.2\%$$

Such an event will occur on average every $1/0.002 = 500$ months, so approximately every 42 years.

We see with this first example that the recurrence time of the event "monthly decrease of 35%" is much more realistic with a 1.7-stable Lévy process than with a Brownian motion.

9.1.4.4 Computation of VaR on the USD/FRF Forex Rate

We consider the daily variations of the USD/FRF Forex rate between January 2, 1986 and September 1, 1995 ($n = 2,522$ quotes in total). We chose this old data set voluntarily in order to show that jumps and extreme risks are not specific to the agitated equity markets of the past decade.

Table 9.1 Counting exceedances beyond VaR

Level c	99%	98.5%	98%	95%
$q_c \sigma_1 = \text{VaR}^c(1)$	−1.66%	−1.55%	−1.47%	−1.18%
Theoretical Nb $< \text{VaR}^c(1)$	25	38	50	126
Observed Nb $< \text{VaR}^c(1)$	40	51	59	127
Exceedance (Nb)	15	13	9	1
Exceedance (Percent)	59%	35%	17%	1%

Under the Gaussian assumption, the estimation of parameters yields $\mu_1 = -0.02\%$ and $\sigma_1 = 0.71\%$ (so an annualized volatility equal to $\sigma_{250} = 0.71 \times \sqrt{250} \simeq 11.20\%$). Important results on VaR under the Gaussian assumption are displayed in Table 9.1. The observed number of exceedances is superior to the theoretical number predicted by the Gaussian model. This implies an underevaluation of risk with Brownian motion. We observe that the difference between the observed and theoretical exceedances increases with the value of c: this phenomenon is more pronounced in the distribution tail.

9.1.4.5 Computation of TCE with Two MDA Assumptions

Assume that X follows a Fréchet distribution beyond a certain threshold x_0 and write this relationship with the parameterization of a Pareto distribution of type I:

$$M(x_0) = \frac{\alpha}{\alpha - 1} x_0 \qquad (9.18)$$

and:

$$e(x_0) = M(x_0) - x_0 = \frac{1}{\alpha - 1} x_0$$

We see that the quantity $M(x_0)$ is equal to the product of x_0 by a term $\alpha/(\alpha - 1)$. For example, if $\alpha = 1.5$, then $M(x_0) = 3x_0$ and $e(x_0) = 2x_0$. This means that, for each threshold x_0, the average of the losses superior to x_0 is equal to $3x_0$. Put differently, if x_0 is exceeded, then the conditional mean is equal to three times this level, and the amount of reserves to constitute ($e(x_0)$) is equal to twice this level. Moreover, the smaller α, the larger $M(x_0)$ and $e(x_0)$.

When $\alpha \to 1$, $\alpha/(\alpha - 1) \to \infty$ and the expected loss superior to x_0 can become infinite. The choice of the protection level $e(x_0)$ becomes extremely difficult in this context. If this protection is computed using the empirical mean $\mu_n(x_0)$, then, depending on the chosen value x_0, it will not be efficient enough in the presence of rare events, and too expensive in their absence. In this situation, the mean value of "normal" decreases gives no indication on the amplitude of large decreases that constitute the principal source of risk.

Conversely, if $\alpha \to \infty$, then $M(x_0)$ tends to x_0, which corresponds to the case of exponential distribution tails. We can overcome the insufficiencies of these tails by artificially increasing the volatility of markets. Indeed, for distributions with exponential tails, with $\xi = 0$, we have from equation (9.15) that $e(a) = \sigma$ where σ is the standard deviation of the exponential distribution. This means that the mean excess loss occurring when the threshold a is exceeded is equal to the standard deviation. So, increasing σ allows us to increase $e(a)$ in the case of the Gumbel MDA.

However, this increase is artificial if the true distribution is Paretian. Considering that $e(a) = e(a, \xi)$ from Eq. (9.15), where the value of ξ determines the nature of the distribution (its shape), we see that forcing the volatility is equivalent to compensating ξ by σ, i.e. compensating the absence of a shape by an increase in the size of fluctuations.

We make this more precise using Eq. (9.15). Suppose that the true distribution is Paretian with the tail index $\xi_0 \neq 0$, $\xi_0 < 1$, and the scale parameter σ_0. In this case, using Eq. (9.15), the true value of $e(a)$, denoted arbitrarily by M, is:

$$M = \frac{\xi_0}{1-\xi_0} a + \frac{\sigma_0}{1-\xi_0}$$

If we force the standard deviation to the level σ_1 in the case $\xi = 0$, then, using Eq. (9.15), we need $\sigma_1 = M$, so that:

$$\sigma_1 = \frac{\xi_0}{1-\xi_0} a + \frac{\sigma_0}{1-\xi_0}$$

which gives the value of the forced standard deviation. For example, if $\xi_0 = 2/3$, we have $M = 2a + 3\sigma_0$, and we must force the standard deviation to $\sigma_1 = 2a + 3\sigma_0$.

It is indeed not possible to substitute a normal shape, even if expanded by a scale operation, to a non-normal distribution shape: these are two distinct statistical representations. We insist on the following point: it is not possible to replace the absence of a shape parameter by an increase of a size parameter.

For example, we act on a distribution pertaining to the Gumbel MDA by increasing the number of average-size variations, which are not desirable because they are of no use for risk management. Consequently, forcing the volatility increases the discrepancy between the theoretical distribution used, which entails too many average-size variations, and the real market distribution, bearing few average-size variations.

These three examples show the importance of the choice of c.d.f. $F(.)$ when computing risk budgets.

9.2 Computation of Risk Budgets

Let us now turn to the computation of risk budgets for assets whose returns are modeled by Lévy processes.

9.2.1 *Numerical Method*

We expose here the general principle allowing us to compute VaR by Fourier transform in the case where the underlying distributions are infinitely divisible, so correspond to dynamics of the Lévy process type. From a numerical point of view, the goal is to compute the cumulative distribution function for all the values of the considered distribution, then to perform a root search on it. The only specific operation here is the computation of the c.d.f., because the root search operation is completely generic and can be readily implemented.

9.2.1.1 *Computation Procedure*

A Lévy process being generally constructed from the specification of a Lévy measure density (assuming it exists), the question that arises here is that of the computation of the c.d.f. of marginal distributions as an indirect function of the latter density. In most classic models, the Lévy measure density can be easily integrated. The Lévy–Khintchine formula allows us to move directly from the Lévy measure to the characteristic function. The only missing step is thus to move from the characteristic function to the cumulative distribution function. In practice, this step can be achieved using the following Fourier transform:

$$F(x) = \frac{1}{2} - \frac{1}{2\pi} \int_{-\infty}^{+\infty} e^{-iux} \frac{\Phi(u)}{iu} du \qquad (9.19)$$

so, 1/2 minus the Fourier transform of the characteristic function divided by the term iu.

To summarize, the computation of VaR for underlyings modeled by exponentials of Lévy processes requires the following procedure:

$$\nu \to \Phi \to F \to \text{VaR}$$

where we assume that the density ν of the Lévy measure is perfectly specified. In practice, if we wish to calibrate the parameters of the measure ν, we can adopt the following procedure:

$$F \to \Phi \to \nu$$

or:

$$\Phi \to \nu$$

or also:

$$f \to \Phi \to \nu$$

where the distribution density f is the inverse Fourier transform of the characteristic function:

$$f(x) = \frac{1}{2\pi} \int_{-\infty}^{+\infty} e^{-iux} \Phi(u) du$$

We start by presenting below the detailed computation of VaR. Then, we move on to the computation of TCE, which can be conducted in a similar way as that of VaR (for more details, see Le Courtois and Walter (2011)).

9.2.1.2 Computation of VaR

Let f be the density function of the infinitely divisible distribution studied, and Φ the associated characteristic function that we express as:

$$\Phi(u) = \int_{-\infty}^{+\infty} e^{iux} f(x) dx$$

We wish to obtain a simple Fourier formula for the cumulative distribution function F with respect to Φ. The formula that we propose below allows us to compute VaR as a root search performed over a single Fast Fourier Transform:

$$F(x) = \frac{e^{ax}}{2\pi} \int_{-\infty}^{+\infty} e^{iux} \frac{\Phi(ia - u)}{a + iu} du \qquad (9.20)$$

for any positive real number a. The general procedure remains the same: $\Phi \to$ Inverse Fourier Transform $\to F$. We look for a root of F to compute VaR. The interest of formula (9.20) compared to formula (9.19) lies in the fact that it does not display a pole at $u = 0$, so it will not explode around this point.

For a strictly positive, we introduce the function:

$$g(x) = e^{-ax} F(x) = e^{-ax} \int_{-\infty}^{x} f(s) ds$$

The Fourier transform of this function is:

$$\Lambda(u) = \int_{-\infty}^{+\infty} e^{iux} g(x) dx = \int_{-\infty}^{+\infty} e^{iux} \left(e^{-ax} \int_{-\infty}^{x} f(s) ds \right) dx$$

so:

$$\Lambda(u) = \int_{-\infty}^{+\infty} \int_{-\infty}^{x} e^{iux} e^{-ax} f(s) ds dx$$

Noting that $-\infty < s < x < +\infty$, we can swap integrals:

$$\Lambda(u) = \int_{-\infty}^{+\infty} \int_{s}^{+\infty} e^{iux} e^{-ax} f(s) dx ds$$

so that:

$$\Lambda(u) = \int_{-\infty}^{+\infty} f(s) \left(\int_{s}^{+\infty} e^{iux} e^{-ax} dx \right) ds = \int_{-\infty}^{+\infty} f(s) \left[\frac{e^{-(a-iu)x}}{-(a-iu)} \right]_{s}^{+\infty} ds$$

and, because $|e^{-(a-iu)x}| = e^{-ax}$ tends to zero when x tends to plus infinity:

$$\Lambda(u) = \int_{-\infty}^{+\infty} f(s) \left(-\frac{e^{-(a-iu)s}}{-(a-iu)} \right) ds = \frac{1}{a-iu} \int_{-\infty}^{+\infty} f(s) e^{-(a-iu)s} ds$$

Finally:

$$\Lambda(u) = \frac{1}{a-iu} \int_{-\infty}^{+\infty} f(s) e^{i(u+ia)s} ds$$

allows us to write:

$$\Lambda(u) = \frac{\Phi(u+ia)}{a-iu}$$

The function g being the inverse Fourier transform of Λ:

$$g(x) = \frac{1}{2\pi} \int_{-\infty}^{+\infty} e^{-iux} \Lambda(u) du$$

we obtain:
$$e^{-ax}F(x) = \frac{1}{2\pi}\int_{-\infty}^{+\infty} e^{-iux}\frac{\Phi(u+ia)}{a-iu}du$$

and finally:
$$F(x) = \frac{e^{ax}}{2\pi}\int_{-\infty}^{+\infty} e^{-iux}\frac{\Phi(u+ia)}{a-iu}du \qquad (9.21)$$

A simple change of variable ($u = -u$) yields the desired formula:
$$F(x) = \frac{e^{ax}}{2\pi}\int_{-\infty}^{+\infty} e^{iux}\frac{\Phi(-u+ia)}{a+iu}du \qquad (9.22)$$

We can use both formulas (9.21) or (9.22) in which a choice of characteristic function Φ must be made.

9.2.1.3 Computation of TCE

Now, we want to compute:
$$H(x) = \mathbb{E}[X|X \le x] = \frac{\mathbb{E}[X1_{X\le x}]}{\Pr(X \le x)} = \frac{G(x)}{F(x)}$$

We will show how to extend the computation of F to that of G (thus enabling us to obtain the TCE). By analogy with the approach used for F, we show, with a strictly positive:
$$h(x) = e^{-ax}G(x) = e^{-ax}\int_{-\infty}^{x} sf(s)ds$$

The Fourier transform of this function is:
$$\zeta(u) = \int_{-\infty}^{+\infty} e^{iux}h(x)dx = \int_{-\infty}^{+\infty} e^{iux}\left(e^{-ax}\int_{-\infty}^{x} sf(s)ds\right)dx$$

so:
$$\zeta(u) = \int_{-\infty}^{+\infty}\int_{-\infty}^{x} e^{iux}e^{-ax}sf(s)\,ds\,dx$$

Noting that $-\infty < s < x < +\infty$, we swap integrals:
$$\zeta(u) = \int_{-\infty}^{+\infty}\int_{s}^{+\infty} e^{iux}e^{-ax}sf(s)\,dx\,ds$$

so that:
$$\zeta(u) = \int_{-\infty}^{+\infty} sf(s)\left(\int_{s}^{+\infty} e^{iux}e^{-ax}dx\right)ds = \int_{-\infty}^{+\infty} f(s)\left[\frac{e^{-(a-iu)x}}{-(a-iu)}\right]_{s}^{+\infty}ds$$

and, because $|e^{-(a-iu)x}| = e^{-ax}$ tends to zero when x tends to plus infinity:

$$\zeta(u) = \int_{-\infty}^{+\infty} sf(s)\left(-\frac{e^{-(a-iu)s}}{-(a-iu)}\right)ds = \frac{1}{a-iu}\int_{-\infty}^{+\infty} sf(s)e^{-(a-iv)s}ds$$

Finally:

$$\zeta(u) = \frac{1}{a-iu}\int_{-\infty}^{+\infty} sf(s)\,e^{i(u+ia)s}ds$$

Recall that (with $u \in \mathbb{C}$ in full generality):

$$\Phi(u) = \int_{-\infty}^{+\infty} f(s)e^{ius}ds$$

Differentiating, and assuming that the conditions under which we can swap integral and derivative signs are satisfied, we obtain:

$$\frac{\partial \Phi(u)}{\partial u} = \frac{\partial\left(\int_{-\infty}^{+\infty} f(s)e^{ius}ds\right)}{\partial u} = \int_{-\infty}^{+\infty} f(s)\frac{\partial e^{ius}}{\partial u}ds = \int_{-\infty}^{+\infty} f(s)ise^{ius}ds$$

so:

$$\frac{\partial \Phi(u)}{\partial u} = i\int_{-\infty}^{+\infty} sf(s)e^{ius}ds$$

and:

$$\zeta(u) = \frac{1}{a-iu}\frac{\Phi'(u+ia)}{i} = \frac{\Phi'(u+ia)}{u+ia}$$

We only need to invert this expression:

$$h(x) = \frac{1}{2\pi}\int_{-\infty}^{+\infty} e^{-ixu}\zeta(u)dv$$

so that:

$$e^{-ax}G(x) = \frac{1}{2\pi}\int_{-\infty}^{+\infty} e^{-ixu}\frac{\Phi'(u+ia)}{u+ia}du$$

and finally:

$$G(x) = \frac{e^{ax}}{2\pi}\int_{-\infty}^{+\infty} e^{-ixu}\frac{\Phi'(u+ia)}{u+ia}du \qquad (9.23)$$

Using the results obtained in (9.22) and (9.23), we can then compute $H(x) = G(x)/F(x)$.

9.2.2 Semi-Heavy Distribution Tails

We illustrate the use of formulas (9.21) and (9.23) for the computation of VaR and TCE, first with Laplace processes (whose marginal distributions have semi-heavy tails) in calendar time, then with more general processes, such as Laplace processes in deformed time.

9.2.2.1 Risk Budgets with Laplace Processes

We concentrate first on the Fourier-type formula yielding VaR, which is valid for all types of Lévy processes (and beyond, as mentioned above), and we assume that asset returns are modeled by Laplace processes (Variance Gamma processes with drift (VGD), see Table 7.4, p. 167). We compute VaR at various confidence levels and over various temporal horizons.

We compute the VaR that is related to an investment in the CAC40 index. We calibrate a Laplace process on this index for the period ranging from January 3, 2001 to April 15, 2009. Once this stochastic process has been calibrated, we can compute VaR at various confidence levels and over various temporal horizons using formula (9.21) and the formula that gives the spectrum of a Laplace process (constructed from the generalized distribution $\mathcal{L}'(\theta, c, \delta)$). Recall that:

$$\Phi(u) = \frac{e^{i\theta u}}{1 + \frac{1}{2}\sigma^2 u^2 - i\delta u}$$

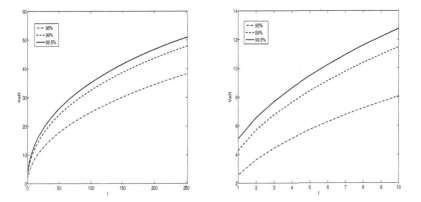

Figure 9.1 Long-term and short-term Laplace VaR (in %)

The left panel of Fig. 9.1 represents VaR computed respectively at the 95, 99, and 99.5% confidence levels, for an investment horizon of 252 work days (one calendar year). The following stylized features can be observed. First, and not surprisingly, the longer the investment horizon, the larger the VaR (and the corresponding protection amount). Then, the higher the confidence level, the larger the VaR. Furthermore, for high confidence levels, further increasing the confidence level has little effect on the VaR. Finally, a change in confidence level has a very small effect compared to a change in time horizon.

This last feature is very important and sheds light on many debates on the optimal choice of confidence level. In Fig. 9.1, it clearly appears that this is not the most important point and that the real difficulty lies in the choice of time horizon. Consider for instance a very illiquid corporate bond, owned by a bank, with an investment period assumed to be of a few days. Due to liquidity problems, it is not easy to determine the holding period of the bond. Because a 30-day VaR is twice as large as a ten-day VaR, we see indeed that very different levels of VaR can be retained, in a very arbitrary way.

In this last example also, the confidence level does not appear to be the key factor for the computation of VaR. The right panel of Fig. 9.1 presents an enlargement of the left panel of the same figure, for confidence levels equal to 95, 99, and 99.5%, and for investment horizons not exceeding ten open days. For very short investment horizons also, we observe that the choice of investment horizon has more effect on VaR than the choice of confidence level. The ten-day, 99% VaR is important in finance because it is the key indicator of the Basel II agreement. We see that in the case of an investment in the CAC40 index, it is equal to 11.5% of the total investment.

Now, looking at the middle-long term (see again the left panel of Fig. 9.1), the Laplace model predicts a 99.5%, one-year VaR equal to approximately 51% of the total investment. This is notably larger than the 32% that is put forward by the QIS4 report of the Solvency II agreement for insurance companies. It is clear from the figure that reserves equal to 32% of an investment in stocks do not correspond to a 99.5%, one-year VaR. However, we do not suggest creating reserves equal to approximately half the investment in stocks. Indeed, as observed above, the sensitivity to the investment horizon is huge and it remains to be proven that a horizon of one year is the best choice for insurance companies. To conclude, charging 32%, as suggested in the QIS4 report, corresponds to a 99%, four-month VaR, or to a 95%, nine-month VaR, on the French market and under the

Laplace process hypothesis.

Finally, we represent values of the TCE in Fig. 9.2 for the CAC40 index. We use the same calibration of the index by a Laplace process over the period ranging from January 3, 2001 to April 15, 2009 as for the preceding VaR computations. TCE can then be computed at different confidence levels and over different time horizons using formula (9.23) and the following formula, which gives the derivative of the spectrum of a Laplace process:

$$\Phi'(u) = \frac{e^{i\theta u}\left(i(\theta + \delta) + (\delta\theta - \sigma^2)u + \frac{i\theta\sigma^2}{2}u^2\right)}{\left(1 + \frac{1}{2}\sigma^2 u^2 - i\delta u\right)^2}$$

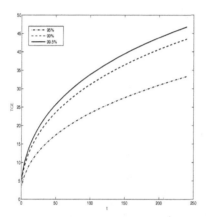

Figure 9.2 TCE under the Laplace process hypothesis

An analysis similar to that of Fig. 9.1 can be conducted on the values displayed in Fig. 9.2. Let us just note here that, *ceteris paribus*, the values of TCE are close, and slightly inferior, to those of VaR.

9.2.2.2 *Risk Budgets with Laplace Processes in Deformed Time*

We now illustrate the changes that occur when we challenge the Lévy process assumption for the modeling of asset returns. To do so, we calibrate Laplace processes both in calendar time and in a time deformed by an ICIR process (whose spectrum is obtained by the method presented in Chapter 7), on the logarithmic return of BNP stock, between December 10, 2009 and June 22, 2010. The parameters of the ICIR clock of the Laplace process are calibrated on the volumes of BNP stock exchanged, while the other parameters of the Laplace process are calibrated in a standard fashion on the logarithmic return of the stock.

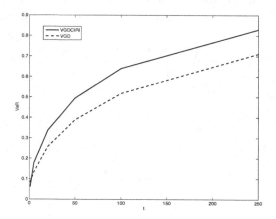

Figure 9.3 Comparison of VaR for BNP stock with a Laplace process in calendar time (dashed line) and in deformed time (solid line)

We display in Fig. 9.3 VaR with respect to the time horizon, for a confidence level equal to 99%, with two Laplace processes: in calendar time and in a time deformed by an ICIR clock. The levels of VaR obtained with the latter assumption are greater than those obtained in calendar time. It is therefore interesting to note that, even though working with Lévy processes allows us to take into account several stylized market features, it is not sufficient to model all the risks affecting markets. Here, using an ICIR clock allows us to keep the Fourier transform methods and to illustrate the computation of a risk premium that is not predicted by the use of Lévy processes.

9.2.3 *Heavy Distribution Tails*

Let us now study the situation where asset return dynamics are best modeled by stable processes, so show marginal distributions with heavy tails.

9.2.3.1 VaR with an Alpha-Stable Motion

We develop here a study on VaR under symmetric alpha-stable distributions (a simple sub-class of stable distributions). The computation of the one-day VaR can be performed by generalizing Eq. (9.17) in the following way:

$$\text{VaR}_c^\alpha(1) = S(0) q_c^\alpha \gamma_1 \qquad (9.24)$$

where γ_1 is the daily dispersion of asset S, measured by the scale parameter of alpha-stable distributions. The estimation of γ_1 can be performed using *ad hoc* statistical methods. For the ten-day VaR, we use the scaling law of alpha-stable motion: $\gamma_{10} = \gamma_1 10^{1/\alpha}$, so that: $\text{VaR}_c^\alpha(10) = S(0) q_c^\alpha \gamma_1 10^{1/\alpha}$. The high quantiles q_c^α are given in Table 5.1 (p. 80) for the $S\alpha S$.

9.2.3.2 Model Risk and Factor 3 of the Basel Agreement

For a chosen confidence level c, the ratio q_c^α / q_c^2 represents a measure of the effect of non-Gaussianity on VaR. Indeed:

$$\frac{\text{VaR}_c^\alpha(1)}{\text{VaR}_c^2(1)} = \frac{S(0) q_c^\alpha \gamma_1}{S(0) q_c^2 \gamma_1} = \frac{q_c^\alpha}{q_c^2} \qquad (9.25)$$

Model risk, in the sense of proposition 9.1 seen above on p. 237, can be assessed with this measure. We can interpret the multiplicative coefficient 3 that has to be applied to VaR, in accordance with the Basel agreements, in terms of the ratio q_c^α / q_c^2, and we can look for the implicit value of α that corresponds to this requirement.

Let VaR^B be the VaR required by the Basel agreement:

$$\text{VaR}_c^B = \text{VaR}_c^2 \times 3$$

Setting $\text{VaR}_c^B = \text{VaR}_c^\alpha$, we have from Eq. (9.25) that $q_c^\alpha / q_c^2 = 3$. For instance, when $c = 99\%$, we require that $q_{0.99}^\alpha / q_{0.99}^2 = 3$, yielding $q_{0.99}^\alpha \simeq 9.87$, or $\alpha \simeq 1.4$. This means that, using a level of 99% and applying the coefficient 3 is equivalent to a regulator assuming that real-world dynamics follow a 1.4-stable motion.

This value would naturally be different if we changed the confidence level. When $c = 97\%$, $q_{0.97}^\alpha \simeq 7.98$, and we have $\alpha \simeq 1.1$. The regulator would then perceive market dynamics as 1.1-stable processes. This shows that model risk cannot be solved by the sole introduction of the factor 3.

We show in Table 9.2 the loss probability $\Pr(X > x) = 1 - F_\alpha(x)$ as a function of x and α. When α is precisely equal to 1 or 2, the corresponding stable distributions have a density function with an explicit analytical form. These are the Cauchy ($\alpha = 1$) and Gaussian ($\alpha = 2$) distributions. We

Table 9.2 Loss probability with alpha-stable distributions

α x	1 Cauchy	1.2	1.4	1.6	1.8	2 Gauss	$N(0,1)$
−1	24.92%	24.66%	24.45%	24.28%	24.13%	23.98%	−0.71
−2	14.76%	12.82%	11.21%	9.87%	8.77%	7.86%	−1.41
−3	10.24%	7.95%	6.02%	4.36%	2.93%	1.69%	−2.12
−4	7.80%	5.58%	3.81%	2.38%	1.21%	0.23%	−2.83
−5	6.28%	4.23%	2.69%	1.53%	0.66%	0.02%	−3.54
−6	5.26%	3.37%	2.03%	1.09%	0.44%	0.00%	−4.24
−7	4.52%	2.79%	1.61%	0.82%	0.31%	0.00%	−4.95
−8	3.96%	2.36%	1.32%	0.65%	0.24%		−5.66
−9	3.52%	2.04%	1.11%	0.53%	0.19%		−6.36
−10	3.17%	1.80%	0.95%	0.44%	0.15%		−7.07
−11	2.89%	1.60%	0.83%	0.38%	0.13%		−7.78
−12	2.65%	1.44%	0.73%	0.33%	0.11%		−8.49
−13	2.44%	1.30%	0.65%	0.29%	0.09%		−9.19
−14	2.27%	1.19%	0.58%	0.25%	0.08%		−9.90
−15	2.12%	1.09%	0.53%	0.23%	0.07%		−10.61
−16	1.99%	1.01%	0.48%	0.20%	0.06%		−11.31
−17	1.87%	0.94%	0.44%	0.18%	0.06%		−12.02
−18	1.77%	0.88%	0.41%	0.17%	0.05%		−12.73
−19	1.67%	0.82%	0.38%	0.15%	0.05%		−13.44
−20	1.59%	0.77%	0.35%	0.14%	0.04%		−14.14

observe that the decrease in loss probability is faster for the values of α. For example, when Pr (loss < −5), we find 0.02% for $\alpha = 2$, and 6.28% for $\alpha = 1$. With the Gaussian distribution, the conversion of quantiles can be performed in the following way:

$$\Pr\left(u_{S_{2,0}(1,0)} \leq -5\right) = \Pr\left(u_{N(0,\sqrt{2})} \leq -5\right)$$
$$= \Pr\left(u_{N(0,1)} \leq -5/\sqrt{2}\right) = \Pr\left(u_{N(0,1)} \leq -3.54\right)$$

Conversely, we can look for the value of the loss x that corresponds to a given probability level. The values underlined in the table give, for a chosen probability level (in this illustration, about 1%), the family of possible couples (α, x), and show the increase of the loss x when α decreases. For instance, we find approximately the same probability with the couples $(1.2; -16)$ and $(1.8; -4)$. This means that a loss of −16 is as likely with a 1.2-stable distribution as a loss of −4 with a 1.8-stable distribution. In all these examples, we assume $X \sim S\alpha S$.

Chapter 10

The Psychology of Risk

Familiarity with and a description of the psychology of financial actors are prerequisites for any theory of portfolio management. The objective of this chapter is therefore to set up the theoretical concepts necessary to derive the ideas presented in the chapters on static and dynamic portfolio management. After having presented the foundations of expected utility by means of the notions of psychological value and certainty equivalent of a random value, an approximation of the risk premium to different orders is performed. These correspond to the psychological attitudes of investors to different moments of distributions. An in-depth analysis of the coefficients of aversion to or taste for the first moments is conducted. The chapter concludes with a study of the Hyperbolic Absolute Risk Aversion (HARA) functions and of their coefficients of aversion expressed to different orders.

10.1 Basic Principles of the Psychology of Risk

We begin by defining the notions of psychological value and of certainty equivalent.

10.1.1 *The Notion of Psychological Value*

It was at the beginning of the 18th century that Daniel Bernoulli (1738) put forward the proposition that, since individuals are not ready to wager the mathematical expectation of a random prospect, it must be that they take into account not the expectations of the gains associated with a risky prospect, but the expectation of some function of the gains. Denoting this function of the gains by $f(X)$, this means that individuals do not consider $\mathbb{E}[X]$ but rather $\mathbb{E}[f(X)]$. This function f works like a psychological filter: it gives the true interest that is held for the random prospect $X(\omega)$, rather than the amount of the gain itself. As Bernoulli himself states (see the translation of his monograph published in Econometrica in 1954, p. 24),

> The determination of an object's value should not rest upon its price, but rather upon the satisfaction that it procures. Whereas the price depends only upon the object itself and remains the same for each, satisfaction depends rather upon the particular circumstances under which the individual makes an evaluation.

In other words, the reference frame that individuals implicitly use is not the expected gain but the expected satisfaction that results from the gain. In contemporary parlance, this satisfaction is termed utility.

We will therefore denote by $u(V)$ the utility resulting from the value V of an object or a lottery. More generally, we will assume that the utility provided to an individual by a given amount is a function $u : \mathbb{R} \times \Omega \to \mathbb{R}$, such that, for any value V, $V \to u(V, \Omega)$ is differentiable and strictly increasing for all $\omega \in \Omega$. We will assume that, for an individual who is naturally not inclined to take risks (*risk averse*), this function is concave. The utility function is increasing to reflect the fact that an individual always prefers "more" to "less". Moreover, the attitude which consists, for an individual who is afraid to take risks, in refusing to invest the mathematical expectation is reflected by the concavity of the utility function.

Indeed, by Jensen's inequality, if X is a random variable of finite expectation and f a concave function such that $f(X)$ is integrable, then $\mathbb{E}[f(X)] \leq f(\mathbb{E}[X])$. The groundwork of the psychology of risk then follows from this concavity. For any utility function u defined as above, and for any random value V, this principle is expressed as follows:

$$\mathbb{E}[u(V)] \leq u\left(\mathbb{E}[V]\right) \qquad (10.1)$$

and states that a typical individual will always prefer the expectation of a gamble to the gamble itself. This means that, if an individual must choose between investing in a random prospect which allows him to obtain a value V providing him with an expected utility of $\mathbb{E}[u(V)]$, and receiving the expectation of the random value V providing the utility $u(\mathbb{E}[V])$, then, to the extent that he seeks out that which satisfies him the most, he will opt for the second possibility and prefer to receive the expectation $\mathbb{E}[V]$ immediately. Bernoulli saw in this an exhortation from nature not to gamble.

However, not all individuals are alike, and for some, winning or losing is less important. These individuals are indifferent with respect to risk (or *risk neutral*). This indifference to changes in their assets is reflected by an equality between certain and expected utilities:

$$\mathbb{E}[u(V)] = u(\mathbb{E}[V]) \qquad (10.2)$$

Mathematically, this equality corresponds to the case where the utility function u is an affine function.

Finally, there is a third type of individual, who specifically seeks out risk (either by addiction, or because of a taste for gambling). In this case, we are confronted with an appetite for risk (*risk taker*), and it is the opposite of formula (10.1) which applies:

$$\mathbb{E}[u(V)] \geq u\left(\mathbb{E}[V]\right) \qquad (10.3)$$

In such cases, an individual will prefer gambling for its own sake, rather than some amount equivalent to the gamble.

10.1.2 *The Notion of Certainty Equivalent*

It is an empirical observation that, because of a fear of loss, individuals are not ready to invest the expected value of a random prospect. Nevertheless, one can ask whether some amount exists such that they are ready to partake of the gamble. According to the intellectual and financial logic underlying the utility criterion, this amount must be such that it equates the utilities resulting from two situations: certainty (without any random draw), and uncertainty (with a random draw). The amount we are looking for is therefore a certain amount providing a utility equivalent to the random amount, i.e. an amount of money providing with certainty the utility that is expected probabilistically. Therefore, let V and u denote a random value and a utility function, respectively. The certainty equivalent to V is a value V^* such that:

$$u(V^*) = \mathbb{E}[u(V)] \qquad (10.4)$$

Observe that if $u(x) = x$, then $V^* = \mathbb{E}[V]$. This situation corresponds to the Pascal–Fermat criterion: the certainty equivalent of a random value is the mathematical expectation of this quantity. Let us recall the fundamental relationship of risk aversion. For a typical individual, i.e. one who has a distaste for risk-taking:

$$\mathbb{E}[u(V)] \leq u(\mathbb{E}[V]) \qquad (10.5)$$

By replacing $\mathbb{E}[u(V)]$ in the relationship (10.4) by the inequality (10.5), we obtain a new inequality:

$$u(V^*) \leq u(\mathbb{E}[V]) \qquad (10.6)$$

Since the utility function u is increasing, we can immediately deduce a very interesting relationship between the certainty equivalent of the random value and its mathematical expectation. Let V be a random value and u the utility function of a risk-averse individual. Let V^* be the certainty equivalent of V. The fundamental relationship describing risk-aversion is:

$$V^* \leq \mathbb{E}[V] \qquad (10.7)$$

and states that, when individuals are afraid of risk, the certainty equivalent of a random value is always smaller than its mathematical expectation. Of course, for individuals who are risk-seeking, it is the opposite relationship that holds.

10.2 The Measurement of Risk Aversion

We now turn to the definition of the risk premium and to its decomposition into several parts reflecting different facets of risk.

10.2.1 *Definitions of the Risk Premium*

The gap between the certainty equivalent V^* of a random value V and the expected value $\mathbb{E}[V]$ shows that individuals require a financial compensation for their risk-taking. Formally, two definitions of the risk premium can be given, which allow us to understand this notion in two different ways: as a difference in evaluation or as a theoretical selling price.

10.2.1.1 The Risk Premium as a Difference in Evaluation

Let V be the random value, with expectation $\mathbb{E}[V]$, and V^* the certainty equivalent to V. The risk premium demanded in order to choose the random value V is the quantity, denoted π, such that:

$$\mathbb{E}[V] - \pi = V^* \tag{10.8}$$

In other words, the risk premium is the amount that must be subtracted from the expected value in order to obtain the certainty equivalent to a random value. By distinguishing the initial financial wealth V_0 in (10.8) from the random prospect X, a new expression for the risk premium π can be obtained:

$$\pi = \mathbb{E}[V] - V^* = \mathbb{E}[V_0 + X] - (V_0 + X^*) = \mathbb{E}[X] - X^*$$

which shows that the risk premium can be interpreted as a difference in evaluation: the difference between the evaluation made by an individual who is indifferent to risk-taking, and another who is not indifferent.

10.2.1.2 The Risk Premium as a Theoretical Selling Price

Let us apply the utility functions to the equality (10.8). This leads to:

$$u(\mathbb{E}[V] - \pi) = u(V^*)$$

and, since by definition $u(V^*) = \mathbb{E}[u(V)]$, a second definition of the risk premium is obtained by means of an equation that uses the utility functions. Thus, let V be a random value and u a utility function. The risk premium required in order to choose the random value V is then the quantity π such that:

$$u(\mathbb{E}[V] - \pi) = \mathbb{E}[u(V)] \tag{10.9}$$

The risk premium π is necessarily positive for a risk-averse individual and negative for a risk-loving individual. Let us now incorporate the initial financial wealth V_0 into the previous equality (10.9):

$$u(V_0 + \mathbb{E}[X] - \pi) = \mathbb{E}[u(V_0 + X)] \tag{10.10}$$

This relationship (10.10) describes the indifference between holding an asset X evaluated at the psychological value described by $\mathbb{E}[u(V_0 + X)]$, and paying π in order to hold the certain amount $V_0 + \mathbb{E}[X]$. The risk premium π can thus be interpreted as the theoretical selling price of a random prospect X for an initial level of financial wealth V_0 and a given utility function u.

10.2.2 Decomposition of the Risk Premium

Here, we perform an expansion of the utility function in the spirit of Pratt (1964), but up to order four. The advantage of this expansion is that it makes explicit the coefficients of aversion or appetite toward the different moments of the distribution of the risk under consideration.

10.2.2.1 Approximation by Taylor Expansion

In his seminal article on the economics of risk, Pratt (1964) suggested using an approximation in the case of small variations in value. The underlying intuition in this approach is to make a certain quantity appear, the coefficient of risk aversion, which measures the psychological reticence experienced by an individual who invests a part of his wealth in a risky asset.

As we have seen above, the risk associated with holding an asset is characterized by a probability distribution corresponding to the possible values that this asset can take at maturity. A probability distribution is generally characterized by its parameters or by its moments. We will therefore characterize the risk associated with an asset, and hence the reluctance to invest in it, by the parameters of its distribution, with each of the parameters representing one of the aspects of the risk. Each coefficient in the expansion thus represents an attitude toward the corresponding asset.

Two types of parameters can be distinguished, each characterizing one specific aspect of the distribution under question. We can first of all consider the scale of risk characteristic. This corresponds to the moment of order 2, of which the variance is the most classic representation. It is associated with the volatility, i.e. with the amplitude of the fluctuations. Next, there are the shape of risk characteristics. There are many of these, of which we will consider only two. First of all, there is the moment of order 3, which describes the asymmetry of the distribution. This involves the risk associated with differences in the probabilities of gains or losses. Second, the moment of order 4, which described the thickness of the tails of the distribution. This corresponds to the risk associated with the likelihood of extreme risks, positive or negative, in the probability distribution under question. In a dynamic setting, this risk is linked to a gap in prices, i.e. to a discontinuity in the trajectory.

In order to make these moments appear, we perform an expansion of the function around the expected value $\mathbb{E}[V]$, up to an order corresponding to the desired moment.

Whereas Arrow (1965) and Pratt (1964) limit their expansion to order 2 (see for instance Eeckhoudt and Gollier (1992)), here we present a more general and more recent development, which includes the moments of order 3 and 4. As with all Taylor expansions, this calculation is, *a priori*, valid only locally around a reference value, and therefore for reasonable fluctuations around this value.

Let us start with Eq. (10.9) for the risk premium:

$$u(\mathbb{E}[V] - \pi) = \mathbb{E}[u(V)] \tag{10.11}$$

and perform an expansion of each term of the equation around $\mathbb{E}[V]$: an expansion limited to to order 1 of $u(\mathbb{E}[V] - \pi)$, as well as an expansion of order 4 of $u(V)$.

We can then rewrite Eq. (10.11), replacing each term by its expansion. The calculation involves the following steps. We first perform an expansion of order 1 of $u(\mathbb{E}[V] - \pi)$:

$$u(\mathbb{E}[V] - \pi) \simeq u(\mathbb{E}[V]) - \pi u'(\mathbb{E}[V]) \tag{10.12}$$

and then an expansion of order 4 of $u(V)$:

$$u(V) \simeq u(\mathbb{E}[V]) + (V - \mathbb{E}[V])u'(\mathbb{E}[V]) + \frac{(V - \mathbb{E}[V])^2}{2!}u''(\mathbb{E}[V])$$
$$+ \frac{(V - \mathbb{E}[V])^3}{3!}u'''(\mathbb{E}[V]) + \frac{(V - \mathbb{E}[V])^4}{4!}u''''(\mathbb{E}[V])$$

We then calculate the expectation of $u(V)$:

$$\mathbb{E}[u(V)] \simeq u(\mathbb{E}[V]) + \mathbb{E}[V - \mathbb{E}[V]]u'(\mathbb{E}[V])$$
$$+ \frac{\mathbb{E}[(V - \mathbb{E}[V])^2]}{2}u''(\mathbb{E}[V]) + \frac{\mathbb{E}[(V - \mathbb{E}[V])^3]}{6}u'''(\mathbb{E}[V])$$
$$+ \frac{\mathbb{E}[(V - \mathbb{E}[V])^4]}{24}u''''(\mathbb{E}[V])$$
$$\simeq u(\mathbb{E}[V]) + \frac{\text{var } V}{2}u''(\mathbb{E}[V]) + \frac{m_3(V)}{6}u'''(\mathbb{E}[V])$$
$$+ \frac{m_4(V)}{24}u''''(\mathbb{E}[V])$$

which will allow us to rewrite Eq. (10.11). In order to do this, let us replace each term in Eq. (10.11) by its expansion. We obtain:

$$u(\mathbb{E}[V]) - \pi u'(\mathbb{E}[V]) \simeq u(\mathbb{E}[V]) + \frac{\text{var}(V)}{2}u''(\mathbb{E}[V])$$
$$+ \frac{m_3(V)}{6}u'''(\mathbb{E}[V]) + \frac{m_4(V)}{24}u''''(\mathbb{E}[V])$$

and infer the expression of π:

$$\pi \simeq \left[-\frac{u''(\mathbb{E}[V])}{u'(\mathbb{E}[V])}\right]\frac{\operatorname{var} V}{2} - \left[\frac{u'''(\mathbb{E}[V])}{u'(\mathbb{E}[V])}\right]\frac{m_3(V)}{6} + \left[-\frac{u''''(\mathbb{E}[V])}{u'(\mathbb{E}[V])}\right]\frac{m_4(V)}{24}$$
(10.13)

Therefore, we have expressed the risk premium as a linear combination of the moments of order 2, 3, and 4, the coefficients of these moments representing exactly the quantification of the corresponding aversion to or taste for risk.

10.2.2.2 Definition of the Coefficients of the Psychology of Risk

We introduce here a number of fundamental definitions that will be used throughout the following chapters.

Coefficient of aversion for volatility risk.

Let us begin with the coefficient of risk aversion, which in practice describes a psychological stance with respect to volatility: the coefficient of absolute risk aversion can be interpreted locally as a coefficient of rejection of the volatility of a random value around its expected value $\mathbb{E}[V]$. Here, the risk is reduced to its moment of order 2. This coefficient is defined by the quantity denoted λ, and defined by:

$$\lambda = -\frac{u''(\mathbb{E}[V])}{u'(\mathbb{E}[V])}$$
(10.14)

It is possible to rank individuals in terms of their local rejection of volatility by means of the coefficient λ by means of the following theorem, which relates λ, π and u:

Ranking of individuals.

Given two individuals 1 and 2 with utility functions u_1 and u_2, with coefficients λ_1 and λ_2, and who require the risk premia π_1 and π_2, the following conditions are equivalent:

(1) $\lambda_1 \geq \lambda_2$
(2) $\pi_1 \geq \pi_2$
(3) u_1 is more concave than u_2.

Conversely, it is possible to define a coefficient which is the opposite of the risk aversion coefficient, and which is defined as follows.

Coefficient of tolerance for volatility risk.

The coefficient for risk tolerance, which is also locally a coefficient of tolerance for the volatility of a random value around its expected value $\mathbb{E}[V]$, is the inverse of the risk aversion coefficient, i.e. $\tau = 1/\lambda$. The fear of seeing a value fluctuate can also depend on the expected value that is used as a reference. In other words, one may normalize the risk premium by this expected value in order to obtain a relative risk premium $\pi/\mathbb{E}[V]$. The new random value to consider is then no longer the random value V, but the relative random value $V/\mathbb{E}[V]$. From the following trivial relation:

$$\mathrm{var}\left(\frac{V}{\mathbb{E}[V]}\right) = \frac{\mathrm{var}\, V}{(\mathbb{E}[V])^2}$$

and Eq. (10.13) taken to order 2:

$$\pi \simeq 0.5\,\lambda\,\mathrm{var}\, V$$

it is possible to express the so-called relative risk aversion:

$$\frac{\pi}{\mathbb{E}[V]} \simeq 0.5\,\lambda\,\mathbb{E}[V]\,\mathrm{var}\left(\frac{V}{\mathbb{E}[V]}\right)$$

which includes a coefficient $\lambda\,\mathbb{E}[V]$, which represents the analog of λ in relative terms.

Coefficient of relative risk aversion.

The coefficient of relative risk aversion of a random value V around its expectation $\mathbb{E}[V]$ is the quantity denoted $\tilde{\lambda}$ defined by:

$$\tilde{\lambda} = -\mathbb{E}[V]\frac{u''(\mathbb{E}[V])}{u'(\mathbb{E}[V])} \qquad (10.15)$$

This coefficient measures the fear of a relative variation (*relative risk aversion*) of a random value V as a share of its expected value $\mathbb{E}[V]$.

We can now proceed with a discussion of the third-order term in Eq. (10.13).

Coefficient of taste for asymmetry.

The coefficient of taste for asymmetry of a random value around its expected value $\mathbb{E}[V]$ is the quantity denoted ψ and defined by:

$$\psi = \frac{u'''(\mathbb{E}[V])}{u'(\mathbb{E}[V])} \qquad (10.16)$$

This coefficient, which appears directly in front of the moment of order three in Eq. (10.13) is different from the coefficient of prudence defined below and usually studied in the economics of risk.

Coefficient of absolute prudence (Kimball (1990)).

The coefficient of absolute prudence for a random value V is the quantity denoted ζ and defined by:

$$\zeta = -\frac{u'''(\mathbb{E}[V])}{u''(\mathbb{E}[V])} \qquad (10.17)$$

We can observe the following link between ψ and ζ:

$$\psi = \frac{u'''(\mathbb{E}[V])}{u'(\mathbb{E}[V])} = \left(-\frac{u'''(\mathbb{E}[V])}{u''(\mathbb{E}[V])}\right)\left(-\frac{u''(\mathbb{E}[V])}{u'(\mathbb{E}[V])}\right)$$

or, more concisely:

$$\psi = \zeta\,\lambda \tag{10.18}$$

which expresses that the coefficient of preference for asymmetry is the prudence multiplied by the risk aversion.

Let us now turn to the moment of order four in Eq. (10.13).

Coefficient of aversion to tail thickness.

The coefficient of aversion to thick tails of a random value about its expected value $\mathbb{E}[V]$ is the quantity denoted φ and defined by:

$$\varphi = -\frac{u''''(\mathbb{E}[V])}{u'(\mathbb{E}[V])} \tag{10.19}$$

Once again, the coefficient is distinct from that usually considered in the literature, and originally defined by Kimball (1992).

Coefficient of absolute temperance (Kimball (1992)).

The coefficient of absolute temperance of a random value V is the quantity denoted χ and defined by:

$$\chi = -\frac{u''''(\mathbb{E}[V])}{u'''(\mathbb{E}[V])} \tag{10.20}$$

We observe the following link between φ and χ:

$$\varphi = -\frac{u''''(\mathbb{E}[V])}{u'(\mathbb{E}[V])} = \left(-\frac{u''''(\mathbb{E}[V])}{u'''(\mathbb{E}[V])}\right)\left(-\frac{u'''(\mathbb{E}[V])}{u''(\mathbb{E}[V])}\right)\left(-\frac{u''(\mathbb{E}[V])}{u'(\mathbb{E}[V])}\right)$$

or, more compactly:

$$\varphi = \chi\,\zeta\,\lambda = \chi\,\psi \tag{10.21}$$

which states that the aversion to tail thickness is the product of risk aversion, prudence, and temperance.

10.2.2.3 Analysis of the Coefficients of the Psychology of Risk

Let us begin by rewriting the risk premium given in Eq. (10.13) as a function of the coefficients λ, ψ and φ:

$$\pi \simeq \frac{1}{2}\lambda \operatorname{var} V - \frac{1}{6}\psi\, m_3(V) + \frac{1}{24}\varphi\, m_4(V) \tag{10.22}$$

Thus, it is apparent that, along each of the dimensions of risk, the risk premium is equal to a product of two factors. A first component is due to the random nature of the future value, characterized by the moments of its distribution, and measures the amount of risk associated with this random value. As we have seen, the risk is decomposed into its size, measured by the moment of order 2 (the variance) of V, and the shape of the risk, namely the asymmetry and the thickness of the tails, respectively measured by the moments of orders 3 and 4 of V. This component can by viewed as independent of the individual, non-psychological, or objective.

The other component is due to the utility function u, and denoted λ for the moment of order 2, ψ for the moment of order 3, and φ for the moment of order 4, and depends only on the individual in question and his level of wealth. In contrast with the preceding expression, this component is psychological, or subjective. In this component, the coefficient $\lambda > 0$ represents a weighting of the volatility. It represents the investor's propensity to accept risks of larger or smaller amplitude. The greater this coefficient is, the more reluctant the investor is to accept a volatility risk. Conversely, the lower λ is, the more willing the investor is to accept a greater volatility. When $\lambda = 0$, the investor examines only the expected gain, regardless of the volatility. The coefficient ψ represents a weighting of the asymmetry of chances of a gain or a loss. It expresses the investor's preferences for unequal odds. The greater this coefficient, the more interested the investor is by such an inequality of the odds (he will seek out assets whose return distribution is highly asymmetric, such as the lotto). Conversely, when $\psi \to 0$, the investor becomes progressively indifferent to this asymmetry and, leaving the order 4 aside, reasons only in terms of the expectation–variance pair: we thus find ourselves back in the classic case. Positive values of ψ will result in a taste for positive values of m_3, and similarly for negative values.

Finally, the coefficient φ represents a weighting of the thickness of the tails of the distribution. It expresses the investor's taste for flattening out and thus, all other things being equal, the willingness to avoid extreme risks, independently of their sign. The greater this coefficient, the more the investor seeks out a flattening of the distribution. Conversely, when $\varphi \to 0$, the investor becomes progressively indifferent to the thickness of the tails.

Equations (10.18) and (10.21) allow us to rewrite the risk premium (10.22) with the coefficients ζ and χ:

$$\pi \simeq \frac{1}{2} \lambda \left(\operatorname{var} V - \frac{1}{3} \zeta \left(m_3(V) - \frac{1}{4} \chi \, m_4(V) \right) \right) \qquad (10.23)$$

We will first analyze this equation taken up to order 3:

$$\pi \simeq \frac{1}{2}\lambda \left(\operatorname{var} V - \frac{1}{3} \zeta\, m_3(V) \right) \tag{10.24}$$

This last expression shows that the risk premium may, locally, be written as the product of the risk aversion and a risk term. The stronger the prudence, the smaller the risk term. Here, prudence measures how the investor will compare the variance and the moment of order 3. The greater the prudence, the greater the weight the investor gives to the moment of order 3, for a given risk aversion.

It is now easy to interpret Eq. (10.23) as a generalization of Eq. (10.24). At each order, a moment is compensated by a psychological term multiplied by another moment. Thus, the moment of order 3, which itself weights the moment of order 2, is weighted by the product of temperance and of the moment of order 4. Temperance may thus be interpreted as the strength of the aversion toward the moment of order 4 allowing us to weight the moment of order 3, for given coefficients of risk aversion and prudence.

Let us now turn to the concept of certainty equivalent V^* to a random value V. Since:

$$V^* = \mathbb{E}[V] - \pi$$

It is also possible to express the certainty equivalent in terms of the previous coefficients:

$$V^* \simeq \mathbb{E}[V] - \frac{1}{2}\lambda \operatorname{var} V + \frac{1}{6}\psi\, m_3(V) - \frac{1}{24}\varphi\, m_4(V) \tag{10.25}$$

The certainty equivalent can therefore be rewritten as a function of the coefficients ζ and χ:

$$V^* \simeq \mathbb{E}[V] + \frac{1}{2}\lambda \left(-\operatorname{var} V + \frac{1}{3}\zeta \left(m_3(V) - \frac{1}{4}\chi\, m_4(V) \right) \right) \tag{10.26}$$

For more details on these approximations and on the coefficients of the psychology of risk, see Le Courtois (2012). Let us now turn to a simple illustration of the concepts of risk aversion, prudence, and temperance.

10.2.3 Illustration

Here, we present some simple tests involving lotteries that will allow the reader to determine his/her attitude toward risk, and in particular whether he/she is prudent or tempered (for a theoretical presentation, see the article of Eeckhoudt and Schlesinger (2006)). In the first of these tests, presented

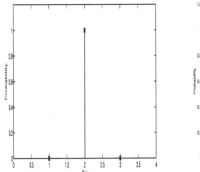 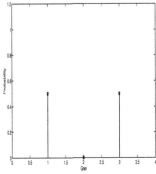

Figure 10.1   Psychological attitude toward volatility: aversion for volatility risk

in Fig. 10.1, a player may choose between winning two euros with certainty, or a lottery which returns one euro with probability 1/2 and three euros with the same probability.

As the calculations of the moments associated with each of these two possibilities confirm:

$$\begin{cases} \mathbb{E}[X^1] = 1 \times 2 = 2 \\ \text{var}(X^1) = 1 \times (2-2)^2 = 0 \end{cases}$$

and:

$$\begin{cases} \mathbb{E}[X^2] = 0.5 \times 1 + 0.5 \times 3 = 2 \\ \text{var}(X^2) = 0.5 \times (1-2)^2 + 0.5 \times (3-2)^2 = 1 \end{cases}$$

where superscript $i = 1, 2$ refers to the lottery for which this expression is evaluated. These lotteries have the same expected value (equal to two, trivially) but have distinct moments of order 2. The first is not risky (in the classic sense of risk which is associated with the moment of order two) whereas the second is. Thus, this test allows us to distinguish between individuals who are risk averse (who choose the first lottery) and those who have a taste for risk (who choose the second lottery).

Let us now consider the lottery choice presented in Fig. 10.2. In this case, as the following expressions confirm:

$$\begin{cases} \mathbb{E}[X^1] = 0.75 \times 1 + 0.25 \times 3 = 1.5 \\ \text{var}(X^1) = 0.75 \times (1-1.5)^2 + 0.25 \times (3-1.5)^2 = 0.75 \\ m_3(X^1) = 0.75 \times (1-1.5)^3 + 0.25 \times (3-1.5)^3 = 0.75 \end{cases}$$

and:

$$\begin{cases} \mathbb{E}[X^2] = 0.25 \times 0 + 0.75 \times 2 = 1.5 \\ \mathrm{var}(X^2) = 0.25 \times (0-1.5)^2 + 0.75 \times (2-1.5)^2 = 0.75 \\ m_3(X^2) = 0.25 \times (0-1.5)^3 + 0.75 \times (2-1.5)^3 = -0.75 \end{cases}$$

both lotteries have the same mean, and the same moment of order 2. However, they differ with respect to their moment of order 3. The individuals who choose the first lottery (a gain of one or three euros with respective probabilities three quarters and one quarter) have a preference for positive asymmetry. In the language of the economics of decision, they are prudent. Most people make this choice, when asked. This explains the existence of the lotto, in its typical form: a probability of a small loss that is virtually equal to one, combined with a very small probability of winning a very significant amount (the reader can easily check that the probability distribution of the lotto has the same shape as the distribution of the first lottery in Fig. 10.2). In contrast, individuals who prefer the second lottery (gains of zero or two euros with probabilities of one quarter and three quarters respectively) have a preference for negative asymmetry. In economic terms, they are imprudent.

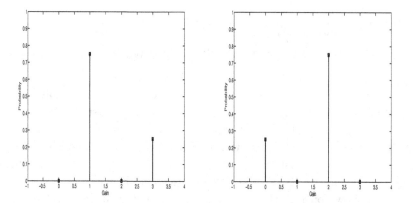

Figure 10.2   Psychological attitude toward asymmetry: prudence

Let us finally turn to Fig. 10.3, which compares two lotteries with the same mean and moments of orders 2 and 3, but whose moments of order 4

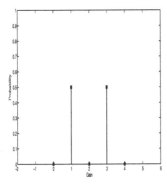 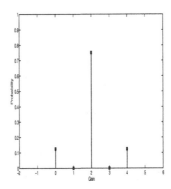

Figure 10.3 Psychological attitude toward kurtosis: the virtue of temperance

differ, as the following calculations show:

$$\begin{cases} \mathbb{E}[X^1] = 0.5 \times 1 + 0.5 \times 3 = 2 \\ \text{var}(X^1) = 0.5 \times (1-2)^2 + 0.5 \times (3-2)^2 = 1 \\ m_3(X^1) = 0.5 \times (1-2)^3 + 0.5 \times (3-2)^3 = 0 \\ m_4(X^1) = 0.5 \times (1-2)^4 + 0.5 \times (3-2)^4 = 1 \end{cases}$$

and:

$$\begin{cases} \mathbb{E}[X^2] = 0.125 \times 0 + 0.75 \times 2 + 0.125 \times 4 = 2 \\ \text{var}(X^2) = 0.125 \times (0-2)^2 + 0.75 \times (2-2)^2 + 0.125 \times (4-2)^2 = 1 \\ m_3(X^2) = 0.125 \times (0-2)^3 + 0.75 \times (2-2)^3 + 0.125 \times (4-2)^3 = 0 \\ m_4(X^2) = 0.125 \times (0-2)^4 + 0.75 \times (2-2)^4 + 0.125 \times (4-2)^4 = 4 \end{cases}$$

An individual who prefers the first lottery (gains of one and three euros with equiprobability) seeks to avoid, all other things being equal, a moment of order four that is too large. In the language of the economics of decision-making, he is temperant. On the other hand, an individual who prefers the second lottery (gains of zero, two, or four euros with probabilities of one eighth, three quarters, and one eighth respectively) has a strong preference for moments of order 4. He will then be described as an intemperant.

10.3 Typology of Risk Aversion

In this last section, we will focus on a particular class of utility functions (the HARA functions) that is very often used in wealth management problems. Let us begin by examining the main types of attitude toward financial risk.

10.3.1 Attitude with Respect to Financial Risk

We now highlight three broad types of individual behavior, which are characterized by different forms – increasing, decreasing, or constant – of risk aversion as a function of financial wealth.

10.3.1.1 Risk Aversion Increasing with Wealth

We consider here the following type of psychological attitude: the greater the wealth, the more troublesome the idea of risk-taking. An increase in financial wealth appears to have the psychological effect of creating a greater apprehension of the risk-taking decision.

Quadratic functions are one example of functions that can represent this psychological attitude:

$$u(x) = x - bx^2 \qquad b > 0$$

where b is a parameter that depends on the individual's psychology.

The utility functions for which risk aversion increases with wealth are known as IARA functions (for *Increasing Absolute Risk Aversion*).

10.3.1.2 Risk Aversion Decreasing with Wealth

The idea here is the opposite of the previous one. This function corresponds to an attitude of the following kind: the greater the wealth, the less fearsome the idea of risk-taking. The individual becomes increasingly inclined to invest, as if he is reassured by the increase in wealth. In this situation, it is as if money provided a form of satisfaction that progressively annihilates the fear of fluctuations in wealth.

Among the utility functions that correspond to this attitude are the logarithm and the power function:

$$u(x) = \frac{1}{a} x^a \qquad 0 < a < 1$$

where a is a parameter that depends on the individual's psychology.

The utility functions for which risk aversion decreases with wealth are known as DARA functions (for *Decreasing Absolute Risk Aversion*). This assumption is the most commonly made in decision economics.

10.3.1.3 Risk Aversion Constant with Respect to Wealth

In contrast with the two preceding mental dispositions in which risk aversion varies with wealth, a third attitude is that in which risk aversion does not change: whatever the individual's level of wealth, be it small or large, the same apprehension of the idea of risk-taking applies.

From a practical standpoint, this constant risk aversion implies that the investor will determine his portfolio allocation independently of the daily value of his portfolio. In economic terms, this is the same as saying that the demand for risky assets is independent of the level of financial wealth.

One consequence of this independence is the possibility of aggregating individual demands into a collective demand. Indeed, to the extent that individuals do not choose the amount of risky assets based on their wealth, it is possible to eliminate wealth from the characterization of individuals, and thus to obtain an average individual who is representative of the group. Denoting this individual's coefficient of risk aversion by λ_M, the representative individual's tolerance for volatility risk $1/\lambda_M$ is then the sum of the risk tolerances $1/\lambda_i$ of all the individuals of the group:

$$\frac{1}{\lambda_M} = \sum_{i=1}^{n} \frac{1}{\lambda_i}$$

which shows that the coefficient of risk aversion λ_M associated with the representative individual is equal to the harmonic mean of the coefficients λ_i of the n individuals. Thus, the distribution of wealth between the different individuals, or actors, present in the market has no influence on the determination of the equilibrium price of the asset in question.

Among the utility functions corresponding to this attitude are the negative exponential functions:

$$u(x) = -c \exp(-\lambda x/c)$$

The utility functions with constant risk aversion are known as CARA (for *Constant Absolute Risk Aversion*). The advantage of this type of utility function becomes apparent when we add the assumption of normality of the potential fluctuations of wealth. This economic framework is then equivalent to the mean-variance setting.

10.3.2 The Family of HARA Functions

The different forms of risk attitudes represented by the preceding utility functions can be viewed as variants of a single generic form in which the risk aversion coefficient $\lambda(x)$ is a function of $1/x$ or, in other words, a hyperbolic function of wealth, with the exception of the negative exponential (a degenerate case). Conversely, in this framework the tolerance for risk (the opposite attitude of aversion) is a linear function of wealth. These functions are called HARA, for *Hyperbolic Absolute Risk Aversion*.

The linearity of the tolerance for volatility risk τ as a function of wealth V can be expressed as follows:

$$\tau(V) = \tau(0) + kV$$

Let us set $\tau(0) = c/b$ and $k = 1/(1-a)$. The tolerance for risk can then be expressed as:

$$\tau(V) = \frac{c}{b} + \frac{V}{1-a} \qquad (10.27)$$

The parameter c/b describes the risk tolerance for a wealth level of zero, and Rubinstein (1973) described the parameter $1/(1-a)$ as the prudence parameter of individuals. In this text, we reserve the term prudence for the attitude with respect to the moment of order 3, as defined by Kimball (1990). The smaller k, the lower the tolerance, which corresponds to large values of a ($k \to 0$ when $a \to \infty$). The risk tolerance and k vary in the same direction. The previous equation can thus be written:

$$-\frac{u'(V)}{u''(V)} = \frac{c}{b} + \frac{1}{1-a}V$$

i.e.:

$$\left(\frac{c}{b} + \frac{1}{1-a}V\right)u''(V) + u'(V) = 0$$

Let us set $v = u'$ and $V = x$. The previous equation becomes:

$$\left(\frac{c}{b} + \frac{1}{1-a}x\right)v' + v = 0 \qquad (10.28)$$

which is a differential equation of order 1 that is separable in its variables. The solutions to this equation form the set of HARA utility functions. In practice:

$$v(x) = b\left(c + \frac{b}{1-a}x\right)^{a-1}$$

from which the general form of HARA utility functions follows:

$$u(x) = \frac{1-a}{a}\left(\frac{b}{1-a}x + c\right)^a \qquad (10.29)$$

Depending on the different parameter values, we can then derive the utility functions defined in the previous subsection.

The class of power utility functions.
Let us set $c = 0$. Equation (10.28) takes the form:
$$\frac{1}{1-a}xv' + v = 0$$
Assume $0 < a < 1$, then $v(x) = x^{a-1}$. We obtain the power functions:
$$u(x) = \frac{1}{a}x^a$$
The class of logarithmic utility functions.
The logarithmic utility function is a degenerate case of the power function. It is obtained by setting $a \to 0$ and $c = 0$. Equation (10.28) can be written as:
$$xv' + v = 0$$
so $v(x) = 1/x$. We therefore obtain the logarithmic function:
$$u(x) = \ln x$$
The class of exponential utility functions.
By taking $a \to \infty$ and arbitrarily setting $b = 1$, we can write (10.28) as:
$$cv' + v = 0$$
Assume that $c \neq 0$, so $v(x) = e^{-x/c}$. We obtain the exponential functions:
$$u(x) = -c\, e^{-x/c}$$
The class of quadratic utility functions.
Let us set $a = 2$ and $c = 1/2$. Equation (10.28) then becomes:
$$\left(-x + \frac{1}{2b}\right)v' + v = 0$$
Assume $b \neq 0$, then $v(x) = 1 - 2bx$. We obtain the quadratic functions:
$$u(x) = x - bx^2$$

Table 10.1 summarizes these results and provides the expressions of the coefficients of risk aversion, appetite for asymmetry, and aversion for tail thickness for the four utility functions mentioned above.

It follows, by inverting Eq. (10.27), that, for the class of HARA utility functions, the coefficient of risk aversion can be expressed as:
$$\lambda(x) = \frac{1}{\frac{1}{1-a}x + \frac{c}{b}}$$

Table 10.1 Class of HARA utility functions

Function	Log	Power	Exp	Quad
a	0	$0 < a < 1$	∞	2
b	–	–	1	$\neq 0$
c	0	0	$\neq 0$	$1/2$
$u(x)$	$\ln x$	$\frac{x^a}{a}$	$-ce^{-x/c}$	$x - bx^2$
$u(0)$	$-\infty$	0	$-c$	0
$u(+\infty)$	$+\infty$	$+\infty$	0	$-\infty$
$u'(x)$	$\frac{1}{x}$	x^{a-1}	$e^{-x/c}$	$1 - 2bx$
$u''(x)$	$-\frac{1}{x^2}$	$-(1-a)x^{a-2}$	$-\frac{e^{-x/c}}{c}$	$2b$
$u'''(x)$	$\frac{2}{x^3}$	$(1-a)(2-a)x^{a-3}$	$\frac{e^{-x/c}}{c^2}$	0
$u''''(x)$	$-\frac{6}{x^4}$	$-(1-a)(2-a)(3-a)x^{a-4}$	$-\frac{e^{-x/c}}{c^3}$	0
$\lambda(x)$	$\frac{1}{x}$	$\frac{1-a}{x}$	$\frac{1}{c}$	$\frac{1}{\frac{1}{2b}-x}$
	↘ CRRA	↘ CRRA	→ CARA	↗
$\tilde{\lambda}(x)$	1	$1 - a$	$\frac{x}{c}$	$\frac{x}{\frac{1}{2b}-x}$
$\frac{1}{\lambda(x)}$	x	$\frac{x}{1-a}$	c	$-x + \frac{1}{2b}$
$\psi(x)$	$\frac{2}{x^2}$	$\frac{(1-a)(2-a)}{x^2}$	$\frac{1}{c^2}$	0
	↘	↘	→	→
$\varphi(x)$	$\frac{6}{x^3}$	$\frac{(1-a)(2-a)(3-a)}{x^3}$	$\frac{1}{c^3}$	0
	↘	↘	→	→

which can be expressed asymptotically as:
$$\lambda(x) \stackrel{+\infty}{\sim} \frac{1-a}{x}$$

In other words, the coefficient of risk aversion varies hyperbolically, explaining the HARA denomination. By calculating the derivatives of orders one and three of HARA functions, we can also obtain the coefficients of taste for asymmetry:
$$\psi(x) = \frac{b^2(2-a)}{(1-a)\left(\frac{b}{1-a}x + c\right)^2}$$

which can be expressed asymptotically as:
$$\psi(x) \stackrel{+\infty}{\sim} \frac{(1-a)(2-a)}{x^2}$$

Finally, in the HARA case the coefficient of aversion to thick tails can be expressed as:
$$\varphi(x) = \frac{b^3(a-2)(a-3)}{(1-a)^2\left(\frac{b}{1-a}x + c\right)^3}$$

and can be expressed asymptotically as:

$$\varphi(x) \overset{+\infty}{\sim} \frac{(1-a)(2-a)(3-a)}{x^3}$$

Figure 10.4 shows the behavior of λ, ψ, and φ as a function of wealth.

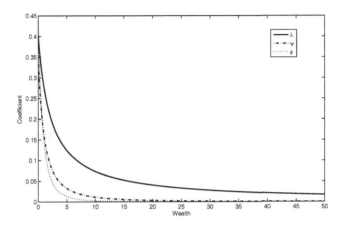

Figure 10.4 Behavior of the coefficients λ, ψ, and φ in the case of HARA functions for the parameters $a = 1/2$, $b = 2$, and $c = 1$

The class of CRRA utility functions is a subclass of the HARA utility functions, and contains the exponential and logarithmic functions, such as those presented in Table 10.1. A CRRA utility function retains the hyperbolic decrease that characterizes its generic form but differentiates itself by virtue of having a constant coefficient of relative risk aversion. In mathematical terms, this function can be expressed as:

$$u(x) = \frac{1-a}{a}\left(\frac{b}{1-a}x\right)^a$$

or more simply, taking $b = (1-a)^{\frac{a-1}{a}}$, as:

$$u(x) = \frac{x^a}{a}$$

so the coefficient of relative risk aversion has the value:

$$\tilde{\lambda}(x) = x\lambda(x) = 1 - a$$

Figure 10.5 shows (solid curve) the variation of $\tilde{\lambda}$ for a utility function of the HARA type. The increase of this coefficient shows that as the wealth of the investor increases, he becomes more psychologically disposed to take risks that are proportional to his wealth. A limiting relative risk aversion exists: beyond a threshold value of wealth, the investor will only slightly alter his choice of investments.

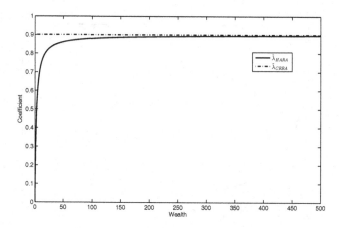

Figure 10.5 Behavior of the coefficient of relative risk aversion in the case of HARA and CRRA functions

In the case of CRRA functions, which are the dotted curves in Fig. 10.5, the agent's relative risk aversion is, by definition, a constant function of his wealth. Two agents with this type of utility function can be compared by their coefficients a. Finally, the constant value $\tilde{\lambda}$ reflects the fact that an agent who fears taking risks will not alter his behavior, even if for instance his revenues are doubled from one day to the next. This results in a constant relative risk aversion, and in the use of CRRA (*Constant Relative Risk Aversion*) functions.

Chapter 11

Monoperiodic Portfolio Choice

The portfolio framework that is adopted here is the monoperiodic one. This is a static framework: the portfolio cannot be modified between the beginning and the end of the period. If the portfolio composition needs to be changed, because of sale or purchase decisions for example, it is necessary to introduce intermediate dates, so as to move to a multiperiodic framework in discrete time (or to a continuous-time framework). Only arbitrary dates will be considered: 0 and t, between which the portfolio composition will not change. Time t represents the investment horizon for decisions taken at time 0. This horizon can be very short (one week) or very long (ten years). For monoperiodic models, which do not take into account the passage of time, the time horizon has no effect on the composition method for the portfolio. However, this horizon t inevitably has an indirect effect on the result, in so far as the values of the parameters $\lambda(t)$ depend on the horizon itself.

Two types of analysis can be conducted. First, a comparative analysis of the results that stem from the use of different models, such as those of Markowitz (1952) or Mitton and Vorkink (2007), for a given set of parameters $\lambda(t)$, for any time t. Second, a comparative statics of the sensitivity of the results of an allocation to a change of value of parameters, as the change from $\lambda(t)$ to $\lambda(at)$. This comparative statics can be interpreted as the influence of time on the allocation. However, such an interpretation is not strictly necessary: we need only to estimate the modification of a portfolio based on the values inputed in the optimization program. This comparative statics does not require the introduction of a time horizon, if even this quantity is linked to the changes of values of parameters. In other words, it is important not to mix two aspects of the tests of portfolios: the comparison of models and the sensitivity of a model to the variations of parameters.

This chapter starts with a general presentation of the monoperiodic optimization program, in the presence of a risk-free asset and of one, or several, risky asset(s). Solutions to this program are then found taking into account several distribution moments, going from simple to complicated settings. Hence, the mean-variance framework is first recalled, then we study a mean-variance-asymmetry framework, and finally a mean-variance-asymmetry-leptokurticity framework, each time in the presence of one, or several, risky asset(s). To conclude, simplifications to the monoperiodic optimization program are suggested. An appendix deals with the treatment of uncertainty in a classic setting.

11.1 The Optimization Program

Consider a portfolio whose value is denoted by $V(t)$ at any time t. In the Bernoulli representation, the satisfaction attached to the portfolio is $u(V(t))$ and the expected satisfaction is $\mathbb{E}[u(V_t)]$. According to this representation, the goal of any rational investor is to maximize satisfaction, which leads to the optimization program:

$$\max \mathbb{E}[u(V(t))] \tag{11.1}$$

For the sake of simplicity, and because we have set ourselves in a monoperiodic framework, we will drop time t when presenting the equations that follow.

Recall that, by definition of the certain amount V^* equivalent to the random value V, $\mathbb{E}[u(V)] = u(V^*)$. Hence:

$$\max \mathbb{E}[u(V)] = \max u(V^*)$$

Assuming that the utility function u is continuous and strictly increasing (and therefore invertible), it is possible to write:

$$\arg\max u(V^*) = \arg\max V^*$$

In other words, looking for a portfolio composition that maximizes $\mathbb{E}[u(V)]$ amounts to trying to find a portfolio composition that maximizes the certain amount V^* equivalent to V. Recall that this amount V^* is defined as a function of the risk premium as follows:

$$V^* = \mathbb{E}[V] - \pi \tag{11.2}$$

The optimization program (11.1) is thus equivalent to:

$$\max (\mathbb{E}[V] - \pi) \tag{11.3}$$

Recall in addition (see Eq. (10.22)) that the risk premium π can be expressed locally and at the fourth order as:

$$\pi \simeq \frac{1}{2} \lambda \operatorname{var} V - \frac{1}{6} \psi\, m_3(V) + \frac{1}{24} \varphi\, m_4(V) \tag{11.4}$$

In the following sections, we concentrate on optimization programs of the type of Eq. (11.3) where π is replaced by its approximate expression given in Eq. (11.4) and stopped at order 2, 3, or 4. It is then necessary to be able to compute the various moments of V.

Let us set ourselves first in dimension 1, investing in the risk-free asset with return r and in the risky asset with return R. Let x be the proportion of funds invested in the risky asset. If the initial total investment is equal to V_0, decomposed into $(V_0 - xV_0) + xV_0$, the final portfolio value is equal to:

$$V = (V_0 - xV_0)(1 + r) + xV_0(1 + R) = V_0(1 + r) + xV_0(R - r)$$

This formula allows us to implicitly incorporate a constraint of the type "the sum of the weights is equal to one". In this chapter, we do not use any other constraint, such as a constraint on the positivity of weights. We define the return of the portfolio as follows:

$$R_p = \frac{V - V_0}{V_0} = \frac{V_0 r + xV_0(R - r)}{V_0} = r + x(R - r)$$

The moments of $V = V_0 + V_0 R_p$ can be expressed as a function of the moments of R_p as follows:

$$\mathbb{E}[V] = V_0 + V_0 \, \mathbb{E}[R_p] \tag{11.5}$$
$$\text{var}(V) = V_0^2 \, \text{var}(R_p) \tag{11.6}$$
$$m_3(V) = V_0^3 \, m_3(R_p) \tag{11.7}$$
$$m_4(V) = V_0^4 \, m_4(R_p) \tag{11.8}$$

The moments of the return R_p satisfy:

$$\mathbb{E}[R_p] = r + x(\mathbb{E}[R] - r) \tag{11.9}$$
$$\text{var}(R_p) = x^2 \, \text{var}(R) \tag{11.10}$$
$$m_3(R_p) = x^3 \, m_3(R) \tag{11.11}$$
$$m_4(R_p) = x^4 \, m_4(R) \tag{11.12}$$

Obtaining the moments of $R(t)$ based on those of $X(t)$ can be performed using Eq. (2.42).

We now set ourselves in dimension d: we consider d risky assets having the vector of returns $\mathbf{R}$ and the vector of weights $\mathbf{x}$. The value of the portfolio can then be expressed as:

$$V = V_0(1 - \langle \mathbf{x}, \mathbf{1} \rangle)(1 + r) + V_0 \langle \mathbf{x}, \mathbf{1} + \mathbf{R} \rangle = V_0(1 + r) + V_0 \langle \mathbf{x}, \mathbf{R} - r\mathbf{1} \rangle$$

and the return R_p of the portfolio becomes:

$$R_p = \frac{V - V_0}{V_0} = \frac{V_0 r + V_0 \langle \mathbf{x}, \mathbf{R} - r\mathbf{1} \rangle}{V_0} = r + \langle \mathbf{x}, \mathbf{R} - r\mathbf{1} \rangle$$

Equations (11.5) to (11.8) are unchanged in dimension d, whereas Eqs (11.9) to (11.12) become:

$$\mathbb{E}[R_p] = r + \langle \mathbf{x}, \mathbb{E}[\mathbf{R}] - r\mathbf{1} \rangle = r + \sum_{i=1}^{d} x_i(\mathbb{E}[R_i] - r) \tag{11.13}$$

and:
$$\text{var}(R_p) = \mathbf{x}' M_2(\mathbf{R}) \mathbf{x} \qquad (11.14)$$
$$m_3(R_p) = \mathbf{x}' M_3(\mathbf{R}) \mathbf{x} \otimes \mathbf{x} \qquad (11.15)$$
$$m_4(R_p) = \mathbf{x}' M_4(\mathbf{R}) \mathbf{x} \otimes \mathbf{x} \otimes \mathbf{x} \qquad (11.16)$$

where $\otimes$ is the tensorial product of two vectors. For example:
$$(x_1, x_2, ..., x_d)' \otimes (x_1, x_2, ..., x_d)' =$$
$$(x_1 x_1, x_1 x_2, ..., x_1 x_d, x_2 x_1, x_2 x_2, ..., x_2 x_d, ..., x_d x_1, x_d x_2, ..., x_d x_d)'$$

In formulas (11.14) to (11.16) we first use $M_2(\mathbf{R})$, the variance-covariance matrix of $\mathbf{R}$. Then comes $M_3(\mathbf{R})$, the asymmetry-coasymmetry matrix of $\mathbf{R}$. This matrix is of dimension $d \times d^2$ and can be written as follows:
$$M_3(\mathbf{R}) = [M_3^1, M_3^2, ..., M_3^d]$$
where each M_3^i is a matrix of terms:
$$m_{j,k}^i = \mathbb{E}\left[(R_i - \mathbb{E}[R_i])(R_j - \mathbb{E}[R_j])(R_k - \mathbb{E}[R_k])\right]$$

Finally, $M_4(\mathbf{R})$ is the leptokurticity-coleptokurticity matrix of $\mathbf{R}$. This matrix is of dimension $d \times d^3$ and can be expressed as follows:
$$M_4(\mathbf{R}) = [M_4^{1,1}, ..., M_4^{1,d}, M_4^{2,1}, ..., M_4^{2,d}, ..., M_4^{d,1}, ..., M_4^{d,d}]$$
where each $M_4^{i,j}$ is a matrix having the terms:
$$m_{k,l}^{i,j} = \mathbb{E}\left[(R_i - \mathbb{E}[R_i])(R_j - \mathbb{E}[R_j])(R_k - \mathbb{E}[R_k])(R_l - \mathbb{E}[R_l])\right]$$

In the absence of coasymmetry and coleptokurticity terms, i.e. when all the terms $m_{j,k}^i$ and $m_{k,l}^{i,j}$ that cannot be cast under the forms $m_{i,i}^i$ and $m_{i,i}^{i,i}$ are zero, then Eqs (11.14) to (11.16) can be expressed more simply as:

$$\text{var}(R_p) = \sum_{i=1}^{d} x_i^2 \text{var}(R_i) \qquad (11.17)$$

$$m_3(R_p) = \sum_{i=1}^{d} x_i^3 m_3(R_i) \qquad (11.18)$$

$$m_4(R_p) = \sum_{i=1}^{d} x_i^4 m_4(R_i) \qquad (11.19)$$

11.2 Optimizing with Two Moments

We start by examining the mean-variance framework in the case of one, and then several, risky asset(s).

11.2.1 One Risky Asset

We consider the optimization program (11.3) where the incorporated risk premium is a cutoff at order 2 of Eq. (11.4). This is merely the classic approach, where the risk premium is expressed with respect to the variance only:

$$\pi = \frac{\lambda}{2} \operatorname{var}(V) \tag{11.20}$$

yielding the classic optimization program (producing the so-called Markowitz solution):

$$\max \left(\mathbb{E}[V] - \frac{\lambda}{2} \operatorname{var}(V) \right) \tag{11.21}$$

It is important to understand that this is not a quadratic utility optimization program: $\operatorname{var}(V) = \mathbb{E}[V^2] - \mathbb{E}[V]^2$ cannot be written under the form $\mathbb{E}[f(V)]$ for any function (say any utility function) f, because of the presence of the term $\mathbb{E}[V]^2$. Indeed, an investor can optimize a criterion, here the mean-variance criterion, and later the mean-variance-asymmetry and the mean-variance-asymmetry-leptokurticity criteria, without optimizing an expected utility function. This stems in particular from the fact that the retained expressions of π are approximations of the "true" π. However, the mean-variance framework yields the same results (the same portfolio compositions) as the Markowitz (1952) framework, without being equivalent to it. Indeed, minimizing $\mathbb{E}[V^2] - \mathbb{E}[V]^2$ with $\mathbb{E}[V]$ constant is equivalent to minimizing $\mathbb{E}[V^2]$ with $\mathbb{E}[V]$ constant.

The program (11.21) can be rewritten, using Eqs (11.5) and (11.6), in a way that displays the portfolio return:

$$\max \left(V_0 + V_0 \mathbb{E}[R_p] - \frac{\lambda V_0^2}{2} \operatorname{var}(R_p) \right)$$

and this can be simplified as follows:

$$\max \left(\mathbb{E}[R_p] - \frac{\lambda V_0}{2} \operatorname{var}(R_p) \right) \tag{11.22}$$

which is merely:

$$\max \left(\mathbb{E}[R_p] - \frac{\tilde{\lambda}}{2} \operatorname{var}(R_p) \right) \tag{11.23}$$

expressed with respect to $\tilde{\lambda} = \lambda V_0$ corresponding to the relative aversion parameter in an expected utility framework.

We first study the case of a unique risky asset. The optimization program can be redefined, using Eqs (11.9) and (11.10), as follows:

$$\max_{x} \left(r + x \left(\mathbb{E}[R] - r \right) - \frac{\tilde{\lambda} \, x^2}{2} \, \mathrm{var}(R) \right) \qquad (11.24)$$

yielding the first order condition:

$$\mathbb{E}[R] - r - \tilde{\lambda} x \, \mathrm{var}(R) = 0$$

and then the value of x:

$$x = \frac{1}{\tilde{\lambda}} \frac{\mathbb{E}[R] - r}{\mathrm{var}(R)} \qquad (11.25)$$

Example (CAC40 and BTF):
Consider a monoperiodic optimization where the period is equal to one year, so all the parameters are expressed in the same time unit equal to one year. The risky asset is represented by the French CAC40 stock index and we traditionally assume $\tilde{\lambda} = 2.5$. The logarithmic return of the CAC40 index was equal to $\mathbb{E}[R] = 4.64\%$ per year between December 31, 1990 and December 31, 2010, whereas its standard deviation during the same period was equal to $\sigma(R) = 23.43\%$. The risk-free asset is set to $r = 4.37\%$ as the mean return over the same period of French zero-coupon Government bonds ("Bons du Trésor Français", or BTF) of maturity equal to one year (their standard deviation, equal to 2.6% on the period, is disregarded). We obtain an estimation for the investment weight in the CAC40 index over a one-year horizon:

$$x = \frac{1}{2.5} \frac{0.0464 - 0.0437}{0.2343^2} = 2\%$$

where the investment weight in BTF is therefore:

$$1 - x = 98\%$$

The data incorporating the recent financial crisis are clearly not conducive to investing in the risky asset. How about an investor making his decisions based on data ranging from December 31, 1990 and December 31, 2007? Over this period, the logarithmic return of the CAC40 index was on average equal to $\mathbb{E}[R] = 7.74\%$ per year, whereas its standard deviation was equal to $\sigma(R) = 20.11\%$. The risk-free rate was equal to $r = 4.76\%$. We obtain an alternative estimation of the investment weight in the CAC40 over a one-year horizon:

$$x = \frac{1}{2.5} \frac{0.0774 - 0.0476}{0.2011^2} = 29.5\%$$

and the investment weight in the BTF:

$$1 - x = 70.5\%$$

The investor who did not predict the crisis invested more heavily in the risky asset.

11.2.2 Several Risky Assets

In dimension d, Eqs (11.21) and (11.23) are unchanged whereas the matrix equivalent to Eq. (11.24) can be expressed as follows:

$$\max_{x} \left(r + \langle \mathbf{x}, \mathbb{E}[\mathbf{R}] - r\mathbf{1} \rangle - \frac{\tilde{\lambda}}{2} \mathbf{x}' \, M_2(\mathbf{R}) \, \mathbf{x} \right) \quad (11.26)$$

The first order condition becomes in this situation:

$$\mathbb{E}[\mathbf{R}] - r\mathbf{1} - \tilde{\lambda} M_2(\mathbf{R}) \, \mathbf{x} = 0$$

The optimal weight has a closed-form expression:

$$\mathbf{x} = \frac{1}{\tilde{\lambda}} M_2(\mathbf{R})^{-1} (\mathbb{E}[\mathbf{R}] - r\mathbf{1}) \quad (11.27)$$

Example (Société Générale, Carrefour, and BTF):
We provide an illustration using two risky assets. We consider the stocks of Société Générale (a bank, having a correlation coefficient with the CAC40 index equal to 82.5% between December 31, 1990 and December 31, 2010) and of Carrefour (a consortium of supermarkets, which is supposedly contracyclical but which is not, presenting a correlation coefficient with the CAC40 index equal to 76.5% between December 31, 1990 and December 31, 2010). The risk-free asset is still comprised of the one-year maturity BTF. The return of Société Générale stock was on average equal to $\mathbb{E}[R_1] = 5.75\%$ per year between the years 1990 and 2010, whereas its standard deviation was equal to $\sigma(R_1) = 31.73\%$ over the same period. The return of Carrefour stock was equal to $\mathbb{E}[R_2] = 7.46\%$ per year on average and its standard deviation was $\sigma(R_2) = 31.08\%$. The correlation between the two stocks was equal to $\rho(R_1, R_2) = 56.49\%$, corresponding to a covariance of $\text{covar}(R_1, R_2) = 5.57\%$. The variance-covariance matrix of these two assets was therefore equal to:

$$M_2(R_1, R_2) = \begin{bmatrix} 0.3173^2 & 0.0557 \\ 0.0557 & 0.3108^2 \end{bmatrix} = \begin{bmatrix} 0.1007 & 0.0557 \\ 0.0557 & 0.0966 \end{bmatrix}$$

Using formula (11.27), we obtain the investment weight for the Société Générale stock:
$$x_1 = -2\%$$
the investment weight for the Carrefour stock:
$$x_2 = 14\%$$
and the investment weight for the BTF:
$$1 - x_1 - x_2 = 88\%$$

The conclusion of this experience is not pleasant reading for Société Générale: an investor taking into account the data of the past 20 years should short sale the bank's stocks in order to buy Carrefour stock. Based on the past 20 years' data, this latter company appears more attractive than the CAC40 index for a standard investor (this is nothing less than an inducement to under-diversify).

Let us perform this experiment again, this time based on stock data ranging from December 31, 1990 to December 31, 2007, therefore, not incorporating data from the latest crisis. Over this period, the return of Société Générale stock was on average equal to $\mathbb{E}[R_1] = 11.35\%$ per year, whereas its standard deviation was equal to $\sigma(R_1) = 21.68\%$. The return of Carrefour stock was on average equal to $\mathbb{E}[R_2] = 11.99\%$ and its standard deviation to $\sigma(R_2) = 27.82\%$. The correlation between these two assets was $\rho(R_1, R_2) = 24.98\%$, corresponding to a covariance of $\text{covar}(R_1, R_2) = 1.51\%$. Over this period, the variance-covariance matrix of these two assets was therefore equal to:

$$M_2(R_1, R_2) = \begin{bmatrix} 0.2168^2 & 0.0151 \\ 0.0151 & 0.2782^2 \end{bmatrix} = \begin{bmatrix} 0.0470 & 0.0151 \\ 0.0151 & 0.0774 \end{bmatrix}$$

Computing formula (11.27), we obtain the investment weight for the Société Générale stock:
$$x_1 = 47\%$$
the investment weight for the Carrefour stock:
$$x_2 = 28\%$$
and the investment weight for the BTF:
$$1 - x_1 - x_2 = 25\%$$

If we do not take into account the data related to the recent financial crisis, we see that we should invest half of our wealth in the Société Générale stock, a quarter in the Carrefour stock, and (only) the remaining quarter in the risk-free asset.

11.3 Optimizing with Three Moments

We now extend the mean-variance framework by taking into account the centered moment of order 3. Again, we start by examining the case of a unique risky asset, and then the case of several risky assets.

11.3.1 One Risky Asset

We extend the program (11.21) using a truncation up to order 3 of Eq. (11.4) giving the risk premium:

$$\max \left(\mathbb{E}[V] - \frac{\lambda}{2} \text{var}(V) + \frac{\psi}{6} m_3(V) \right) \quad (11.28)$$

This is a rewriting of the program solved by Mitton and Vorkink (2007).

Let us write this optimization program as a function of R_p using Eqs (11.5) to (11.7). We then need to solve:

$$\max \left(V_0 + V_0 \mathbb{E}[R_p] - \frac{\lambda V_0^2}{2} \text{var}(R_p) + \frac{\psi V_0^3}{6} m_3(R_p) \right) \quad (11.29)$$

which is equivalent to:

$$\max \left(\mathbb{E}[R_p] - \frac{\lambda V_0}{2} \text{var}(R_p) + \frac{\psi V_0^2}{6} m_3(R_p) \right) \quad (11.30)$$

so to:

$$\max \left(\mathbb{E}[R_p] - \frac{\tilde{\lambda}}{2} \text{var}(R_p) + \frac{\tilde{\psi}}{6} m_3(R_p) \right) \quad (11.31)$$

where $\tilde{\lambda} = \lambda V_0$ and $\tilde{\psi} = \psi V_0^2$ correspond respectively to the relative risk aversion and the relative asymmetry attraction coefficients.

Let us now express the program (11.31) with respect to R. Using Eqs (11.9) to (11.11), we can write:

$$\max_x \left(r + x(\mathbb{E}[R] - r) - \frac{x^2 \tilde{\lambda}}{2} \text{var}(R) + \frac{x^3 \tilde{\psi}}{6} m_3(R) \right) \quad (11.32)$$

leading to the first order condition:

$$\mathbb{E}[R] - r - x\tilde{\lambda} \text{var}(R) + x^2 \frac{\tilde{\psi}}{2} m_3(R) = 0 \quad (11.33)$$

which can also be expressed as follows:

$$x^2 - \frac{2\tilde{\lambda} \text{var}(R)}{\tilde{\psi} m_3(R)} x + \frac{2(\mathbb{E}[R] - r)}{\tilde{\psi} m_3(R)} = 0$$

This equation has the two solutions:

$$x = \frac{\frac{2\tilde{\lambda}\operatorname{var}(R)}{\tilde{\psi}m_3(R)} \pm \sqrt{\left(\frac{2\tilde{\lambda}\operatorname{var}(R)}{\tilde{\psi}m_3(R)}\right)^2 - \frac{8(\mathbb{E}[R]-r)}{\tilde{\psi}m_3(R)}}}{2}$$

which can be simplified as:

$$x = \frac{\tilde{\lambda}\operatorname{var}(R)}{\tilde{\psi}m_3(R)}\left(1 \pm \sqrt{1 - \frac{2(\mathbb{E}[R]-r)\tilde{\psi}m_3(R)}{\tilde{\lambda}^2 \operatorname{var}(R)^2}}\right) \tag{11.34}$$

These solutions can also be expressed under the form:

$$x = \frac{\tilde{\lambda}\operatorname{var}(R)}{\tilde{\psi}m_3(R)}\left(1 \pm \sqrt{1 - \frac{2\tilde{\psi}m_3(R)}{\tilde{\lambda}\operatorname{var}(R)}x^*}\right) \tag{11.35}$$

where:

$$x^* = \frac{\mathbb{E}[R]-r}{\tilde{\lambda}\operatorname{var}(R)}$$

is the optimal amount according to the mean-variance criterion obtained in formula (11.25). To simplify matters, we set:

$$A = \frac{\tilde{\psi}m_3(R)}{2\tilde{\lambda}\operatorname{var}(R)}$$

and obtain:

$$x = \frac{1}{2A}\left(1 \pm \sqrt{1 - 4Ax^*}\right)$$

Let us first perform a limited development up to order 2 of the preceding expression. Recall first that:

$$\sqrt{1+x} \simeq 1 + \frac{x}{2} - \frac{x^2}{8}$$

so that:

$$x \simeq \frac{1}{2A}\left[1 \pm \left(1 - 2Ax^* - 2A^2x^{*2}\right)\right]$$

yields the two approximate solutions:

$$x_1 \simeq x^*(1 + Ax^*) \tag{11.36}$$

and:

$$x_2 \simeq \frac{1}{A} - x^*(1 + Ax^*) \tag{11.37}$$

The second solution is unstable. In particular, when $m_3(R)$ becomes zero, $\frac{1}{A}$ diverges, and x_2 also. When $m_3(R) = 0$, equation (11.32) can be expressed as a maximization on an upside-down parabola. The solution x_2 of an infinite investment in the risky asset is therefore impossible. We shall retain only the first solution, which we show in the form:

$$x \simeq x^*(1 + km_3(R)x^*) \qquad (11.38)$$

where $k = \frac{\tilde{\psi}}{2\tilde{\lambda}\operatorname{var}(R)}$ is positive as long as the investor is assumed to be attracted by positive asymmetry ($\tilde{\psi} > 0$).

Three situations arise. If $m_3(R) = 0$, i.e. in the absence of an asymmetry term, then $x = x^*$ and the optimal weight is naturally that of the mean-variance framework. If $m_3(R) > 0$, i.e. in the presence of a thick right tail, then $x > x^*$: the amount invested in the risky asset increases, representing a concentration on this asset. Finally, if $m_3(R) < 0$, so in the presence of a thick left tail, then $x < x^*$: the amount invested in the risky asset decreases, representing a concentration on the risk-free asset.

Thus, it appears that with (11.38), directly and unsurprisingly, and under the hypothesis of an investor attracted by asymmetry, a positive $m_3(R)$ increases the investment weight in the risky asset, whereas a negative $m_3(R)$ diminishes it.

Example (CAC40 and BTF):

We extend the previously studied example to the case of an investor who takes into account the moment of order 3 in his choice of portfolio. We need two new parameters for this illustration. First, we need to estimate $\tilde{\psi}$. In the absence of experimental economics research allowing us to know with a reasonable degree of precision the order of magnitude of $\tilde{\psi}$ for a standard investor, we assume that the investor is endowed with a power-type utility function (see Table 10.1 in Chapter 10). Under this hypothesis $\tilde{\lambda} = 1 - a = 2.5$ so that $a = -1.5$ and because $\tilde{\psi} = (1-a)(2-a)$ then $\tilde{\psi} = 2.5 * 3.5 = 8.75$. Next, we need to compute the third moment of the return of the CAC40 index using annual data between December 31, 1990 and December 31, 2010. Our estimation yields $m_3(R) = -0.0132$, corresponding to a skewness coefficient $S = -1.02$. Using equation (11.34), we obtain the estimation for the investment weight in the CAC40 index over a one-year horizon:

$$x = 1.95\%$$

and the investment weight in the BTF:

$$1 - x = 98.05\%$$

which are nearly the same as the values obtained in the mean-variance framework. We conduct this experience again, but based on the data between December 31, 1990 and December 31, 2007. In this case $m_3(R) =$

-0.0066 and the skewness coefficient is equal to $S = -0.81$. The effect of negative asymmetry is stronger, where the weights are respectively equal to:

$$x = 27.4\%$$

and:

$$1 - x = 72.6\%.$$

so approximately half a percent less is invested in the risky asset and half a percent more in the risk-free asset, compared to the results of the mean-variance framework.

Note also that it is possible to express the optimal weight given in formula (11.34) with respect to $S(R)$ instead of $m_3(R)$. We first write:

$$x = \frac{\tilde{\lambda}\sigma(R)^3}{\tilde{\psi}m_3(R)\sigma(R)} \left(1 \pm \sqrt{1 - \frac{2(\mathbb{E}[R] - r)\tilde{\psi}m_3(R)}{\tilde{\lambda}^2\sigma(R)^3\sigma(R)}}\right)$$

which displays the Pearson–Fisher skewness coefficient $S = m_3/\sigma^3$. Therefore:

$$x = \frac{\tilde{\lambda}}{\tilde{\psi}S(R)\sigma(R)} \left(1 \pm \sqrt{1 - \frac{2(\mathbb{E}[R] - r)\tilde{\psi}S(R)}{\tilde{\lambda}^2\sigma(R)}}\right)$$

We define the relative prudence coefficient (see Le Courtois (2012) for a general overview on the notions of prudence and temperance) as $\tilde{\nu} = \tilde{\psi}/\tilde{\lambda}$. We obtain:

$$x = \frac{1}{\tilde{\nu}S(R)\sigma(R)} \left(1 \pm \sqrt{1 - \frac{2(\mathbb{E}[R] - r)\tilde{\nu}S(R)}{\tilde{\lambda}\sigma(R)}}\right) \quad (11.39)$$

which can also be expressed as a function of the optimal investment weight in the risky asset x^* from the mean-variance framework as follows:

$$x = \frac{1}{\tilde{\nu}S(R)\sigma(R)} \left(1 \pm \sqrt{1 - 2x^*\tilde{\nu}S(R)\sigma(R)}\right)$$

so, for x^* small:

$$x \simeq x^* \left(1 + \frac{\tilde{\nu}S(R)\sigma(R)}{2}\right)$$

whose interpretation as a function of $S(R)$ is quite similar to that of (11.38) as a function of $m_3(R)$.

11.3.2 Several Risky Assets

In dimension d, the program (11.32) can be generalized, using Eqs (11.13) to (11.15), as follows:

$$\max_x \left(r + \langle \mathbf{x}, \mathbb{E}[\mathbf{R}] - r\mathbf{1} \rangle - \frac{\tilde{\lambda}}{2} \mathbf{x}' M_2(\mathbf{R}) \mathbf{x} + \frac{\tilde{\psi}}{6} \mathbf{x}' M_3(\mathbf{R}) \mathbf{x} \otimes \mathbf{x} \right)$$
(11.40)

Because $\nabla (\mathbf{x}' M_3(\mathbf{R}) \mathbf{x} \otimes \mathbf{x}) = 3 M_3(\mathbf{R}) \mathbf{x} \otimes \mathbf{x}$, the first order condition becomes:

$$\mathbb{E}[\mathbf{R}] - r\mathbf{1} - \tilde{\lambda} M_2(\mathbf{R}) \mathbf{x} + \frac{\tilde{\psi}}{2} M_3(\mathbf{R}) \mathbf{x} \otimes \mathbf{x} = 0$$

This problem cannot be solved analytically, but by using a numerical approach, as in the example that follows.

Example (Société Générale, Carrefour, and BTF):

Let us come back to our illustration using two risky assets, where the investor can choose between stocks of Société Générale or of Carrefour, and can purchase BTF. The estimation of the matrices M_3^1 and M_3^2 for the centered moments and comoments of order 3 linking the returns of the stocks of Société Générale and Carrefour yields, over the period ranging from December 31, 1990 to December 31, 2010:

$$M_3^1 = \begin{bmatrix} m_3^{1,1,1} & m_3^{1,1,2} \\ m_3^{1,2,1} & m_3^{1,2,2} \end{bmatrix} = \begin{bmatrix} -0.0454 & -0.0322 \\ -0.0322 & -0.0215 \end{bmatrix}$$

and:

$$M_3^2 = \begin{bmatrix} m_3^{2,1,1} & m_3^{2,1,2} \\ m_3^{2,2,1} & m_3^{2,2,2} \end{bmatrix} = \begin{bmatrix} -0.0322 & -0.0215 \\ -0.0215 & -0.0087 \end{bmatrix}$$

where $M_3(R_1, R_2) = [M_3^1, M_3^2]$. Solving numerically the program (11.27) yields the weight for the Société Générale stock:

$$x_1 = -2.8\%$$

the weight for the Carrefour stock:

$$x_2 = 14.4\%$$

and the investment weight for the BTF:

$$1 - x_1 - x_2 = 88.4\%$$

For the shorter period ranging from December 31, 1990 to December 31, 2007, the estimated moments and comoments of order 3 become:

$$M_3^1 = \begin{bmatrix} m_3^{1,1,1} & m_3^{1,1,2} \\ m_3^{1,2,1} & m_3^{1,2,2} \end{bmatrix} = \begin{bmatrix} -0.0030 & 0.0015 \\ 0.0015 & 0.0009 \end{bmatrix}$$

and:

$$M_3^2 = \begin{bmatrix} m_3^{2,1,1} & m_3^{2,1,2} \\ m_3^{2,2,1} & m_3^{2,2,2} \end{bmatrix} = \begin{bmatrix} 0.0015 & 0.0009 \\ 0.0009 & 0.0026 \end{bmatrix}$$

Solving numerically the program (11.27), we deduce the investment weight for the Société Générale stock:

$$x_1 = 45.9\%$$

the investment weight for the Carrefour stock:

$$x_2 = 30.2\%$$

and the investment weight for the BTF:

$$1 - x_1 - x_2 = 23.9\%$$

where it appears that a decision-maker taking asymmetry into account will reduce his investment in Société Générale stock by about 1% and increase his investment in Carrefour stock by about 2% compared to an individual presenting a simple mean-variance criterion.

11.4 Optimizing with Four Moments

We continue our study on multi-moment optimization with a framework taking into account the mean, the variance, the centered moment of order 3, and the centered moment of order 4. As in the preceding developments, we consider first a unique risky asset, then the case of several risky assets.

11.4.1 *One Risky Asset*

We extend the program (11.28) using all the terms of the risk premium given by formula (11.4):

$$\max \left(\mathbb{E}[V] - \frac{\lambda}{2} \text{var}(V) + \frac{\psi}{6} m_3(V) - \frac{\varphi}{24} m_4(V) \right) \quad (11.41)$$

keeping in mind that the risk premium is approximated at order 4 and that this is an arbitrary criterion that does not correspond to the optimization of a polynomial utility function of order 4. For more details on the programs of order 4, see Le Courtois (2012).

We adapt the optimization program by writing it in terms of R_p using Eqs (11.5) to (11.8). We then need to solve:

$$\max \left(V_0 + V_0 \mathbb{E}[R_p] - \frac{\lambda V_0^2}{2} \text{var}(R_p) + \frac{\psi V_0^3}{6} m_3(R_p) - \frac{\varphi V_0^4}{24} m_4(R_p) \right) \tag{11.42}$$

which is equivalent to:

$$\max \left(\mathbb{E}[R_p] - \frac{\lambda V_0}{2} \text{var}(R_p) + \frac{\psi V_0^2}{6} m_3(R_p) - \frac{\varphi V_0^3}{24} m_4(R_p) \right) \tag{11.43}$$

so to:

$$\max \left(\mathbb{E}[R_p] - \frac{\tilde{\lambda}}{2} \text{var}(R_p) + \frac{\tilde{\psi}}{6} m_3(R_p) - \frac{\tilde{\varphi}}{24} m_4(R_p) \right) \tag{11.44}$$

where $\tilde{\lambda} = \lambda V_0$, $\tilde{\psi} = \psi V_0^2$, $\tilde{\varphi} = \varphi V_0^3$ correspond respectively to the relative risk aversion, relative asymmetry attraction, and relative tail thickness aversion coefficients.

Let us express program (11.44) in terms of R. Using Eqs (11.9)–(11.12), we can write:

$$\max_x \left(r + x(\mathbb{E}[R] - r) - \frac{x^2 \tilde{\lambda}}{2} \text{var}(R) + \frac{x^3 \tilde{\psi}}{6} m_3(R) - \frac{x^4 \tilde{\varphi}}{24} m_4(R) \right) \tag{11.45}$$

which yields the first order condition:

$$\mathbb{E}[R] - r - x \tilde{\lambda} \text{var}(R) + x^2 \frac{\tilde{\psi}}{2} m_3(R) - x^3 \frac{\tilde{\varphi}}{6} m_4(R) = 0 \tag{11.46}$$

This is a third degree equation that can be rewritten as follows:

$$\alpha x^3 + \beta x^2 + \gamma x + \delta = 0 \tag{11.47}$$

where $\alpha = -\frac{\tilde{\varphi}}{6} m_4(R)$, $\beta = \frac{\tilde{\psi}}{2} m_3(R)$, $\gamma = -\tilde{\lambda} \text{var}(R)$ and $\delta = \mathbb{E}[R] - r$.

The discriminant of this third degree equation is:

$$\Delta_3 = 18\alpha\beta\gamma\delta - 4\beta^3\delta + \beta^2\gamma^2 - 4\alpha\gamma^3 - 27\alpha^2\delta^2 \tag{11.48}$$

For standard values of the parameters (see the example below), this discriminant is negative, which implies the existence of a unique real solution and of two complex solutions. We are only interested in the real solution, which can be expressed as follows:

$$x = -\frac{\beta}{3\alpha} - \frac{1}{3\alpha} \left(\frac{1}{2} \left(\Lambda_1 + \sqrt{\Lambda_1^2 - 4\Lambda_2^3} \right) \right)^{\frac{1}{3}} - \frac{1}{3\alpha} \left(\frac{1}{2} \left(\Lambda_1 - \sqrt{\Lambda_1^2 - 4\Lambda_2^3} \right) \right)^{\frac{1}{3}}$$

where:
$$\Lambda_1 = 2\beta^3 - 9\alpha\beta\gamma + 27\alpha^2\delta$$
and:
$$\Lambda_2 = \beta^2 - 3\alpha\gamma$$
so that:
$$x = \frac{\tilde{\psi}m_3(R)}{\tilde{\varphi}m_4(R)} + \frac{2}{\tilde{\varphi}m_4(R)}\left(\frac{1}{2}\left(\Lambda_1(R) + \sqrt{\Lambda_1(R)^2 - 4\Lambda_2(R)^3}\right)\right)^{\frac{1}{3}}$$
$$+ \frac{2}{\tilde{\varphi}m_4(R)}\left(\frac{1}{2}\left(\Lambda_1(R) - \sqrt{\Lambda_1(R)^2 - 4\Lambda_2(R)^3}\right)\right)^{\frac{1}{3}} \quad (11.49)$$
where:
$$\Lambda_1(R) = \frac{\tilde{\psi}^3}{4}m_3(R)^3 - \frac{3\tilde{\varphi}\tilde{\psi}\tilde{\lambda}}{4}m_4(R)m_3(R)\operatorname{var}(R) + \frac{3\tilde{\varphi}^2}{4}m_4(R)^2(\mathbb{E}[R] - r)$$
and:
$$\Lambda_2(R) = \frac{\tilde{\psi}^2}{4}m_3(R)^2 - \frac{\tilde{\varphi}\tilde{\lambda}}{2}m_4(R)\operatorname{var}(R)$$

Example (CAC40 and BTF):
We end our illustration on the CAC40 and the BTF by additionally taking into account the moment of order 4. We need two supplementary parameters. First, we must estimate $\tilde{\varphi}$. In the absence of experimental economics research giving with a reasonable degree of precision the value of $\tilde{\varphi}$ for a standard investor (we made the same remark for $\tilde{\psi}$), we assume that the decision-maker is endowed with a power-type utility function (see Table 10.1 in Chapter 10). Under this hypothesis $\tilde{\lambda} = 1 - a = 2.5$ so that $a = -1.5$, and because $\tilde{\varphi} = (1-a)(2-a)(3-a)$ then $\tilde{\varphi} = 2.5 \times 3.5 \times 4.5 = 39.375$. Further, $m_4(R) = 0.0106$ is the estimation of the fourth centered moment of the CAC40 return between December 31, 1990 and December 31, 2010 (corresponding to a kurtosis of $K = 3.53$). We first compute the discriminant Δ_3 as in formula (11.48):
$$\Delta_3 = -0.0007$$

This discriminant is negative, confirming the uniqueness of the real solution. We can therefore use formula (11.49) in order to estimate the investment weight of the CAC40 index over a one-year horizon:
$$x = 0.0195$$
and the investment weight for the BTF:
$$1 - x = 0.9805$$

These weights are unchanged compared to those of the portfolio optimization program at order 3. Let us now concentrate on the period ranging from December 31, 1990 to December 31, 2007. The centered moment of order 4 is then equal to $m_4(R) = 0.0053$, corresponding to a kurtosis of $K = 3.23$. The discriminant Δ_3 is equal to:

$$\Delta_3 = -0.0002$$

and is negative, confirming the uniqueness of the real root. We therefore use Eq. (11.49) to estimate the investment weight for the CAC40 index over a one-year horizon:

$$x = 26.8\%$$

and the investment weight for the BTF:

$$1 - x = 73.2\%$$

where taking into account the moment of order 4 beyond the moments of 1, 2, and 3 implies a reduction of half a percent of the investment in the risky asset and an increase of half a percent of the investment in the risk-free asset.

11.4.2 Several Risky Assets

In dimension d, the program (11.45) can be generalized, using Eqs (11.13)–(11.16), as follows:

$$\max_x \left(r + \langle \mathbf{x}, \mathbb{E}[\mathbf{R}] - r\mathbf{1} \rangle - \frac{\tilde{\lambda}}{2} \mathbf{x}' M_2(\mathbf{R}) \mathbf{x} + \frac{\tilde{\psi}}{6} \mathbf{x}' M_3(\mathbf{R}) \mathbf{x} \otimes \mathbf{x} \right.$$
$$\left. - \frac{\tilde{\varphi}}{24} \mathbf{x}' M_4(\mathbf{R}) \mathbf{x} \otimes \mathbf{x} \otimes \mathbf{x} \right) \tag{11.50}$$

Because $\nabla \left(\mathbf{x}' M_4(\mathbf{R}) \mathbf{x} \otimes \mathbf{x} \right) = 4 M_4(\mathbf{R}) \mathbf{x} \otimes \mathbf{x} \otimes \mathbf{x}$, the first order condition becomes:

$$\mathbb{E}[\mathbf{R}] - r\mathbf{1} - \tilde{\lambda} M_2(\mathbf{R}) \mathbf{x} + \frac{\tilde{\psi}}{2} M_3(\mathbf{R}) \mathbf{x} \otimes \mathbf{x} - \frac{\tilde{\varphi}}{6} M_4(\mathbf{R}) \mathbf{x} \otimes \mathbf{x} \otimes \mathbf{x} = 0$$

This problem cannot be solved using a simple analytical approach. In full generality, a numerical method is necessary, as in the following example.

Example (Société Générale, Carrefour, and BTF):
Let us return to the simple case with two risky assets, so to an investor who buys only stocks of Société Générale and Carrefour, and can also purchase BTF. The estimation of the matrices $M_4^{1,1}$, $M_4^{1,2}$, $M_4^{2,1}$, and $M_4^{2,2}$

of the centered moments and comoments of order 4 linking the returns of the stocks of Société Générale and Carrefour yields, for the period ranging from December 31, 1990 to December 31, 2010:

$$M_4^{1,1} = \begin{bmatrix} m_4^{1,1,1,1} & m_4^{1,1,1,2} \\ m_4^{1,1,2,1} & m_4^{1,1,2,2} \end{bmatrix} = \begin{bmatrix} 0.0559 & 0.0395 \\ 0.0395 & 0.0307 \end{bmatrix}$$

and:

$$M_4^{1,2} = \begin{bmatrix} m_4^{1,2,1,1} & m_4^{1,2,1,2} \\ m_4^{1,2,2,1} & m_4^{1,2,2,2} \end{bmatrix} = \begin{bmatrix} 0.0395 & 0.0307 \\ 0.0307 & 0.0233 \end{bmatrix}$$

and also:

$$M_4^{2,1} = \begin{bmatrix} m_4^{2,1,1,1} & m_4^{2,1,1,2} \\ m_4^{2,1,2,1} & m_4^{2,1,2,2} \end{bmatrix} = \begin{bmatrix} 0.0395 & 0.0307 \\ 0.0307 & 0.0233 \end{bmatrix}$$

and:

$$M_4^{2,2} = \begin{bmatrix} m_4^{2,2,1,1} & m_4^{2,2,1,2} \\ m_4^{2,2,2,1} & m_4^{2,2,2,2} \end{bmatrix} = \begin{bmatrix} 0.0307 & 0.0233 \\ 0.0233 & 0.0263 \end{bmatrix}$$

where $M_4(R_1, R_2) = [M_4^{1,1}, M_4^{1,2}, M_4^{2,1}, M_4^{2,2}]$. Solving numerically the program (11.27) allows us to compute the investment weight for the Société Générale stock:

$$x_1 = -2.6\%$$

the investment weight for the Carrefour stock:

$$x_2 = 11.8\%$$

and the investment weight for the BTF:

$$1 - x_1 - x_2 = 90.8\%$$

Taking into account the moment of order 4 in addition to the first three moments leads to a decrease in investment in Carrefour stock of about 2.5% whereas the short position on Société Générale is almost unchanged.

Over the period ranging from December 31, 1990 to December 31, 2007, our estimation of the moments and comoments of order 4 yields:

$$M_4^{1,1} = \begin{bmatrix} m_4^{1,1,1,1} & m_4^{1,1,1,2} \\ m_4^{1,1,2,1} & m_4^{1,1,2,2} \end{bmatrix} = \begin{bmatrix} 0.0054 & 0.0020 \\ 0.0020 & 0.0032 \end{bmatrix}$$

and:

$$M_4^{1,2} = \begin{bmatrix} m_4^{1,2,1,1} & m_4^{1,2,1,2} \\ m_4^{1,2,2,1} & m_4^{1,2,2,2} \end{bmatrix} = \begin{bmatrix} 0.0020 & 0.0032 \\ 0.0032 & 0.0029 \end{bmatrix}$$

and also:
$$M_4^{2,1} = \begin{bmatrix} m_4^{2,1,1,1} & m_4^{2,1,1,2} \\ m_4^{2,1,2,1} & m_4^{2,1,2,2} \end{bmatrix} = \begin{bmatrix} 0.0020 & 0.0032 \\ 0.0032 & 0.0029 \end{bmatrix}$$

and:
$$M_4^{2,2} = \begin{bmatrix} m_4^{2,2,1,1} & m_4^{2,2,1,2} \\ m_4^{2,2,2,1} & m_4^{2,2,2,2} \end{bmatrix} = \begin{bmatrix} 0.0032 & 0.0029 \\ 0.0029 & 0.0122 \end{bmatrix}$$

where $M_4(R_1, R_2) = [M_4^{1,1}, M_4^{1,2}, M_4^{2,1}, M_4^{2,2}]$. Solving numerically the program (11.27) allows us to compute the investment weight for the Société Générale stock:

$$x_1 = 21.8\%$$

the investment weight for the Carrefour stock:

$$x_2 = 14.2\%$$

and the investment weight for the BTF:

$$1 - x_1 - x_2 = 64\%$$

This latter calculation is interesting because it shows that solving the optimization program containing the fourth moment of distributions implies a reduction by more than two of the exposures to Société Générale and Carrefour stocks compared to what is obtained with a program of the mean-variance-asymmetry type, based on data finishing in the year 2007. It is therefore extremely important not to stop at the third moment when conducting a monoperiodic portfolio optimization.

11.5 Other Problems

We have just exposed criteria allowing us to construct portfolios in a monoperiodic framework and taking into account the mean and moments up to order 4. Maintaining this line of thought, we can find other types of criteria, having analogous characteristics, but based on slightly different quantities. We present two examples below.

11.5.1 Giving up Comoments

In the preceding sections, we made the hypothesis that individuals, when constructing multidimensional portfolios, take into account not only the asymmetry and tail thickness of asset return distributions, but also the comoments between these return distributions, which complicates solving optimization programs in high dimensions (making them more difficult to calibrate and more time-consuming to compute). In practice, the vast majority of investors, typically mutual funds, banks, and insurance companies, consider only asymmetry and leptokurticity in their management and do not take into account comoments. By way of illustration, we consider here only asymmetry and leptokurticity terms in monoperiodic portfolio optimization programs with several risky assets.

It is then possible to generalize the optimization program that is given in formula (11.45) and recalled here for the sake of clarity:

$$\max_x \left(r + x(\mathbb{E}[R] - r) - \frac{x^2 \tilde{\lambda}}{2} \text{var}(R) + \frac{x^3 \tilde{\psi}}{6} m_3(R) - \frac{x^4 \tilde{\varphi}}{24} m_4(R) \right)$$

in a dimension d program. Disregarding the comoments of order 3 and 4, we obtain:

$$\max_x \left(r + \langle \mathbf{x}, \mathbb{E}[R] - r\mathbf{1} \rangle - \frac{\tilde{\lambda}}{2} \mathbf{x}' \, M_2(\mathbf{R}) \, \mathbf{x} + \frac{\tilde{\psi}}{6} \sum_{i=1}^{d} x_i^3 m_3(R_i) \right.$$
$$\left. - \frac{\tilde{\varphi}}{24} \sum_{i=1}^{d} x_i^4 m_4(R_i) \right) \tag{11.51}$$

which is also a simplified form of Eq. (11.50).

This type of problem presents two main advantages and one main drawback. Among the advantages, we can cite first the speed of computation. It is no longer necessary to estimate numerous coasymmetry and coleptokurticity terms for several hundreds or thousands of assets (the size of the coasymmetry matrix being $d \times d^2$ and the size of the coleptokurticity matrix being $d \times d^3$ in the presence of d risky assets), and then to launch time-consuming root searches on the first order condition. The second advantage is the simplicity of the approach for portfolio managers: for a practitioner, constructing an intuitive process based on the moments of order 3 and 4 is possible with little experience, but becomes a lot more problematic when including the corresponding comoments. Therefore, all the parameters of the optimization program (11.51) correspond to simple intuitions. These parameters are on the one hand the three parameters of the utility function – although, we saw above that choosing a power-type utility function allows us to contract these three parameters into a single one, and on the other hand the mean, variance, asymmetry, and leptokurticity parameters. The optimization problem then rests only those inputs over which managers have a direct and sensitive catch. The main drawback associated with this program lies in its simplifying assumption. Not taking into account the comoments of order 3 and 4 amounts to strongly approximating the joint distribution of risky asset returns.

11.5.2 Perturbative Approach and Normalized Moments

We start from the standard deviation that is the centered moment of order 2 elevated to the power of $1/2$:

$$m_2^{1/2} = \mathbb{E}\left[(R-\mu)^2\right]^{1/2} \qquad (11.52)$$

and we consider by extension the other centered moments of order k elevated to the power of $1/k$. These moments are homogeneous with respect to the quantity to which they are related (here a stock return):

$$s_k = m_k^{1/k} = \mathbb{E}\left[(R-\mu)^k\right]^{1/k} \qquad (11.53)$$

Equipped with these geometrically normalized moments, we can construct new optimization criteria where asymmetry and leptokurticity come into play, not through the coefficients S or K or the centered third and fourth moments, but through the third and fourth moments elevated respectively to the powers of $1/3$ and $1/4$. Kraus and Litzenberger (1976) constructed for instance a three-moment CAPM where asymmetry is expressed in this fashion.

In the one-dimensional case, we have the following relations:

$$s_3(R_p) = x \; s_3(R)$$

and:

$$s_4(R_p) = x \; s_4(R)$$

which extend naturally the relations (11.11) and (11.12).

We use these quantities inside a new portfolio choice criterion. Starting from the expression (11.45), recalled below:

$$\max_x \left(r + x(\mathbb{E}[R] - r) - \frac{x^2 \tilde{\lambda}}{2} \mathrm{var}(R) + \frac{x^3 \tilde{\psi}}{6} m_3(R) - \frac{x^4 \tilde{\varphi}}{24} m_4(R) \right)$$

we replace the moments m_3 and m_4 by the expressions s_3 and s_4^2, and use a new set of multiplicative constants. This yields the optimization program:

$$\max_x \left(r + x(\mathbb{E}[R] - r) - \frac{x^2 \zeta}{2} \mathrm{var}(R) + x\chi s_3(R) - \frac{x^2 \xi}{2} s_4^2(R) \right) \qquad (11.54)$$

having the first order condition:

$$\mathbb{E}[R] - r - x\zeta \operatorname{var}(R) + \chi s_3(R) - x\xi s_4^2(R) = 0$$

which gives the investment weight for the risky asset:

$$x = \frac{\mathbb{E}[R] - r + \chi s_3(R)}{\zeta \operatorname{var}(R) + \xi s_4^2(R)}$$

This program is interesting because it considers that asymmetry is a perturbation of the mean (these two terms being here proportional to x) and that tail thickness is proportional to the variance (these two terms being here proportional to x^2). Written differently, the third moment induces an effect similar to that of the mean for the investor, whereas the fourth moment induces an effect comparable to that of the variance.

The multidimensional generalization of problem (11.54) is not direct. However, if we make the additional hypothesis that the investor is indifferent to the coasymmetry and coleptokurticity terms, then we obtain in any dimension d:

$$\max\left\{V_0(1+r) + \langle \mathbf{x}, \mathbb{E}[\mathbf{R}] - r\mathbf{1}\rangle - \frac{\zeta}{2}\mathbf{x}'\Gamma\mathbf{x} + \chi\mathbf{x}'s_3(\mathbf{R}) - \frac{\xi}{2}\mathbf{x}'s_4(\mathbf{R})\mathbf{x}\right\} \tag{11.55}$$

where $s_3(\mathbf{R})$ is the vector of the moments of order 3 normalized geometrically and $s_4(\mathbf{R})$ is a diagonal matrix that contains the moments of order 4 normalized geometrically.

Solving the criterion defined in (11.55) yields the vector of weights:

$$\mathbf{x} = (\zeta\,\Gamma + \xi\,s_4(\mathbf{R}))^{-1}\left(\mathbb{E}[\mathbf{R}] - r + \chi\,s_3(\mathbf{R})\right)$$

This way of introducing a criterion of the fourth order is different from the approach starting from a limited development at order 4 of an expected utility function.

Appendix: Dealing with Uncertainty

We present an approach that leads to a smaller diversification of portfolios and that takes into account the uncertainty that affects returns and volatilities. This approach is that of the model of Van Nieuwerburgh and Veldkamp (2007). For the sake of simplicity, we develop the treatment of uncertainty only in the mean-variance framework.

Conceptual Framework of the Model

Consider a collection of assets producing a (vector of) flux **F** of income (for example an annual flow). Assume this flow is characterized by a mean-variance couple, with the mean μ (a vector) and a variance-covariance matrix Γ, so that $\mathbf{F} \sim \mathcal{N}(\mu, \Gamma)$. The quantities μ and Γ therefore represent the *ex ante* or *a priori* values of the means, and variances and covariances. Then, after a length of time has passed, we obtain the *ex post* estimations of these quantities, which we denote by the classic conventions of $\hat{\mu}$ and $\hat{\Gamma}$.

The question is therefore, at the time of constructing a portfolio, knowing how to combine the *a priori* values with our knowledge of companies. Or, specifically, to best determine $\hat{\mu}$ and $\hat{\Gamma}$ in function of μ and Γ and of the information available. If we can obtain good information on companies, and if we can obtain from their current situation an indication of their future performance, then we should incorporate these signals into the quantities μ and Γ. Denote by s this informational signal and by $\mu(s)$ the expectation of the resulting flow. This signal is itself uncertain and we can consider that the uncertainty of s is quantified by its variance matrix denoted by $\Gamma(s)$. The goal is then to combine the *a priori* estimations and the information s in order to obtain the values μ and Γ.

Van Nieuwerburgh and Veldkamp (2007) formulate the hypothesis that:

$$\hat{\mu} = \frac{\Gamma^{-1}\mu + \Gamma(s)^{-1}\mu(s)}{\Gamma^{-1} + \Gamma(s)^{-1}}$$

and:

$$\hat{\Gamma} = \left(\Gamma^{-1} + \Gamma(s)^{-1}\right)^{-1}$$

Note that the *ex post* variance satisfies:

$$\frac{1}{\hat{\Gamma}} = \frac{1}{\Gamma} + \frac{1}{\Gamma(s)}$$

so that it is the harmonic mean of the *a priori* and informational variances. This expression is convenient for interpreting $\hat{\Gamma}$: if, for instance, the informational signal appears very safe, then $|\Gamma(s)|$ is very small, and therefore $|\Gamma(s)|^{-1}$ is very large (tending to infinity). In this case, we have from the preceding expression that $|\hat{\Gamma}|$ becomes very small (tending to zero).

Written differently, if the signal obtained by the investor displays a small variance-covariance matrix $\Gamma(s)$, then the variance of assets such as it appeared *ex post* to the investor is diminished. From there stems the interest of the maximal reduction of uncertainty as modeled by $|\hat{\Gamma}|$. Note that the contribution of Van Nieuwerburgh and Veldkamp (2007) lies in the characterization of portfolios for which utility is optimized and for which it has been possible to reduce $|\hat{\Gamma}|$ as much as possible. However, these authors say nothing about the nature of the information and the way in which it can be obtained.

The uncertainty framework studied here rests on two constraints.

(1) First, the variance-covariance matrix of assets must be positive definite. This is a standard constraint.
(2) The second constraint is more innovative: the investor is assumed initially endowed with a limited capacity to obtain information. This capacity, expressed by the real number K, induces a maximal reduction of the *ex post* variance of assets according to:

$$\frac{|\Gamma|}{|\hat{\Gamma}|} \leq e^{2K} \tag{11.56}$$

Put differently, the greater the capacity of the investor to obtain information (the larger K), the more it is possible to reduce $|\hat{\Gamma}|$. Equivalently, it is not possible to reduce $|\hat{\Gamma}|$ beyond a factor that depends on K: the relation (11.56) introduces a bound on the capacity to obtain information on individual companies. An incompressible uncertainty component always remains.

Let us make this point more precise. Van Nieuwerburgh and Veldkamp (2007) consider that it is not possible to reduce $\hat{\Gamma}_i$ beyond a certain level for each asset i. An information threshold beyond which variance cannot be reduced therefore exists. This minimal variance is called non-learnable risk. It is a kind of intrinsic volatility for which we can obtain no information, a natural fluctuation that is not accessible by a standard informational learning process and that only diversification can cancel out.

In the presence of such a risk, the nature of portfolios is modified as follows. First, the investor allocates all his information capacity to the asset (or component, if assets are correlated) with the highest Sharpe ratio. From a certain level of capacity, no gain can be obtained in terms of reduction of variance. Additional means are then allocated to the asset with the second highest Sharpe ratio, and so on. We observe the following phenomenon. First, increasing the capacity to get information leads to an increase in concentration. Then, beyond a certain level, concentration diminishes and the portfolio becomes more and more diversified. This corresponds to the fact that a theoretically infinite informational capacity would offer pertinent information for all of the existing assets, leading therefore to a perfect portfolio diversification.

The Van Nieuwerburgh and Veldkamp Solution

We consider the solution of the optimization program first for independent assets, then for dependent assets.

Independent Assets

We consider the simplified case of n independent assets. The matrices Γ and $\hat{\Gamma}$ are diagonal and the optimization program can be written as:

$$\max_{\hat{\Gamma}} \left(\sum_{i=1}^{n} \frac{\Gamma_i}{\hat{\Gamma}_i} \left(1 + \mathrm{SR}_i^2\right) \right)$$

where:

$$\mathrm{SR}_i = \frac{\mu_i - P_i r}{\Gamma_i^{1/2}}$$

is the Sharpe ratio (SR) of asset i with P_i the price of asset i.

Solving this program yields the following result: it is optimal to allocate all of one's information research capacity to the asset with the highest Sharpe ratio. Put differently, we aim to minimize the $\hat{\Gamma}_i$ of the asset i with the highest SR_i. This choice maximizes utility of the mean-variance type. We obtain:

$$\hat{\Gamma}_i = e^{-2K} \Gamma_i$$

for the asset i with the highest Sharpe ratio, and:

$$\hat{\Gamma}_j = \Gamma_j$$

for the other assets. The quantity invested in asset i (denoted by θ_i) is equal to:

$$\theta_i = \tau \hat{\Gamma}_i^{-1} (\mu_i - P_i r)$$

where τ is the risk tolerance coefficient. This quantity increases when $\hat{\Gamma}_i$ decreases.

In the case of independent assets, we therefore obtain a strong concentration on the asset with the highest Sharpe ratio. Furthermore, this asset represents a proportion of the portfolio that is all the more important that we have a high information capacity K for this asset.

Dependent Assets

In the case of correlated assets, any learning process related to the risky characteristics of an asset provides information on the risk of other assets. We can demonstrate risk factors using a principal component analysis (PCA), which decomposes the initial variance-covariance matrix (the *a priori* estimations) as follows:

$$\hat{\Gamma} = \Phi \hat{\Lambda} \Phi'$$

where Φ is the matrix containing the eigenvectors of the PCA and $\hat{\Lambda}$ is the matrix containing the eigenvalues, or variances of each risk factor (we recover the APT framework of Ross).

In this general framework, we can define a Sharpe ratio indicator for each principal component. SR_i therefore becomes:

$$\text{SR}_i = \frac{(\mu_i - P_i r)' \Phi_i}{\Lambda_i^{1/2}}$$

where the notation ' designates the transpose of a vector or matrix. Van Nieuwerburgh and Veldkamp (2007) show that a rational investor allocates all this information research capacity to diminish the variance of the component with the highest Sharpe ratio. We obtain in this case $\hat{\Lambda}_i = e^{-2K} \Lambda_i$ for the component i with the highest Sharpe ratio and $\hat{\Lambda}_j = \Lambda_j$ for the other components.

The concentration of the portfolio on a few assets (if they are correlated) appears in the following way: starting from a maximally diversified portfolio, investors will first increase their capacity for obtaining information on the principal component (linear combination of assets) having the highest generalized Sharpe ratio. They invest massively in this component, particularly given that they invested in information capacity. They do so because the invested quantities are inversely proportional to $\hat{\Lambda}_i$, which is further reduced given that the investment to obtain a high capacity K has been significant.

Chapter 12

Dynamic Portfolio Choice

The principles of portfolio construction in a dynamic setting rest on the modeling of stock fluctuations and on the solution of an optimization program bearing on future consumption and wealth. At the beginning of the research in the field, the modeling of stock fluctuations was Brownian and optimization was performed solving a Hamilton–Jacobi–Bellman equation. This is the classic portfolio choice theory in continuous time, as developed by Merton (1969, 1971). For a detailed presentation, see the books of Merton (1992) and Quittard-Pinon (2003). See also Cox and Huang (1989) for the martingale approach. Other possibilities then appeared: for instance introducing jumps into the dynamics of stocks, or adding uncertainty to probabilities in the case of classic Brownian motions.

We present hereafter these two perspectives, after recalling the general principles of the classic dynamic framework for solving portfolio choice programs. The treatment of dynamic portfolio choice in the presence of jumps follows the presentation of Le Courtois (2011). The uncertainty framework is inspired by the contribution of Uppal and Wang (2003).

12.1 The Optimization Program

We start by constructing the objective function that we aim to compute and we then concentrate on the modeling of the market in which the investor operates.

12.1.1 The Objective Function

We write the objective function with respect to the utility function of consumption and final wealth of the investor. We then transform the problem to be solved into a Hamilton–Jacobi–Bellman equation. Finally, we examine the case of an infinite investment horizon.

12.1.1.1 Utilities of Wealth and Consumption

We consider an investor (individual, institutional) who wants to set up a portfolio with the following two goals: obtaining a regular flow of income and ensuring a final residual value. In the case of an individual, we can think of an investor who wants to obtain an annual income for himself and pass on the residual capital to his heirs. In the case of an institution, such as an insurance company, we can think of an investor who has to compensate for annual flows of claims – and therefore needs annual inflows of cash – while preserving the residual value of his portfolio.

We denote by c_s the financial consumption rate at time s. The consumption between times 0 and T is therefore:

$$\int_0^T c_s ds$$

We assume that the investor's satisfaction criteria are distinct if he wants to evaluate a consumption flow or a value V_T at the terminal date T (for instance 30 years). Furthermore, preferences are assumed to be additively separable, so that it is possible to formally separate the corresponding satisfaction functions. We denote these two satisfaction functions by u_1 and u_2: u_1 is applied to the consumption of income and u_2 to the future value of funds. The global satisfaction function between times t and T is denoted traditionally by U and is defined at time t as follows:

$$U(t, c, V_T) = \int_t^T u_1(c_s) ds + u_2(V_T)$$

This utility function can be generalized assuming the investor does not appreciate the utility of future consumption but the discounted value at a rate ρ of this utility. The goal is then to optimize:

$$U(t, c, V_T) = \int_t^T e^{-\rho s} u_1(c_s) ds + u_2(V_T) \qquad (12.1)$$

where it is enough to set $\rho = 0$ in order to come back to the preceding situation.

Thus, we want to find the optimal asset quantities in which to invest, and the optimal consumption, i.e. a couple (π^*, c^*) such that the utility function U is maximized. Therefore, we want to solve the following optimization program:

$$J(V_t, t) = \sup_{\pi, c} \mathbb{E}_t \left[\int_t^T e^{-\rho s} u_1(c_s) ds + u_2(V_T) \right] \qquad (12.2)$$

We can now rewrite this program under the form of a Hamilton–Jacobi–Bellman equation, whose solution yields, upon verification, the couple (π^*, c^*) of optimal asset quantities and consumption.

12.1.1.2 Hamilton–Jacobi–Bellman Equation

Assume that the optimal utility J defined in (12.2) has the dynamics:

$$dJ(V_s, s) = \mathcal{A}J(V_s, s) ds + dM_s \qquad (12.3)$$

where M is a martingale. We can also write this equation as follows:

$$dJ(V_s, s) = J_s ds + \mathcal{L}J(V_s, s) ds + dM_s \qquad (12.4)$$

where J_s is the explicit derivative of J with respect to time. Because:

$$J(V_T, T) = \sup_{\pi, c} \mathbb{E}_T \left[\int_T^T e^{-\rho s} u_1(c_s) ds + u_2(V_T) \right] = \sup_{\pi, c} \mathbb{E}_T \left[u_2(V_T) \right]$$

and because viewed from time T it is meaningless to establish an optimal investment strategy for a time horizon also equal to T or to take the expectation of quantities known at that time, we have:

$$J(V_T, T) = u_2(V_T)$$

From this equation and from Eq. (12.2), we deduce:

$$J(V_t, t) = \sup_{\pi, c} \mathbb{E}_t \left[\int_t^T e^{-\rho s} u_1(c_s) ds + J(V_T, T) \right] \qquad (12.5)$$

We can now write the integrated form of Eq. (12.3):

$$J(V_T, T) = J(V_t, t) + \int_t^T \mathcal{A}J(V_s, s) ds + \int_t^T dM_s$$

in order to replace $J(V_T, T)$ in Eq. (12.5):

$$J(V_t, t) = \sup_{\pi,c} \mathbb{E}_t \left[\int_t^T e^{-\rho s} u_1(c_s) ds + J(V_t, t) \right.$$
$$\left. + \int_t^T \mathcal{A} J(V_s, s) ds + \int_t^T dM_s \right]$$

Because the term $J(V_t, t)$ inside the right-hand side of the equation is known at time t, it can be factored out of the expectation as well as out of the supremum, on which it has no effect. Therefore, we have:

$$0 = \sup_{\pi,c} \left(\mathbb{E}_t \left[\int_t^T e^{-\rho s} u_1(c_s) ds + \int_t^T \mathcal{A} J(V_s, s) ds \right] + \mathbb{E}_t \left[\int_t^T dM_s \right] \right)$$

Further, because M is by definition a martingale, we can write:

$$\mathbb{E}_t \left[\int_t^T dM_s \right] = \mathbb{E}_t \left[M_T - M_t \right] = \mathbb{E}_t \left[M_T \right] - M_t = M_t - M_t = 0$$

yielding the new form of the optimization program:

$$0 = \sup_{\pi,c} \mathbb{E}_t \left[\int_t^T e^{-\rho s} u_1(c_s) ds + \int_t^T \mathcal{A} J(V_s, s) ds \right]$$

This expression, divided by $T - t$, and letting $T - t$ tend to zero, yields an equation whose terms are expressed at time t and for which there is no need to apply the expectation operator:

$$0 = \sup_{\pi,c} \{ e^{-\rho t} u_1(c_t) + \mathcal{A} J(V_t, t) \} \qquad (12.6)$$

This is the Hamilton–Jacobi–Bellman equation of the considered optimization problem. This equation can be rewritten as follows:

$$0 = \sup_{\pi,c} \{ e^{-\rho t} u_1(c_t) + J_t + \mathcal{L} J(V_t, t) \}$$

or, because J_t does not depend explicitly on c or π (contrary to $\mathcal{L} J(V_t, t)$, as we shall observe later on), as follows:

$$0 = J_t + \sup_{\pi,c} \{ e^{-\rho t} u_1(c_t) + \mathcal{L} J(V_t, t) \} \qquad (12.7)$$

This equation does not mean that no utility is attached to the terminal value of wealth. Indeed, u_2 operates indirectly as an initial condition during the backward solving from times T to 0.

It is important to note that the solution to Eq. (12.7) may not be a maximum solving Eq. (12.2), just like canceling the derivative of a function does not always mean finding a maximum for this function. In order to check that the solution to Eq. (12.7) is indeed a solution to Eq. (12.2), it is necessary to use a so-called verification theorem. For examples of verification theorems in a diffusive context, see Korn and Korn (2001); in the presence of jumps, see Oksendal and Sulem (2010).

This approach, which allows us to obtain, *in fine* and upon verification, the couple (π^*, c^*), comes down in general to solving a partial differential equation (PDE) in the case of assets that follow diffusions, or a partial integro-differential equation (PIDE) in the case of assets that follow jump-diffusion processes. This general PIDE has the expression:

$$0 = J_t + e^{-\rho t} u_1(c_t^*) + \mathcal{L}J(V_t, t)_{\pi^*, c^*} \qquad (12.8)$$

For the moment, we concentrate on the case of an infinite time horizon and show how Eqs (12.7) and (12.8) can be simplified in this setting.

12.1.1.3 *Infinite Time Horizon Case*

In the case where $T = +\infty$, Eq. (12.2) can be rewritten as follows:

$$J(V_t, t) = \sup_{\pi, c} \mathbb{E}_t \left[\int_t^{+\infty} e^{-\rho s} u_1(c_s) ds \right]$$

where we assume that $\lim_{T \to +\infty} \mathbb{E}_t [u_2(V_T)] = 0$. Let us now define:

$$K(V_t, t) = e^{\rho t} J(V_t, t) = \sup_{\pi, c} \mathbb{E}_t \left[\int_t^{+\infty} e^{-\rho(s-t)} u_1(c_s) ds \right]$$

which becomes, after the change of variable $v = s - t$:

$$K(V_t, t) = \sup_{\pi, c} \mathbb{E}_t \left[\int_0^{+\infty} e^{-\rho v} u_1(c_{v+t}) dv \right]$$

Because there is fundamentally no difference between computing this expectation at time t or at time 0, we write:

$$K(V_t, t) = K(V_t) = \sup_{\pi, c} \mathbb{E}_0 \left[\int_0^{+\infty} e^{-\rho v} u_1(c_v) dv \right]$$

so that the function K does not depend explicitly on time. We can now replace J by K in Eq. (12.7), noting that:

$$J_t = \frac{\partial J}{\partial t} = \frac{\partial K(V_t) e^{-\rho t}}{\partial t} = \frac{\partial K(V_t)}{\partial t} e^{-\rho t} + K(V_t) \frac{\partial e^{-\rho t}}{\partial t}$$

so:

$$J_t = 0 \cdot e^{-\rho t} - \rho K(V_t) e^{-\rho t} = -\rho K(V_t) e^{-\rho t}$$

and:

$$\mathcal{L}J(V_t, t) = e^{-\rho t} \mathcal{L}K(V_t)$$

We obtain, by insertion into (12.7):

$$0 = -\rho K(V_t)e^{-\rho t} + \sup_{\pi,c}\{e^{-\rho t}u_1(c_t) + e^{-\rho t}\mathcal{L}K(V_t)\} \quad (12.9)$$

so finally (because $K(V_t)$ does not depend explicitly on π or c):

$$0 = \sup_{\pi,c}\{u_1(c_t) - \rho\, K(V_t) + \mathcal{L}K(V_t)\} \quad (12.10)$$

One of the main interests of Eq. (12.9) is that it boils down to a differential equation and not to a partial differential equation (up to a potential integral term that is related to jumps). Therefore, it is simpler to solve than Eq. (12.7). This differential equation can be presented in the form:

$$0 = u_1(c_t^*) - \rho\, K(V_t)_{\pi^*,c^*} + \mathcal{L}K(V_t)_{\pi^*,c^*} \quad (12.11)$$

Let us now come to the specification of stock dynamics.

12.1.2 Modeling Stock Fluctuations

The goal of the developments that follow is to obtain a general expression for the dynamics V of the investor's portfolio and to compute the infinitesimal generator associated with this stochastic process in order to insert it in the general optimization program (12.6).

12.1.2.1 General Notations

Consider a financial market with d risky assets and one risk-free asset. We use the following notation for the composition of the portfolio:

- $\mathbf{S}(t) = (S_i(t))_{i=1,\cdots,d}$ is the vector of the prices at time t of the d risky assets,
- $\boldsymbol{\theta}(t) = (\theta_i(t))_{i=1,\cdots,d}$ is the vector of the quantities invested in risky assets at time t,
- $\mathbf{w}(t) = (w_i(t))_{i=1,\cdots,d}$ is the vector of the weights of risky assets in the portfolio at time t. The proportion invested in the risk-free asset is consequently:

$$w_0(t) = 1 - \sum_{i=1}^{d} w_i(t)$$

- $\boldsymbol{\pi}(t) = (\pi_i(t))_{i=1,\cdots,d}$ is the vector of the values of the lines of the portfolio at time t. Each line is worth $\pi_i(t) = \theta_i(t) S_i(t)$; this is the amount in euros invested in asset i at time t.
- $V(t)$ is the value of the portfolio at time t:

$$V(t) = \sum_{i=0}^{d} \theta_i(t) S_i(t) = \sum_{i=0}^{d} \pi_i(t) = V(t) \sum_{i=0}^{d} w_i(t)$$

Thus, the composition of the portfolio is given equally by the vectors $\boldsymbol{\theta}$, $\mathbf{w}$, or $\boldsymbol{\pi}$.

12.1.2.2 Risk-Free Asset

The risk-free asset, denoted by $S_0(t)$, has a return $r(t)$ and follows the dynamics:

$$dS_0(t) = S_0(t)\, r(t)\, dt$$

which, by definition, do not depend on an explicit source of randomness. We shall assume in what follows that r is constant. The dynamics of the risk-free asset then become:

$$dS_0(t) = S_0(t)\, r\, dt$$

or, upon integration:

$$S_0(t) = S_0(0)\, e^{rt}$$

12.1.2.3 Risky Assets

We assume that the d risky assets present geometric dynamics. In full generality, we allow each asset to depend on m Brownian motions and on n Lévy processes.

We model Lévy processes here by Poisson random measures. This allows us to present infinitesimal operators (operators $\mathcal{L}$ and $\mathcal{A}$) useful for solving dynamic portfolio problems. Let N be a Poisson random measure associated with a Lévy process. Then, $N(dt, dz)$ describes (counts) the number of jumps of size z to $z + dz$ that occur between times t and $t + dt$. The notation $N(t, dz)$ corresponds to the number of jumps of size z to $z + dz$ that occurred between times 0 and t.

A Lévy process is correctly described by a Poisson random measure if its small jumps are compensated. Let us denote by:

$$\tilde{N}(dt, dz) = N(dt, dz) - \mathbb{E}(N(dt, dz))$$

a compensated Poisson random measure. In the case of a Lévy process, the time and space components are separable in the compensator:

$$\mathbb{E}(N(dt,dz)) = \nu(dz)dt$$

where ν is the Lévy measure. Therefore, we can write:

$$\tilde{N}(dt,dz) = N(dt,dz) - \nu(dz)dt$$

which is a compensated Poisson random measure. The reader may note that by construction $\tilde{N}$ is a martingale.

The pure jump part of a Lévy process can be written in full generality as follows:

$$\int_{|z|<\Lambda} z\tilde{N}(t,dz) + \int_{|z|\geq\Lambda} zN(t,dz)$$

where the truncation parameter Λ is arbitrary. In the following developments, we consider Lévy processes constructed with an infinite truncation parameter, i.e. a compensation that applies to jumps of all sizes. Thus, we use Lévy processes L of the type:

$$\int_{\mathbb{R}} z\tilde{N}(t,dz)$$

which are martingales by construction.

The dynamics S_i of any risky asset can then be modeled in full generality by the stochastic differential equation:

$$dS_i(t) = S_i(t^-)\left(\mu_i\,dt + \sum_{k=1}^{m}\sigma_i^k\,dW_t^k + \sum_{l=1}^{n}\gamma_i^l\int_{\mathbb{R}} z_l\,\tilde{N}^l(dt,dz_l)\right) \quad (12.12)$$

where $W = (W^1, W^2, \cdots, W^m)$ is an m-dimensional standard Brownian motion, with:

$$\sigma_i = \sqrt{\sum_{k=1}^{m}(\sigma_i^k)^2}$$

the standard deviation of the continuous component of the return of S_i and:

$$\sigma_{i,j} = \rho_{i,j}\,\sigma_i\,\sigma_j = \sum_{k=1}^{m}\sigma_i^k\sigma_j^k$$

the covariance between the continuous components of the returns of S_i and S_j ($\rho_{i,j}$ describing their correlation), and where:

$$L(t) = \left(\int_{\mathbb{R}} z_1\tilde{N}^1(t,dz_1), \int_{\mathbb{R}} z_2\tilde{N}^2(t,dz_2), \cdots, \int_{\mathbb{R}} z_n\tilde{N}^n(t,dz_n)\right)$$

expresses the value at time t of an n-dimensional compensated Lévy process.

The jumps of the returns of each S_i are assumed superior to -1, in order to ensure that stock prices remain positive. This implies for all i the relationship that must be satisfied by any jump z_l:

$$\gamma_i^l z_l \geq -1$$

within Eq. (12.12). It this perspective, it is sufficient that the jumps z_l of each Lévy process satisfy:

$$z_l \geq \max\left\{-\frac{1}{\gamma_1^l}, -\frac{1}{\gamma_2^l}, \cdots, -\frac{1}{\gamma_d^l}\right\} := \lambda^l$$

We can therefore rewrite the dynamics (12.12) in the more explicit form:

$$dS_i(t) = S_i(t^-)\left(\mu_i\, dt + \sum_{k=1}^m \sigma_i^k\, dW_t^k + \sum_{l=1}^n \gamma_i^l \int_{\lambda^l}^{+\infty} z_l\, \tilde{N}^l(dt, dz_l)\right) \quad (12.13)$$

We can use a simplified form of Itō's lemma for jump processes (see Oksendal and Sulem (2010), and Jacod and Shiryaev (2002) for several versions of Itō's lemma) in order to express the dynamics in an integrated way. This lemma asserts that for a vectorial stochastic process $\mathbf{X}$ of dimension d and dynamics:

$$d\mathbf{X}(t) = \mathbf{a}(\mathbf{X}_{t^-}, t)\, dt + \mathbf{b}(\mathbf{X}_{t^-}, t)d\mathbf{W}_t + \int \mathbf{c}(\mathbf{X}_{t^-}, \mathbf{z}, t)\, \tilde{\mathbf{N}}(dt, d\mathbf{z}) \quad (12.14)$$

where $\mathbf{a}$, $\mathbf{b}$, $\mathbf{W}$ (Brownian motion), $\mathbf{c}$, and $\tilde{\mathbf{N}}$ (compensated Poisson random measure) are of dimension d, $d \times m$, m, $d \times n$, and n respectively, where each column $\mathbf{c}_{.,l}$ depends only on z_l ($\mathbf{c}$ does not mix the jumps of directing Lévy processes), and where the integral is multiple by construction, then, if we set a sufficiently regular function f that constructs a new univariate process Y as follows:

$$Y(t) = f(\mathbf{X}_t, t)$$

with in particular a potential explicit dependence on time, we obtain that the dynamics of Y can be expressed as follows:

$$dY(t) = \frac{\partial f}{\partial t}dt + \sum_{i=1}^{d}\frac{\partial f}{\partial x_i}(a_i dt + \mathbf{b}_{i,.}d\mathbf{W}_t) + \frac{1}{2}\sum_{i=1}^{d}\sum_{j=1}^{d}(\mathbf{bb}^t)_{i,j}\frac{\partial^2 f}{\partial x_i \partial x_i}dt$$

$$+ \sum_{l=1}^{n}\int [f(\mathbf{X}_{t^-} + \mathbf{c}_{.,l}(\mathbf{X}_{t^-}, z_l, t), t) - f(\mathbf{X}_{t^-}, t)$$

$$- \sum_{i=1}^{d} c_{i,l}(\mathbf{X}_{t^-}, z_l, t)\frac{\partial f}{\partial x_i}]\,\nu^l(dz_l)dt$$

$$+ \sum_{l=1}^{n}\int [f(\mathbf{X}_{t^-} + \mathbf{c}_{.,l}(\mathbf{X}_{t^-}, z_l, t), t) - f(\mathbf{X}_{t^-}, t)]\,\tilde{N}^l(dt, dz_l)$$

(12.15)

where $\mathbf{b}_{i,.}$ is a row vector of $\mathbf{b}$ and $\mathbf{c}_{.,l}$ is a column vector of $\mathbf{c}$.

In order to integrate (12.13) using Itō's lemma, we can set $f = \ln$ and identify in (12.14): $\mathbf{X} = S_i$, of dimension 1, $\mathbf{a}(\mathbf{X}_{t^-}, t) = \mu_i S_i(t^-)$, of dimension 1, $\mathbf{b}(\mathbf{X}_{t^-}, t) = \sigma_i S_i(t^-)$, of dimension $1 \times m$, and $\mathbf{c}(\mathbf{X}_{t^-}, \mathbf{z}, t) = [\gamma_i^l z_l]_{l=1,..,n}\, S_i(t^-)$, of dimension $1 \times n$. Replacing these expressions and the derivatives of the logarithm in Eq. (12.15):

$$\frac{\partial \ln(S_i(t))}{\partial t} = 0,$$

$$\frac{\partial \ln(S_i(t))}{\partial S_i(t)} = \frac{1}{S_i(t)}$$

and:

$$\frac{\partial^2 \ln(S_i(t))}{\partial S_i^2(t)} = -\frac{1}{S_i^2(t)}$$

we obtain:

$$d\ln(S_i(t)) = \left[\mu_i - \frac{1}{2}\sum_{k=1}^{m}(\sigma_i^k)^2\right]dt + \sum_{k=1}^{m}\sigma_i^k\, dW_t^k$$

$$+ \left[\sum_{l=1}^{n}\int_{\lambda^l}^{+\infty}(\ln(1+\gamma_i^l z_l) - \gamma_i^l z_l)\,\nu^l(dz_l)\right]dt$$

$$+ \sum_{l=1}^{n}\int_{\lambda^l}^{+\infty}\ln(1+\gamma_i^l z_l)\,\tilde{N}^l(dt, dz_l)$$

which gives upon integration and exponentiation:

$$S_i(t) = e^{\left[\mu_i - \frac{1}{2}\sum_{k=1}^{m}(\sigma_i^k)^2 + \sum_{l=1}^{n}\int_{\lambda^l}^{+\infty}(\ln(1+\gamma_i^l z_l) - \gamma_i^l z_l)\nu^l(dz_l)\right]t}$$
$$\times\, e^{\sum_{k=1}^{m}\sigma_i^k W_t^k} \times e^{\sum_{l=1}^{n}\int_0^t\int_{\lambda^l}^{+\infty}\ln(1+\gamma_i^l z_l)\tilde{N}^l(ds, dz_l)} \quad (12.16)$$

In the simple case where we consider d Brownian motions, d compensated Lévy processes, and where each asset depends on a unique Brownian motion and a unique compensated Lévy process, each asset then has the dynamics:

$$dS_i(t) = S_i(t^-)\left(\mu_i\, dt + \sigma_i\, dW_t^i + \gamma_i \int_{-\frac{1}{\gamma_i}}^{+\infty} z_i\, \tilde{N}^i(dt, dz_i)\right) \qquad (12.17)$$

which can be integrated, using Itō's lemma, into:

$$S_i(t) = S_i(0) e^{\left[\mu_i - \frac{1}{2}\sigma_i^2 + \int_{-\frac{1}{\gamma_i}}^{+\infty}(\ln(1+\gamma_i z_i) - \gamma_i z_i)\nu^i(dz_i)\right]t + \sigma_i W_t^i} \\ \times e^{\int_0^t \int_{-\frac{1}{\gamma_i}}^{+\infty} \ln(1+\gamma_i z_i)\,\tilde{N}^i(ds, dz_i)} \qquad (12.18)$$

12.1.2.4 Portfolio

The value of the portfolio follows the dynamics:

$$dV_t = \sum_{i=0}^{d} \theta_i(t) dS_i(t) - c_t dt$$

which can also be expressed in a developed form when stocks follow the dynamics (12.13):

$$\begin{aligned} dV_t &= (V_t r - c_t)\, dt + \sum_{i=1}^{d} \pi_i(t)(\mu_i - r)dt \\ &+ \sum_{i=1}^{d}\sum_{k=1}^{m} \pi_i(t)\, \sigma_i^k\, dW_t^k + \sum_{i=1}^{d}\sum_{l=1}^{n} \pi_i(t)\, \gamma_i^l \int_{\lambda^l}^{+\infty} z_l\, \tilde{N}^l(dt, dz_l) \end{aligned} \qquad (12.19)$$

Let us now compute the infinitesimal operator $\mathcal{A}$ applied to any sufficiently regular function f of the value of the portfolio and of explicit time. Recall that according to our definition (12.3) of $\mathcal{A}$:

$$df(V_t, t) = \mathcal{A}f(V_t, t)ds + dM_s$$

where M is a martingale. The expression of df is also given by Eq. (12.15) of Itō's lemma. If we subtract the martingale component from Eq. (12.15), we obtain:

$$Af(\mathbf{X}_t, t) = \frac{\partial f}{\partial t}dt + \sum_{i=1}^{d} \frac{\partial f}{\partial x_i} a_i + \frac{1}{2}\sum_{i=1}^{d}\sum_{j=1}^{d} (\mathbf{bb}^t)_{i,j} \frac{\partial^2 f}{\partial x_i \partial x_i}$$

$$+ \sum_{l=1}^{n} \int \left[f\left(\mathbf{X}_{t^-} + \mathbf{c}_{.,l}(\mathbf{X}_{t^-}, z_l, t), t\right) - f\left(\mathbf{X}_{t^-}, t\right) \right.$$

$$\left. - \sum_{i=1}^{d} c_{i,l}(\mathbf{X}_{t^-}, z_l, t)\frac{\partial f}{\partial x_i} \right] \nu^l(dz_l) \quad (12.20)$$

Comparing Eqs (12.14) and (12.19), we perform the following identification: $\mathbf{X} = V$, of dimension 1, $a(\mathbf{X}_{t^-}, t) = V_t\, r - c_t + \sum_{i=1}^{d} \pi_i(t)(\mu_i - r)$, of dimension 1, $\mathbf{b}(\mathbf{X}_{t^-}, t) = \left[\sum_{i=1}^{d} \pi_i(t)\, \sigma_i^k\right]_{k=1,\cdots,m}$, of dimension $1 \times m$, and $\mathbf{c}(\mathbf{X}_{t^-}, \mathbf{z}, t) = [\sum_{i=1}^{d} \pi_i(t)\gamma_i^l z_l]_{l=1,..,n}$, of dimension $1 \times n$. Replacing in (12.20), we obtain:

$$Af(V_t, t) = \frac{\partial f}{\partial t} + \frac{\partial f}{\partial v}\left(V_t\, r - c_t + \sum_{i=1}^{d} \pi_i(t)(\mu_i - r)\right)$$

$$+ \frac{1}{2}\sum_{k=1}^{m}\left(\sum_{i=1}^{d} \pi_i(t)\, \sigma_i^k\right)^2 \frac{\partial^2 f}{\partial v^2}$$

$$+ \sum_{l=1}^{n} \int \left[f\left(V_{t^-} + \sum_{i=1}^{d} \pi_i(t^-)\gamma_i^l z_l, t\right) - f(V_{t^-}, t) \right.$$

$$\left. - \sum_{i=1}^{d} \pi_i(t)\gamma_i^l z_l \frac{\partial f}{\partial v} \right] \nu^l(dz_l) \quad (12.21)$$

When the stocks have the simplified dynamics (12.17), then the dynamics (12.19) of the portfolio can be simplified into:

$$dV_t = (V_t\, r - c_t)\, dt + \sum_{i=1}^{d} \pi_i(t)\, (\mu_i - r)\, dt$$

$$+ \sum_{i=1}^{d} \pi_i(t)\, \sigma_i\, dW_t^i + \sum_{i=1}^{d} \pi_i(t)\, \gamma_i \int_{-\frac{1}{\gamma_i}}^{+\infty} z_i\, \tilde{N}^i(dt, dz_i) \quad (12.22)$$

and the infinitesimal generator (12.21) can be expressed in the simplified form:

$$\mathcal{A}f(V_t,t) = \frac{\partial f}{\partial t} + \frac{\partial f}{\partial v}\left(V_t\, r - c_t + \sum_{i=1}^{d} \pi_i(t)(\mu_i - r)\right)$$

$$+ \frac{1}{2}\frac{\partial^2 f}{\partial v^2}\sum_{i=1}^{d} \pi_i(t)^2\, \sigma_i^2$$

$$+ \sum_{i=1}^{d}\int \left[f\left(V_{t^-} + \pi_i(t^-)\gamma_i z_i, t\right) - f(V_{t^-}, t)\right.$$

$$\left.- \pi_i(t)\gamma_i z_i \frac{\partial f}{\partial v}\right]\nu^i(dz_i) \qquad (12.23)$$

To conclude this part, dynamic portfolio optimization consists in solving Eq. (12.6) starting from a specification of the infinitesimal operator as in (12.21) or in (12.23). In the general case, and despite a few sub-cases where there exist closed-form formulas, the solution is achieved numerically. In the following sections, we give two simple examples for which the problem is solved in closed-form: first in the case of diffusive lognormal assets and in the presence of a HARA utility function, then in the case of assets modeled as diffusions plus Poisson jumps (with a very specific Lévy measure as we shall observe) and in the presence of a CARA utility function.

12.2 Classic Approach

In order to illustrate how it is possible to arrive at an asset allocation in the dynamic context, we solve the preceding program in dimension $d = 1$, so in the case of a market with a risk-free asset and a unique risky asset. We can imagine that the risky asset represents an index: we look for the optimal distribution of funds between this index and the monetary market.

We set ourselves in the standard lognormal framework where μ and σ do not depend on S. In dimension 1, the dynamics of the index are:

$$dS_t = S_t(\mu dt + \sigma dW_t)$$

The proportion (expressed as a percentage) allocated to the index is θ_t and the portfolio therefore has the composition $(\theta_t, 1 - \theta_t)$. At each time t, the value V_t can be decomposed into $\theta_t V_t + (1 - \theta_t)V_t$ and the consumption is c_t. From this observation, we deduce the stochastic differential equation driving the value of the portfolio:

$$dV_t = \theta_t V_t\,(\mu dt + \sigma dW_t) + (1 - \theta_t)V_t\, r dt - c_t\, dt$$
$$= (\theta_t V_t(\mu - r) + rV_t - c_t)\, dt + \theta_t V_t\, \sigma\, dW_t$$

This expression can also be obtained directly as a sub-case of Eq. (12.22) setting $d = 1$, $\pi_1(t) = \theta_t V_t$, and canceling the jump component. Applying these simplifications to Eq. (12.23), we obtain the infinitesimal generator:

$$\mathcal{A}J(t) = J_t + (\theta_t V_t(\mu - r) + rV_t - c_t) J_v + \frac{1}{2}(\theta_t V_t)^2 \sigma^2 J_{vv}$$

where J_t is the temporal derivative of J, and J_v and J_{vv} are respectively the first and second derivatives of J with respect to the portfolio value. Inserting this expression into Eq. (12.6), we obtain the optimization program:

$$0 = \sup_{\theta,c} \left\{ u(c_t, t) + J_t + (\theta_t V_t(\mu - r) + rV_t - c_t) J_v + \frac{1}{2}(\theta_t V_t)^2 \sigma^2 J_{vv} \right\} \quad (12.24)$$

where $u(c_t, t) = e^{-\rho t} u_1(c_t)$. We can obtain the optimal consumption c_t^* by differentiating with respect to c_t the expression inside the sup of Eq. (12.24):

$$\left(\frac{\partial u}{\partial c}\right)(c_t^*, t) - J_v = 0$$

hence:

$$c_t^* = (u')^{-1}(J_v)_t \quad (12.25)$$

denoting by u' the derivative of u with respect to c. We now compute the optimal weight θ_t^* by differentiating with respect to θ_t the expression inside the supremum of Eq. (12.24):

$$V_t(\mu - r) J_v + \theta_t^* V_t^2 \sigma^2 J_{vv} = 0$$

hence:

$$\theta_t^* = -\frac{1}{V_t} \frac{J_v}{J_{vv}} \frac{\mu - r}{\sigma^2} \quad (12.26)$$

Inserting the expressions (12.25) and (12.26) into the HJB equation (12.24), we obtain a partial differential equation:

$$u\left((u')^{-1}(J_v)_t, t\right) + J_t + \left(-\frac{J_v}{J_{vv}} \frac{(\mu - r)^2}{\sigma^2} + rV_t - (u')^{-1}(J_v)_t\right) J_v$$
$$+ \frac{1}{2}\left(-\frac{1}{V_t} \frac{J_v}{J_{vv}} \frac{\mu - r}{\sigma^2}\right)^2 V_t^2 \sigma^2 J_{vv} = 0$$

so:

$$u\left((u')^{-1}(J_v)_t, t\right) + J_t - \frac{J_v^2}{J_{vv}} \frac{(\mu - r)^2}{\sigma^2} + \left(rV_t - (u')^{-1}(J_v)_t\right) J_v$$
$$+ \frac{1}{2} \frac{J_v^2}{J_{vv}} \frac{(\mu - r)^2}{\sigma^2} = 0$$

and therefore the partial differential equation:
$$u\left((u')^{-1}(J_v)_t, t\right) + J_t + J_v(rV_t - (u')^{-1}(J_v)_t) - \frac{1}{2}\frac{J_v^2}{J_{vv}}\frac{(\mu-r)^2}{\sigma^2} = 0 \quad (12.27)$$
yielding the function J that gives, when inserted through its derivatives in the expressions (12.25) and (12.26), the optimal consumption and weights.

We now present an explicit solution in the case of a HARA utility function defined as in (10.29) and discounted at the continuous rate ρ between times 0 and t, so in the case when:
$$u(c_t, t) = \frac{1-a}{a}\left(\frac{b}{1-a}c_t + e\right)^a e^{-\rho t}$$
This is the classic case explored by Merton (1971). We obtain first the expression of the optimal consumption, using:
$$\frac{\partial u}{\partial c}(c_t^*, t) = b\left(\frac{b}{1-a}c_t^* + e\right)^{a-1} e^{-\rho t} = J_v$$
so that:
$$c_t^* = \frac{1-a}{b}\left(\left(\frac{J_v\, e^{\rho t}}{b}\right)^{\frac{1}{a-1}} - e\right)$$
and we also have:
$$u(c_t^*, t) = \frac{1-a}{a}\left(\frac{J_v\, e^{\rho t}}{b}\right)^{\frac{a}{a-1}} e^{-\rho t}$$
We can insert these two expressions into the partial differential equation (12.27) to obtain:
$$\frac{1-a}{a}\left(\frac{J_v\, e^{\rho t}}{b}\right)^{\frac{a}{a-1}} e^{-\rho t} + J_t + J_v\left(rV_t - \frac{1-a}{b}\left(\left(\frac{J_v\, e^{\rho t}}{b}\right)^{\frac{1}{a-1}} - e\right)\right)$$
$$-\frac{1}{2}\frac{J_v^2}{J_{vv}}\frac{(\mu-r)^2}{\sigma^2} = 0$$
or:
$$\frac{1-a}{a}\left(\frac{J_v\, e^{\rho t}}{b}\right)^{\frac{a}{a-1}} e^{-\rho t} + J_t + J_v\left(rV_t + \frac{(1-a)e}{b}\right)$$
$$- (1-a)\frac{J_v e^{\rho t}}{b}\left(\frac{J_v\, e^{\rho t}}{b}\right)^{\frac{1}{a-1}} e^{-\rho t} - \frac{1}{2}\frac{J_v^2}{J_{vv}}\frac{(\mu-r)^2}{\sigma^2} = 0$$
yielding:
$$\frac{1-a}{a}\left(\frac{J_v\, e^{\rho t}}{b}\right)^{\frac{a}{a-1}} e^{-\rho t} + J_t + J_v\left(rV_t + \frac{(1-a)e}{b}\right)$$
$$- \frac{a-a^2}{a}\left(\frac{J_v\, e^{\rho t}}{b}\right)^{\frac{a}{a-1}} e^{-\rho t} - \frac{1}{2}\frac{J_v^2}{J_{vv}}\frac{(\mu-r)^2}{\sigma^2} = 0$$

so that finally:

$$\frac{(1-a)^2}{a} \left(\frac{J_v \, e^{\rho t}}{b}\right)^{\frac{a}{a-1}} e^{-\rho t} + J_t + J_v \left(rV_t + \frac{(1-a)e}{b}\right)$$
$$-\frac{1}{2} \frac{J_v^2}{J_{vv}} \frac{(\mu-r)^2}{\sigma^2} = 0 \qquad (12.28)$$

We leave it to the reader to check that the expression that follows is a solution of the partial differential equation (12.28):

$$\frac{J(V_t, t)}{e^{-\rho t}} = \frac{1-a}{a} b^a \left[\frac{1}{k}\left(1 - e^{-k(T-t)}\right)\right]^{1-a} \left[\frac{V_t}{1-a} + \frac{e}{b}\frac{1}{r}\left(1 - e^{-r(T-t)}\right)\right]^a \qquad (12.29)$$

where we have set:

$$\nu = r + \frac{1}{2}\left(\frac{\mu-r}{\sigma}\right)^2 \frac{1}{1-a}$$

and:

$$k = \frac{\rho - a\nu}{1-a}$$

and where we can easily check that $J(V_T, T) = 0$, corresponding to the absence, in our example, of a utility attached to the terminal value of the portfolio.

Inserting the expression (12.29) into Eqs (12.25) and (12.26), we obtain the target couple (c^*, θ^*):

$$c_t^* = \frac{k}{1 - e^{-k(T-t)}} \left[V(t) + \frac{e}{b}\frac{1-a}{r}\left(1 - e^{-r(T-t)}\right)\right] - \frac{(1-a)e}{b} \qquad (12.30)$$

and:

$$\theta_t^* = \frac{\mu-r}{\sigma^2}\left(\frac{1}{1-a} + \frac{1}{V(t)}\frac{e}{b}\frac{1}{r}\left(1 - e^{-r(T-t)}\right)\right) \qquad (12.31)$$

Taking the restriction $e = 0$, and specifying $b = (1-a)^{\frac{a-1}{a}}$, we obtain the power function or CRRA utility function:

$$u(x) = \frac{1}{a} x^a \qquad 0 < a < 1$$

and we find the solution couple:

$$c_t^* = \frac{k}{1 - e^{-k(T-t)}} V(t) \qquad (12.32)$$

and:

$$\theta_t^* = \frac{\mu-r}{(1-a)\sigma^2} \qquad (12.33)$$

where this latter equation is identical to Eq. (11.25), giving the optimal weight in a monoperiodic optimization setting with a mean-variance criterion. We use the data for the CAC40 index and the BTF Government bonds of the previous chapter in an illustration of the dynamic framework.

Example:
Let $1 - a = 2.5$, $\mu = 4.64\%$, $\sigma(R) = 23.43\%$, and $r = 4.37\%$. Then, we find again:

$$\theta^* = 2\%$$

which is independent of time and identical to the weight calculated in the preceding chapter. Assume in addition that $\rho = r$. The consumption rate can then be expressed as follows:

$$\frac{c_t^*}{V_t} = \frac{0.0437}{1 - e^{-0.0437(T-t)}}$$

When the investment horizon $T - t$ is large enough, then:

$$\frac{c_t^*}{V_t} \simeq 0.0437$$

which shows that the consumption rate is close to the risk-free rate. However, when we are near to the investment horizon, then:

$$\lim_{t \to T} \left(\frac{c_t^*}{V_t} \right) = +\infty$$

which shows that, in the absence of a terminal utility as in this example, the consumption rate becomes infinite before the investment horizon in order to liquidate residual funds.

12.3 Optimization in the Presence of Jumps

We present here a recent model taking into account jumps (the Aït-Sahalia, Cacho-Diaz, and Hurd (2009) model) and then we illustrate it. The developments that follow will show that many assumptions must be made in the presence of jumps in order to be able to arrive at closed-form formulas.

12.3.1 *Presentation of the Model*

Let us now examine the model proposed by Aït-Sahalia, Cacho-Diaz, and Hurd (2009) that simplifies the question of the jumps of individual assets by assuming that each asset reacts with its own sensitivity to global jumps that are common to all assets. Each asset will amplify or reduce this common shock via a sensitivity coefficient denoted by γ_i. The trajectory of common shocks is modeled by a compound Poisson process. In the case of a market of dimension d, we will therefore consider the following stock dynamics, which have a simplified form with respect to those presented in Eq. (12.12):

$$\frac{dS_i(t)}{S_i(t^-)} = \mu_i \, dt + \sum_{j=1}^{d} \sigma_i^j \, dW_t^j + \gamma_i \int_{\zeta}^{+\infty} z \, N(dt, dz) \qquad (12.34)$$

where we assumed the existence of as many driving Brownian motions as there are risky assets, where N is a Poisson random measure that is not compensated (in order to take into account the hypothesis of Aït-Sahalia, Cacho-Diaz, and Hurd (2009) of a non-compensated compound Poisson process) and that is common to all of the assets, and where we assume for the sake of consistency:

$$\zeta = \max\left\{-\frac{1}{\gamma_1}, -\frac{1}{\gamma_2}, \cdots, -\frac{1}{\gamma_d}\right\}$$

As in (12.12), an implicit correlation structure links each asset to all other assets, through the multi-dimensional Brownian motion that models the diffusive part of the fluctuations of assets.

The dynamics of the portfolio of value V for a consumption c_t at time t can be obtained by simplifying Eq. (12.19) as follows:

$$\begin{aligned} dV_t &= (V_t r - c_t) \, dt + \sum_{i=1}^{d} \pi_i(t) \, (\mu_i - r) dt \\ &+ \sum_{i=1}^{d} \sum_{j=1}^{d} \pi_i(t) \, \sigma_i^j \, dW_t^j + \sum_{i=1}^{d} \pi_i(t) \, \gamma_i \int_{\zeta}^{+\infty} z \, N(dt, dz) \end{aligned} \qquad (12.35)$$

or, with respect to the weights described by $\mathbf{w}(t) = (w_i(t))_{i=1,\cdots,d}$:

$$\begin{aligned} dV_t &= (V_t r - c_t) \, dt + V_t \sum_{i=1}^{d} w_i(t) \, (\mu_i - r) dt \\ &+ V_t \sum_{i=1}^{d} \sum_{j=1}^{d} w_i(t) \, \sigma_i^j \, dW_t^j + V_t \sum_{i=1}^{d} w_i(t) \, \gamma_i \int_{\zeta}^{+\infty} z \, N(dt, dz) \end{aligned}$$

$$(12.36)$$

The infinitesimal generator (12.21) becomes, applied to the maximal expected utility (or indirect utility function) J:

$$\mathcal{A}J(V_t,t) = \frac{\partial J}{\partial t} + \frac{\partial J}{\partial v}\left(V_t\, r - c_t + \sum_{i=1}^{d}\pi_i(t)(\mu_i - r)\right)$$

$$+ \frac{1}{2}\sum_{k=1}^{d}\left(\sum_{i=1}^{d}\pi_i(t)\,\sigma_i^k\right)^2 \frac{\partial^2 J}{\partial v^2}$$

$$+ \int_{\zeta}^{+\infty}\left[J\left(V_{t^-} + \sum_{i=1}^{d}\pi_i(t)\gamma_i z, t\right) - J(V_{t^-},t)\right]\nu(dz) \tag{12.37}$$

or, with respect to the weights described by $\mathbf{w}(t) = (w_i(t))_{i=1,\cdots,d}$:

$$\mathcal{A}J(V_t,t) = J_t + J_v\left(V_t\, r - c_t + V_t\sum_{i=1}^{d}w_i(t)(\mu_i - r)\right)$$

$$+ \frac{J_{vv}}{2}V_t^2\sum_{k=1}^{d}\left(\sum_{i=1}^{d}w_i(t)\,\sigma_i^k\right)^2$$

$$+ \int_{\zeta}^{+\infty}\left[J\left(V_{t^-}\left(1 + \sum_{i=1}^{d}w_i(t)\gamma_i z\right), t\right) - J(V_{t^-},t)\right]\nu(dz) \tag{12.38}$$

or again, in vectorial notation:

$$\mathcal{A}J(V_t,t) = J_t + J_v\left(V_t\, r - c_t + V_t\,\mathbf{w}^t(t)(\boldsymbol{\mu} - \mathbf{r})\right) + \frac{J_{vv}}{2}V_t^2\,\mathbf{w}^t(t)\boldsymbol{\Sigma}\mathbf{w}(t)$$

$$+ \int_{\zeta}^{+\infty}\left[J\left(V_{t^-}(1 + \mathbf{w}^t(t)\boldsymbol{\gamma}z), t\right) - J(V_{t^-},t)\right]\nu(dz) \tag{12.39}$$

where the notation t in the exponent corresponds to a transpose and where we can come back to the variance-covariance matrix $\boldsymbol{\Sigma}$ in the following way:

$$\sum_{k=1}^{d}\left(\sum_{i=1}^{d}w_i(t)\,\sigma_i^k\right)^2 = \sum_{k=1}^{d}\left(\mathbf{w}^t(t)\boldsymbol{\sigma}_{.,k}\right)^2 = \sum_{k=1}^{d}\mathbf{w}^t(t)\boldsymbol{\sigma}_{.,k}\boldsymbol{\sigma}^t_{.,k}\mathbf{w}(t)$$

so:

$$\sum_{k=1}^{d}\left(\sum_{i=1}^{d}w_i(t)\,\sigma_i^k\right)^2 = \sum_{k=1}^{d}\mathbf{w}^t(t)\,[\sigma_{i,k}\sigma_{j,k}]_{i,j}\,\mathbf{w}(t)$$

and:

$$\sum_{k=1}^{d}\left(\sum_{i=1}^{d}w_i(t)\,\sigma_i^k\right)^2 = \mathbf{w}^t(t)\left[\sum_{k=1}^{d}\sigma_{i,k}\sigma_{j,k}\right]_{i,j}\mathbf{w}(t)$$

so that finally:
$$\sum_{k=1}^{d}\left(\sum_{i=1}^{d} w_i(t)\,\sigma_i^k\right)^2 = \mathbf{w}^t(t)\boldsymbol{\sigma}\boldsymbol{\sigma}^t\mathbf{w}(t) = \mathbf{w}^t(t)\boldsymbol{\Sigma}\mathbf{w}(t)$$

and where the goal is to solve the Hamilton–Jacobi–Bellman equation that remains in the form:
$$0 = \sup_{\mathbf{w},c}\left[e^{-\rho t}u_1(c_t) + \mathcal{A}J(V_t,t)\right] \qquad (12.40)$$

In the precise case of the article of Aït-Sahalia, Cacho-Diaz, and Hurd (2009), the investment horizon is infinite and it is therefore the following optimization that is performed (see Eq. (12.10) and the corresponding paragraph):
$$0 = \sup_{\mathbf{w},c}\left[u_1(c_t) - \rho K(V_t) + \mathcal{L}K(V_t)\right] \qquad (12.41)$$

where classically $K(V_t) = e^{\rho t}J(V_t,t)$ does not depend explicitly on time and where $\mathcal{L}K(V_t) = \mathcal{A}K(V_t) - K_t$.

Solving this equation numerically is complex in the general case. This is because adding jumps transforms the partial differential equation into a partial integro-differential equation. We present below a simple setting suggested by Aït-Sahalia, Cacho-Diaz, and Hurd (2009) that yields an explicit solution, unfortunately at the cost of strong simplifying assumptions.

12.3.2 Illustration

We wish to solve Eq. (12.41) in the presence of a unique risky asset. From Eq. (12.39), we deduce:
$$\mathcal{L}K(V_t) = K_v\left(V_t\, r - c + V_t\, w\,(\mu - r)\right) + \frac{K_{vv}}{2} V_t^2\, \sigma^2\, w^2$$
$$+ \int_{-\frac{1}{\gamma}}^{+\infty} \left[K\left(V_{t-}(1 + w\gamma z),t\right) - K\left(V_{t-},t\right)\right] \nu(dz) \qquad (12.42)$$

where w is the investment weight of the risky asset. In order to obtain the optimal weight w^*, we differentiate with respect to w the interior of (12.41) in which we have introduced (12.42). We have:
$$K_v\, V_t\, (\mu - r) + K_{vv}\, V_t^2\, \sigma^2\, w^*$$
$$+ \left(\frac{\partial}{\partial w} \int_{-\frac{1}{\gamma}}^{+\infty} \left[K\left(V_{t-}(1 + w\gamma z),t\right) - K\left(V_{t-},t\right)\right] \nu(dz)\right)_{w=w^*} = 0$$
$$(12.43)$$

In order to simplify this equation, we must obtain more information on the function K. If we assume that the utility function depends on the consumption and is of the CRRA type, i.e. if we suppose that:

$$u_1(c) = \frac{c^a}{a}$$

then, typically, we can also assume that K takes an identical form, allowing the optimization problem to be fully solved in most examples. We consider therefore, up to a multiplicative constant that will not be made explicit here, the following function:

$$K(V_t) = \frac{V_t^a}{a}$$

whose first and second order derivatives are:

$$K_v(V_t) = V_t^{a-1} = a\,\frac{K(V_t)}{V_t}$$

and:

$$K_{vv}(V_t) = (a-1)\,V_t^{a-2} = a\,(a-1)\,\frac{K(V_t)}{V_t^2}$$

We introduce these expressions in (12.43) and we obtain:

$$a\,\frac{K(V_t)}{V_t}\,V_t\,(\mu - r) + a\,(a-1)\,\frac{K(V_t)}{V_t^2}\,V_t^2\,\sigma^2\,w^*$$
$$+ \left(\frac{\partial}{\partial w}\int_{-\frac{1}{\gamma}}^{+\infty}[K(V_{t-})(1+w\gamma z)^a - K(V_{t-})]\,\nu(dz)\right)_{w=w^*} = 0$$

that can be simplified into:

$$a(\mu - r) + a(a-1)\sigma^2 w^* + \left(\frac{\partial}{\partial w}\int_{-\frac{1}{\gamma}}^{+\infty}[(1+w\gamma z)^a - 1]\,\nu(dz)\right)_{w=w^*} = 0$$

so into:

$$(\mu - r) + (a-1)\,\sigma^2\,w^* + \int_{-\frac{1}{\gamma}}^{+\infty}\gamma z\,(1+w^*\gamma z)^{a-1}\,\nu(dz) = 0 \quad (12.44)$$

It appears that Eq. (12.44) has a simple solution only if additional hypotheses are made. We assume that the relative risk aversion is precisely equal to two:

$$1 - a = 2$$

and that the Lévy measure can be expressed in the form:

$$\nu(dz) = \begin{cases} \frac{\lambda^+ dz}{z} & \forall z \in\,]0, \frac{1}{\gamma}[\\ -\frac{\lambda^- dz}{z} & \forall z \in\,]-\frac{1}{\gamma}, 0[\end{cases}$$

Then, Eq. (12.44) can be simplified into:

$$(\mu - r) - 2\sigma^2 w^* - \lambda^- \int_{-\frac{1}{\gamma}}^{0} \frac{\gamma dz}{(1+w^*\gamma z)^2} + \lambda^+ \int_{0}^{\frac{1}{\gamma}} \frac{\gamma dz}{(1+w^*\gamma z)^2} = 0$$

giving, after the change of variable $s = \gamma z$:

$$(\mu - r) - 2\sigma^2 w^* - \lambda^- \int_{-1}^{0} \frac{ds}{(1+w^*s)^2} + \lambda^+ \int_{0}^{1} \frac{ds}{(1+w^*s)^2} = 0$$

so:

$$(\mu - r) - 2\sigma^2 w^* - \lambda^- \left[-\frac{1}{w^*}\frac{1}{1+w^*s}\right]_{-1}^{0} + \lambda^+ \left[-\frac{1}{w^*}\frac{1}{1+w^*s}\right]_{0}^{1} = 0$$

which is identical to:

$$(\mu - r) - 2\sigma^2 w^* + \frac{\lambda^-}{w^*}\left[1 - \frac{1}{1-w^*}\right] - \frac{\lambda^+}{w^*}\left[\frac{1}{1+w^*} - 1\right] = 0$$

so to:

$$(\mu - r) - 2\sigma^2 w^* + \frac{\lambda^-}{w^*}\left[\frac{-w^*}{1-w^*}\right] - \frac{\lambda^+}{w^*}\left[\frac{-w^*}{1+w^*}\right] = 0$$

yielding finally:

$$(\mu - r) - 2\sigma^2 w^* - \frac{\lambda^-}{1-w^*} + \frac{\lambda^+}{1+w^*} = 0 \qquad (12.45)$$

This expression can be easily transformed into a third degree equation. If we multiply it by $(1-w^*)(1+w^*)$, we obtain:

$$(\mu - r)(1 - (w^*)^2) - 2\sigma^2 w^*(1 - (w^*)^2) - \lambda^-(1+w^*) + \lambda^+(1-w^*) = 0$$

so that, indeed:

$$2\sigma^2 (w^*)^3 - (\mu - r)(w^*)^2 - (2\sigma^2 + \lambda^- + \lambda^+)w^* + (\mu - r + \lambda^+ - \lambda^-) = 0 \quad (12.46)$$

Let us solve Eq. (12.45) in two particular cases. First of all, we assume that $\lambda^+ = \lambda^- = 0$, which amounts to assuming that the risky asset is not affected by jumps. Then:

$$(\mu - r) - 2\sigma^2 w^* = 0$$

so that:

$$w^* = \frac{\mu - r}{2\sigma^2}$$

which is indeed the Merton weight when the relative risk aversion coefficient is equal to 2. Let us assume now that $\lambda^+ = 0$: the risky asset is only affected by negative jumps. We have:

$$(\mu - r) - 2\sigma^2 w^* - \frac{\lambda^-}{1-w^*} = 0$$

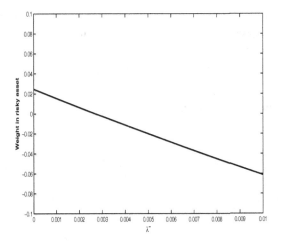

Figure 12.1 Risky asset weight with respect to λ^-

which gives, after multiplying by $(1 - w^*)$:

$$(\mu - r)(1 - w^*) - 2\,\sigma^2\,w^*(1 - w^*) - \lambda^- = 0$$

so:

$$2\sigma^2(w^*)^2 - \left(\mu - r + 2\sigma^2\right)w^* + (\mu - r - \lambda^-)$$

We shall keep the following solution to this second degree equation (this is the only solution that produces consistent results):

$$w^* = \frac{\mu - r + 2\sigma^2 - \sqrt{(\mu - r + 2\sigma^2)^2 - 8\sigma^2(\mu - r - \lambda^-)}}{4\sigma^2} \qquad (12.47)$$

To conclude, we represent w^* in Fig. 12.1 with respect to λ^- for the traditional set of parameters: $\mu = 4.64\%$, $\sigma(R) = 23.43\%$, and $r = 4.37\%$; but here we assume $1 - a = 2$. We observe not surprisingly that adding negative jumps to the modeling leads to a decrease in the investment in the risky asset.

Appendix: Dealing with Uncertainty

As for the previous chapter, we conclude our presentation of dynamic portfolio management with a study of the effect of uncertainty. We present here the particular approach of Uppal and Wang (2003). This approach is close to that of Maenhout (2004), which is itself similar to that of Anderson, Hansen, and Sargent (2000). For more information on robust control theory, see the book of Hansen and Sargent (2007).

The Uppal and Wang Model

The departure point of the Uppal and Wang (2003) approach is a generalized Bellman equation in discrete time, which we present below in a more precise form than in their text:

$$J(t) = su_1(c_t)$$
$$+ e^{-\rho s} \inf_{\xi} \left\{ \psi \left(\mathbb{E}^{Q_\xi}[J_{t+s}] \right) \sum_{i=1}^{n} \phi_i \, \mathbb{E}^{Q_{\xi_i}} \left[\ln \frac{dQ_{\xi_i}}{dP_i} \right] + \mathbb{E}^{Q_\xi}[J_{t+s}] \right\}$$
(12.48)

where the variables have the following meaning:

- P is the historical probability measure, which corresponds to assets in the real world. However, there exists uncertainty regarding P, which prevents a fair description of the dynamics of assets.
- Q_ξ is another probability measure, the relevance of which we want to examine. It corresponds to the behavior of assets in a dual world. Classically for probability changes, we move from P to Q_ξ using the density ξ, which can be understood as the density of uncertainty:

$$\xi = \frac{dQ_\xi}{dP}$$

- This density ξ can in turn be decomposed into several sub-densities denoted by ξ_i, where each sub-density is relative to a given asset.
- Q_ξ corresponds, therefore, to a collection of probability measures Q_{ξ_i}, each of which describes the asset i in the new universe.

To simplify the computations and for the sake of clarity, we did not assume the existence of subsets of $\{J_i\}$ as in the article of Uppal and Wang.

The Bellman equation presented below can be interpreted as follows. Because we want to minimize uncertainty, the goal is to minimize ξ, yielding an infimum on ξ. The terms denoted by L_i and defined as:

$$L_i = \mathbb{E}^{Q_{\xi_i}} \left[\ln \frac{dQ_{\xi_i}}{dP_i} \right]$$

are entropic expressions of the uncertainty for each asset i. Indeed, we have from the definition of mathematical expectation that:

$$L_i = \int \ln \frac{dQ_{\xi_i}}{dP_i} \, dQ_{\xi_i} = \int \frac{dQ_{\xi_i}}{dP_i} \ln \frac{dQ_{\xi_i}}{dP_i} \, dP_i$$

so:

$$L_i = \int \xi_i \ln \xi_i \, dP_i$$

which is precisely the definition of entropy. Each of these terms is weighted by a constant ϕ_i that depends on asset i. Finally, the solution is renormalized by a term $\psi E^{Q_\xi}[J_{t+s}]$.

Then, Eq. (12.48) can be extended to the continuous case in the following way:

$$0 = \inf_\nu \left\{ u(c_t) - \rho J + \mathcal{A} J(t) + \nu^t J_S + \frac{\psi(J)}{2} \nu^t \Phi \nu \right\} \qquad (12.49)$$

We can note an important aspect: uncertainty, which was described by ξ, is now contained in the vector ν. In practice, this amounts to moving from an approach based on a measure change (through ξ) to an approach based on a drift change, where uncertainty affects asset returns through ν. Indeed, the measure change allows us to move from the asset dynamics:

$$dS_t = \mu(S_t, t) \, dt + \sigma(S_t, t) \, dW_t^P$$

to the new dynamics:

$$dS_t = (\mu(S_t, t) + \nu) \, dt + \sigma(S_t, t) \, dW_t^{Q_\xi}$$

where this expression should be understood in the vectorial sense.

The interpretation of Φ in Eq. (12.49) is as follows. Assuming that assets are not correlated and that their return is affected by a specific degree of uncertainty measured by ϕ_i, then Φ is the diagonal matrix comprising the terms:

$$\Phi_{i,i} = \frac{\phi_i}{\sigma_i^2}$$

where σ_i is the volatility of asset i.

Uncertainty and Optimal Portfolio Composition

We consider Eq. (12.49) in a portfolio context where we look for an optimal consumption and optimal weights. This amounts to taking the supremum:

$$0 = \sup_{w,c} \inf_\nu \left\{ u(c_t) - \rho J + \mathcal{A} J + \nu^t J_X + \frac{\psi(J)}{2} \nu^t \Phi \nu \right\} \qquad (12.50)$$

Let us assume that the dynamics the asset are lognormal. We therefore obtain:

$$dV = (V(r + w^t(\mu - r)) - c)dt + w\sigma V dW$$

Assuming in addition a CRRA utility function, so $u(c_t) = \frac{c_t^{1-a}}{1-a}$, we have the equation:

$$0 = \sup_{w,c} \inf_{\nu} \left\{ \frac{c_t^{1-a}}{1-a} - \rho J + \mathcal{A}J + \nu^t J_V + \frac{\psi(J)}{2} \nu^t \Phi \nu \right\} \quad (12.51)$$

with:

$$\mathcal{A}J = J_t + J_V(V(r + w^t(\mu - r)) - c) + \frac{1}{2} J_{VV} w^t \sigma^2 V^2 w$$

where we recall that σ is diagonal and where:

$$\nu^t J_V = \nu_V^t J_V = w^t \nu_S V J_V$$

and:

$$\frac{\psi(J)}{2} \nu^t \Phi \nu = \frac{\psi(J)}{2} \nu_S^t w^t \Phi w \nu_S$$

Solving this equation requires two preliminary steps. First, the minimization with respect to ν_S, which yields an optimal value ν_S^*:

$$\nu_S = -\frac{1}{\psi(J)} \Phi^{-1} J_V V w$$

This value is then introduced into Eq. (12.51). We can then proceed with the minimization with respect to w and c. We obtain:

$$c^* = J_V^{-\frac{1}{a}}$$

and:

$$w^* = \left(1 - \frac{J_V{}^2}{\psi(J) J_{VV}} (\sigma^{-1})^2 \Phi^{-1}\right)^{-1} \frac{1}{a} \frac{\mu - r}{\sigma^2}$$

so:

$$w^* = \left[\frac{1}{1 - \frac{J_V{}^2}{\psi(J) J_{VV}} \frac{1}{\phi_i}}\right] \frac{1}{a} \frac{\mu - r}{\sigma^2}$$

where [.] is a diagonal matrix whose terms are indicated in the brackets. If we make the additional assumptions:

$$\psi(J) = \frac{1-a}{a} J \quad \text{and} \quad J(V) = \frac{V^{1-a}}{1-a}$$

then we observe that:
$$\frac{J_V{}^2}{\psi(J)\, J_{VV}} = -1$$
yielding readily:
$$w^* = \left[\frac{\phi_i}{1+\phi_i}\right] \frac{1}{\gamma} \frac{\mu - r}{\sigma^2} \qquad (12.52)$$
which is the simple formula given by Uppal and Wang at the end of their article.

We present an illustration in dimension $d = 1$, so with a risk-free and a risky asset. In Fig. 12.3.2, we display the proportion to be invested in the risky asset with respect to the prevailing degree of confidence regarding the predictions for the return of this asset. It appears that the higher the degree of confidence, the greater the proportion invested in the risky asset. In particular, we observe that the increase of the weight invested in the risky asset is highly non-linear: it first increases very quickly, then stabilizes to a constant level, equal here to 16%.

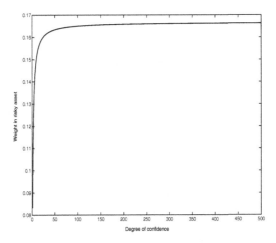

Figure 12.2 Proportion of risky asset w.r.t. degree of confidence of return predictions

Chapter 13

Conclusion

The studies conducted between the 1950s and the 1970s by Markowitz, Sharpe, Samuelson, Black, Scholes, and Merton made possible the construction of a consistent framework for the management of portfolios and the valuation of derivatives. This framework, as well as the 1992 model of Heath, Jarrow, and Morton, which unites numerous interest rate submodels, yields results and predictions that can vary considerably depending on the hypotheses that are made. Most traditional financial models rest on a Brownian representation of dynamics, so pertain to the Leibnizian continuity principle mentioned in the introduction. This principle has been abandoned in physics and genetics, but retained in economics and finance. Contrary to this natural philosophy where things change gradually, we aimed to take into account discontinuities within all financial techniques of portfolio and risk management. Modeling asset returns using Lévy processes was systematically explored and included within the classic methods dating back to the 1970s.

Although it allows for a fruitful re-examination of classic approaches and produces very interesting results, the framework studied here also has its own limitations. The independence and stationarity hypotheses for asset returns allow us to relatively easily undertake a certain number of developments. However, these are strong hypotheses that do not take into account several relevant characteristics of financial markets. Practitioners and researchers may wish to design a larger framework. The seventh chapter of this book proposed a possible extension, through the use of a stochastic clock that is not a Lévy process. Numerous developments in this direction remain to be carried out. Furthermore, we have only briefly considered dependence between financial assets. This topic deserves an ample treatment, as does the use of Lévy processes for interest rate and credit risk modeling.

Appendix A

Concentration vs Diversification

In this appendix, we examine the following question: in order to protect a portfolio against losses, is it better to diversify or to concentrate the portfolio? This question is that of the determination of the adequate level of concentration of a portfolio. This book showed how risk management is affected when smooth randomness, modeled typically by Brownian diffusions, is replaced by irregular randomness, modeled by non-Brownian Lévy processes. Characterizing the problem of portfolio concentration by formulating it in a more mathematical way remains to be addressed. Whether concentration is desirable or not, it should be possible to measure it correctly. It should also be possible to determine which of any two portfolios is the most concentrated.

Risk budget functions will be linked to the degree of concentration by the notion of Schur-convexity. If the risk budget function is Schur-convex, then the risk budget should be increasingly significant as the portfolio concentration increases (classic finance case). Conversely, if this function is Schur-concave, a highly concentrated portfolio will have a lower capital requirement. In this case, the increased concentration reduces the need for capital: it becomes less risky to invest in concentrated portfolios.

Classic Concentration Measures. Measuring concentration amounts to defining an indicator called the "degree of concentration". Portfolios can then be ranked based on their degree of concentration. We consider a portfolio consisting of n lines and we aim to characterize its degree of concentration.

A first intuitive measure is the number of lines: the larger n, the greater the diversification of the portfolio. The ratio $1/n$ can therefore represent a measure of concentration. The larger the number of lines, the closer the concentration to zero: a fully diversified portfolio will have a zero measure of concentration. Let us denote this measure by C_1:

$$C_1 = \frac{1}{n} \qquad (A.1)$$

However, if lines have different weights in the portfolio, this measure might not be relevant. For example, a portfolio comprising two lines in the proportions 99% and 1%, and another portfolio which is evenly distributed (50–50 weighting), will both have the concentration measure $C_1 = 0.5$. However, the risk of these two portfolios is not identical. Therefore, we need to introduce a second measure that takes into account the differences of weights between the lines.

A way to introduce this measure is to follow the proposition of Friend and Blume (1975), which consists in comparing the distribution of the assets in the portfolio with the distribution of a market index representing by construction a well-diversified portfolio. Let w_i^M be the weight of the asset i in the market index M. We compute the following quantity:

$$C_2 = \sum_{i=1}^{n} \left(w_i - w_i^M\right)^2$$

Assuming that the market index is well-diversified and that the weights w_i^M are weak, it is possible to make the approximation $w_i^M = 1/n$, yielding:

$$C_2 = \sum_{i=1}^{n} \left(w_i - \frac{1}{n}\right)^2 \simeq \sum_{i=1}^{n} w_i^2$$

We therefore obtain:

$$C_2 = \sum_{i=1}^{n} w_i^2 \qquad (A.2)$$

which is the Herfindhal index of the portfolio. For example, if we use these two measures to compare the two portfolios of type 99/1 and 50/50 (as well as some additional allocations), we obtain:

Allocation (in %)	C1	C2
2 lines		
50/50	0.50	0.50
75/25	0.50	0.63
80/20	0.50	0.68
90/10	0.50	0.82
99/01	0.50	0.98
3 lines		
33/33/33	0.33	0.33
90/08/02	0.33	0.82

We observe that C2 displays an increase in concentration when the weights become unbalanced, whereas all portfolios appear identical when viewed from the perspective of C1.

However, these two concentration measures do not take into consideration possible comovements of asset prices, which can have a sharp effect on the risk of the portfolio. For instance, C2 gives an identical result for two 80/20 portfolios, whether their assets belong to the same industrial sector or to different sectors. In order to include a sector effect, Mitton and Vorkink propose a third concentration measure by introducing the mean correlation of the n lines of the portfolio as a function of weights. This correlation is denoted by $\overline{\rho}$ and defined as:

$$\overline{\rho} = \frac{\sum_{i \neq j} w_i w_j \rho_{i,j}}{\sum_{i \neq j} w_i w_j}$$

With this mean correlation, the new concentration measure is:

$$C_3 = \sum_{i=1}^{n} w_i^2 + \left(1 - \sum_{i=1}^{n} w_i^2\right) \overline{\rho} \qquad (A.3)$$

From there, we deduce the following relationship between C2 and C3:

$$C_3 = C_2(1 - \overline{\rho}) + \overline{\rho}$$

When the correlation is zero, $C_3 = C_2$: we come back to the previous concentration measure. In other situations, the existence of a positive correlation strengthens risk, corresponding therefore to an increase in the concentration measure C3, and *vice versa* for a negative correlation. The table below illustrates the results obtained for various choices of the concentration measure.

Allocation (in %)	C1	C2	C3
2 lines			$\rho = 0.50$
75/25	0.50	0.63	0.81
80/20	0.50	0.68	0.84
2 lines			$\rho = -0.50$
75/25	0.50	0.63	0.44
80/20	0.50	0.68	0.52

Ranking of portfolios by concentration degree. Let us now come to a more recent ranking system for the degree of portfolio concentration. Consider two portfolios A and B parameterized in weights, i.e. two vectors of weights $w^A = (w_1^A, \cdots, w_n^A)$ where w_j^A is the investment weighting of line j for portfolio A, and $w^B = (w_1^B, \cdots, w_n^B)$. Which of these two vectors describes the most concentrated portfolio? To answer this question, we use the work of Marshall and Olkin (1979) as revisited by Ibragimov (2009).

We start by ordering the portfolio lines by decreasing weights. We keep the conventions introduced in Chapter 2: $w_{(1)} = \max(w_j)$ and $w_{(n)} = \min(w_j)$. The portfolio ordered by decreasing weight is the ordered vector $(w_{(1)}, \cdots, w_{(n)})$.

We can now introduce a ranking relation between portfolios A and B using their degree of concentration. We say that A is less concentrated (more diversified) than B, and we denote $A \prec B$ when:

$$\begin{cases} \sum_{i=1}^k w_{(i)}^A \leq \sum_{i=1}^k w_{(i)}^B & k = 1, \cdots, n-1 \\ \sum_{i=1}^n w_{(i)}^A = \sum_{i=1}^n w_{(i)}^B & k = n \end{cases} \quad (A.4)$$

It is possible to write the relation $\prec$ ("is less concentrated than") in a more compact way than (A.4), using the notion of a doubly stochastic matrix, i.e. using a matrix whose components are all positive and whose lines and columns all sum up to 1. If M is a doubly stochastic matrix, then A is less concentrated than B if $w^A = M w^B$.

Let us try to reach an intuitive understanding of this definition taking a simple example. We assume that portfolio A is composed of n iso-weighted lines and that portfolio B is composed of $n-1$ iso-weighted lines. Intuitively, portfolio A should be less concentrated than portfolio B: $A \prec B$. We will verify this using the definition previously introduced.

The vectors of the weights of the two portfolios are respectively:

$$w^A = \left(\underbrace{\frac{1}{n}, \cdots, \frac{1}{n}}_{n \text{ lines}} \right) \quad w^B = \left(\underbrace{\frac{1}{n-1}, \cdots, \frac{1}{n-1}}_{n-1 \text{ lines}}, 0 \right)$$

We compute first the partial sums for all $k = 1, \cdots, n-1$, and we check that we have indeed:

$$\sum_{i=1}^k \frac{1}{n} \leq \sum_{i=1}^k \frac{1}{n-1}$$

before verifying that:

$$\sum_{i=1}^n \frac{1}{n} = \sum_{i=1}^{n-1} \frac{1}{n-1} + 0 = 1$$

which validates the order relation between A and B.

The relation $\prec$ ("is less concentrated than") is a preorder. This means that it is both reflexive:

$$\forall w \quad w \prec w$$

and transitive:

$$\forall\ w^A, w^B, w^C \quad w^A \prec w^B \text{ et } w^B \prec w^C \Rightarrow w^A \prec w^C$$

Thus, if portfolio A is less concentrated than portfolio B, and if portfolio B is less concentrated than portfolio C, then portfolio A is less concentrated than portfolio C.

However, the relation $\prec$ is not an order relation because antisymmetry is not satisfied:

$$\forall\ w^A, w^B \quad w^A \prec w^B \text{ et } w^B \prec w^A \not\Rightarrow w^A = w^B$$

If A is less concentrated than B and if B is less concentrated than A, this does not imply that A and B have the same asset compositions. This implies the same weight compositions, which are not necessarily applied to the same assets. In other words, the vector of weights of portfolio A is defined by a permutation of the components of the vector of weights of portfolio B.

Let us introduce the notion of a permutation matrix: this is a matrix that contains one and only one non-zero coefficient equal to 1 in each line and in each column. We denote by P this permutation matrix and write:

$$\forall\ w^A, w^B \quad w^A \prec w^B \text{ et } w^B \prec w^A \Rightarrow w^A = P w^B$$

Concentration and risk budgets. We now look at the link between the concentration degree of a portfolio and the risk budget that should be allocated to this portfolio. This amounts to establishing a quantification of the relation $\prec$ ("is less concentrated than"): we need to find a mathematical tool for this purpose. We introduce now the notion of a Schur-convex function.

Let A and B two portfolios ranked according to $A \prec B$. We denote by ϕ the risk budget to be allocated to a portfolio: ϕ is a function from $\mathbb{R}^d$ to $\mathbb{R}$. This function is Schur-convex when:

$$w^A \prec w^B \quad \Rightarrow \quad \phi(w^A) \leq \phi(w^B)$$

Conversely, ϕ is Schur-concave when:

$$w^A \prec w^B \quad \Rightarrow \quad \phi(w^A) \geq \phi(w^B)$$

If the risk budget function is Schur-convex, then the risk budget will be more significant when the portfolio concentration is high. Conversely, if the risk budget function is Schur-concave, then a very concentrated portfolio will require little capital. The first case corresponds implicitly to a classic, or "orthodox" conception of finance, in accordance with the ideas of Markowitz–Tobin–Sharpe, where diversification is a protection against risk. Thus, a well-diversified portfolio does not require an important risk budget in this context. In the second case, everything is reversed because an increase in concentration reduces the need for capital: it becomes less risky to invest in concentrated portfolios. This is an approach that validates the criticisms of many practitioners: a "heterodox" conception that finds its roots in the theories of Williams and Keynes. We see how Schur-convexity allows us to establish a link between risk management and portfolio management.

An example with alpha-stable distributions. Let us illustrate these properties with a portfolio whose returns follow a stable distribution and let us choose a very general risk measure, assumed positively homogeneous. This property is satisfied by VaR.

We consider two portfolios A and B comprising assets of returns X_i and we assume that for all i we have $X_i \sim \mathcal{S}(\alpha, \gamma, 0, 0)$. The risk budgets of these portfolios are respectively equal to, in the case of VaR:

$$\phi(w^A) = \text{VaR}\left(\sum_{i=1}^n w_i^A X_i\right) \qquad \phi(w^B) = \text{VaR}\left(\sum_{i=1}^n w_i^B X_i\right)$$

If returns follow stable distributions $\mathcal{S}(\alpha, \gamma, 0, 0)$, then it is possible to show that:

$$\sum_{i=1}^n w_i^A X_i \stackrel{d}{=} \epsilon^A X \qquad \sum_{i=1}^n w_i^B X_i \stackrel{d}{=} \epsilon^B X$$

where $\epsilon^j = \left(\sum_{i=1}^n \left(w_i^j\right)^\alpha\right)^{\frac{1}{\alpha}}$ for $j = A, B$. By the positive homogeneity property of the risk measure, we have:

$$\phi(w^A) = \phi((1, 0, ..., 0))\, \epsilon^A \qquad \phi(w^B) = \phi((1, 0, ..., 0))\, \epsilon^B$$

It is thus sufficient to be able to compare ϵ^A and ϵ^B in order to determine which of the two portfolios requires the smallest amount of capital.

Set $(w_i^j)^\alpha = g(w_i^j)$. With this convention, we have $\epsilon^j = (\sum_{i=1}^n g(w_i^j))^{\frac{1}{\alpha}}$. The function $\sum_{i=1}^n g(w_i^j)$ is Schur-convex when g is convex and it is Schur-concave when g is concave (see Marshall and Olkin (1979), Chapter 3).

However, the function $g(w_i^j)$ is convex when $\alpha > 1$ and concave when $\alpha < 1$. Equipped with these results, we can now deduce the ranking of portfolios A and B depending on their required level of capital. Two situations arise.

First possibility. Stock fluctuations remain relatively regular. This corresponds to the case $\alpha > 1$:

$$w^A \prec w^B \Rightarrow \sum_{i=1}^n (w_i^A)^\alpha \leq \sum_{i=1}^n (w_i^B)^\alpha \Rightarrow \left(\sum_{i=1}^n (w_i^A)^\alpha\right)^{\frac{1}{\alpha}} \leq \left(\sum_{i=1}^n (w_i^B)^\alpha\right)^{\frac{1}{\alpha}}$$

We deduce the following relationship between risk budgets:

$$w^A \prec w^B \quad \Rightarrow \quad \epsilon^A \leq \epsilon^B \quad \Rightarrow \quad \phi(w^A) \leq \phi(w^B)$$

Second possibility. Stock fluctuations display violent moves. This corresponds to the case $\alpha < 1$:

$$w^A \prec w^B \Rightarrow \sum_{i=1}^n (w_i^A)^\alpha \geq \sum_{i=1}^n (w_i^B)^\alpha \Rightarrow \left(\sum_{i=1}^n (w_i^A)^\alpha\right)^{\frac{1}{\alpha}} \geq \left(\sum_{i=1}^n (w_i^B)^\alpha\right)^{\frac{1}{\alpha}}$$

We deduce the following relationship between risk budgets:

$$w^A \prec w^B \quad \Rightarrow \quad \epsilon^A \geq \epsilon^B \quad \Rightarrow \quad \phi(w^A) \geq \phi(w^B)$$

Let us summarize our conclusions. In the case where stock fluctuations are sufficiently regular ($\alpha > 1$), the optimal management approach consists in diversifying one's portfolio, whereas in the case where stock fluctuations are very important ($\alpha < 1$), the optimal management approach consists, on the contrary, in concentrating one's portfolio. Let us now come back to the ideas of Keynes, this time equipped with practical technical tools.

The Keynesian point of view. The historical investigation of the justifications of the practice of diversification for the management of portfolios and risks is well documented in the history of financial theories. For more details, we refer the reader to Walter (1996, 1999, 2004a, 2005). The main conclusions are as follows: the diversification that is put forward in the Markowitz–Tobin–Sharpe framework led to an important shift in the ways of perceiving the investment decisions of practitioners. This move replaced the classic preliminary thought process on the selection of individual assets with the consideration of the total portfolio of assets (strategic asset allocation). This represented a radical change with regard to the previous conceptions, specifically these defined in the classic manual of Williams (1938). Markowitz stated it extremely clearly: "my theory was in contradiction with the previous dominant theory, the one of J. B. Williams in The Theory of Investment Value" (Markowitz (1992), p. 16).

Beyond the contradiction between the "new" management approach of Markowitz and the practices of professional investors, this theory was in stark opposition to the ideas of Keynes (1942) in which investing on the stock exchange is precisely trying to make good choices of assets irrespective of any other superfluous considerations:

> I am in favor of having as large unit as market conditions will allow To suppose that safety first consists in having a small gamble in a large number of different companies where I have no information to reach a good judgment, as compared with a substantial stake in a company where one's information is adequate, strikes me as a travesty of investment policy.

"A travesty of investment policy". This is how Keynes (see p. 81 in his collected writings published in (1983)) describes the strategic asset allocation that results from the use of the linear model in finance, through the works of Markowitz, Tobin, and Sharpe.

Instead of multiple investments of small quantities of iso-weighted assets in a well-diversified portfolio, Keynes suggests holding large quantities of a small number of individual assets. We recognize here the presence of distributions of the 80/20 type, which, as explained in the previous chapters, stem from the non-Brownian nature of stock fluctuations. Indeed, the Brownian representation as employed for the modeling of financial markets can only lead to the use of maximally diversified indexes by investors (see Walter (2005)).

Therefore, it seems possible to establish a link between the probabilistic debate on the choice of representation (regular or irregular randomness, Brownian motion or non-Brownian Lévy process) and the professional debate on the pros and cons of portfolio diversification and on the risks associated with long-term protection.

Bibliography

Abramowitz, M. and Stegun, I. A. (1965). *Handbook of Mathematical Functions: with Formulas, Graphs, and Mathematical Tables* (Dover Publications, New York).

Aït-Sahalia, Y., Cacho-Diaz, J., and Hurd, T. R. (2009). Portfolio choice with jumps: a closed form solution, *Annals of Applied Probability* **19**, 2, pp. 556–584.

Aït-Sahalia, Y. and Jacod, J. (2009). Testing for jumps in a discretely observed process, *Annals of Statistics* **37**, 1, pp. 184–222.

Aït-Sahalia, Y. and Jacod, J. (2010). Is Brownian motion necessary to model high-frequency data? *Annals of Statistics* **38**, 5, pp. 3093–3128.

Alami, A. and Renault, E. (2000). Risque de modèle de volatilité, *Journal de la Société Française de Statistique* **141**, 1–2, pp. 103–136.

Albin, J. M. P. and Sundén, M. (2009). On the asymptotic behaviour of Lévy processes, part I: subexponential and exponential processes, *Stochastic Processes and their Applications* **119**, 1, pp. 281–304.

Anderson, E., Hansen, L. P., and Sargent, T. J. (2000). Robustness, detection and the price of risk, *Working paper*.

Ané, T. and Geman, H. (2000). Order flow, transaction clock, and normality of asset returns, *Journal of Finance* **55**, 5, pp. 2259–2284.

Applebaum, D. (2009). *Lévy Processes and Stochastic Calculus* (Cambridge University Press, Cambridge, 2nd edition).

Armatte, M. (1995). *Histoire du modèle linéaire*, Ph.D. thesis, EHESS.

Arrow, K. (1965). *Aspects of the Theory of Risk-bearing* (Yrjs Jahnsson foundation).

Artzner, P., Delbaen, F., Eber, J.-M., and Heath, D. (1999). Coherent measures of risk, *Mathematical Finance* **9**, 3, pp. 203–228.

Bachelier, L. (1900). *Théorie de la Spéculation* (Gauthier-Villars, Paris).

Balkema, A. and De Haan, L. (1974). Residual lifetime at great age, *Annals of Probability* **2**, 5, pp. 792–804.

Barbut, M. (1989). Distributions de type paretien et représentation des inégalités, *Mathématiques, Informatique, et Sciences Humaines* **106**, pp. 53–69.

Barbut, M. (1998). Une famille de distributions: des paretiennes aux antiparetiennes. Applications à l'étude de la concentration urbaine et de son évolution, *Mathématiques, Informatique, et Sciences Humaines* **141**, pp. 43–72.

Barndorff-Nielsen, O. E. (1997). Normal inverse gaussian distributions and stochastic volatility modeling, *Scandinavian Journal of Statistics* **24**, 1, pp. 1–13.

Barndorff-Nielsen, O. E. and Shephard, N. (2006). Econometrics of testing for jumps in financial economics using bipower variation, *Journal of Financial Econometrics* **4**, 1, pp. 1–30.

Bernoulli, D. (1738). Specimen theoriae novae de mensura sortis, *Commentarii Academiae Scientiarum Imperialis Petropolitanae* **5**, pp. 175–192.

Bernoulli, D. (1954). Exposition of a new theory on the measurement of risk, *Econometrica* **22**, 1, pp. 23–36.

Bertoin, J. (1996). *Lévy Processes* (Cambridge University Press, Cambridge).

Bienaymé, I.-J. (1867). Sur la probabilité des erreurs d'après la méthode des moindres carrés, *Journal de Mathématiques Pures et Appliquées* **17**, pp. 33–78.

Bouchaud, J.-P. and Potters, M. (2003). *Theory of Financial Risk and Derivative Pricing* (Cambridge University Press, Cambridge, 2nd edition).

Box, G. E. P. and Muller, M. E. (1958). A note on the generation of random normal deviates, *Annals of Mathematical Statistics* **29**, 2, pp. 610–611.

Boyarchenko, S. I. and Levendorskii, S. (2002). *Non-Gaussian Merton–Black–Scholes Theory* (World Scientific Publishing, Singapore).

Brown, G. W. and Tukey, J. W. (1946). Some distributions of sample means, *Annals of Mathematical Statistics* **17**, 1, pp. 1–12.

Carr, P., Geman, H., Madan, D. B., and Yor, M. (2002). The fine structure of asset returns: an empirical investigation, *Journal of Business* **75**, 2, pp. 305–332.

Carr, P., Geman, H., Madan, D. B., and Yor, M. (2003). Stochastic volatility for Lévy processes, *Mathematical Finance* **13**, 3, pp. 345–382.

Carr, P. and Wu, L. (2003). The finite moment log-stable process and option pricing, *Journal of Finance* **58**, 2, pp. 753–777.

Cauchy, A.-L. (1853). Sur les résultats moyens d'observations de même nature, et sur les résultats les plus probables, *Comptes-rendus de l'Académie des Sciences* **37**, pp. 94–104.

Chambers, J. M., Mallows, C., and Stuck, B. W. (1976). A method for simulating stable random variables, *Journal of the American Statistical Association* **7**, 354, pp. 340–344.

Clark, P. K. (1973). A subordinated stochastic process model with finite variance for speculative prices, *Econometrica* **41**, 1, pp. 135–156.

Cont, R. and Tankov, P. (2004). *Financial Modelling with Jump Processes* (Chapman and Hall, London).

Cox, J. C. and Huang, C.-F. (1989). Optimal consumption and portfolio policies when asset prices follow a diffusion process, *Journal of Economic Theory* **49**, 1, pp. 33–83.

Cox, J. C., Ingersoll, J. E., and Ross, S. A. (1985). A theory of the term structure of interest rates, *Econometrica* **53**, 2, pp. 385–407.
Cox, J. C. and Ross, S. A. (1976). The valuation of options for alternative stochastic processes, *Journal of Financial Economics* **3**, pp. 145–166.
Crum, W. L. (1923). The use of the median in determining seasonal variation, *Journal of the American Statistical Association* **18**, 141, pp. 607–614.
Drees, H., De Haan, L., and Resnick, S. (2000). How to make a Hill plot, *Annals of Statistics* **28**, 1, pp. 254–274.
Eberlein, E. and Keller, U. (1995). Hyperbolic distributions in finance, *Bernoulli* **1**, 3, pp. 281–299.
Eberlein, E., Keller, U., and Prause, K. (1998). New insights into smile, mispricing and Value at Risk: the hyperbolic model, *Journal of Business* **71**, 3, pp. 371–405.
Eeckhoudt, L. and Gollier, C. (1992). *Les Risques Financiers. Évaluation, Gestion, Partage* (Édiscience International, Paris).
Eeckhoudt, L. and Schlesinger, H. (2006). Putting risk in its proper place, *American Economic Review* **96**, 1, pp. 280–289.
Embrechts, P., Klüppelberg, C., and Mikosch, T. (1997). *Modelling Extremal Events for Insurance and Finance* (Springer, Berlin).
Engle, R. F. and Russell, J. R. (1994). Forecasting transaction rates: the autoregressive conditional duration model, *NBER working paper*.
Föllmer, H. and Schied, A. (2011). *Stochastic Finance. An Introduction in Discrete Time* (De Gruyter, Berlin, 3rd edition).
Friend, I. and Blume, M. E. (1975). The demand for risky assets, *American Economic Review* **65**, 5, pp. 900–922.
Gauss, J. C. F. (1809). *Theoria Motus Corporum Coelestium in Sectionibus Conicis Solem Ambientium* (Hamburg, F. Perthes, and I. H. Besser).
Geman, G. (2008). Stochastic clock and financial markets, in Marc Yor (ed.), *Aspects of mathematical finance* (Springer, Berlin, Heidelberg), pp. 37–52.
Gilli, M. and Këllezi, E. (2003). An application of extreme value theory for measuring risk, *Computational Economics* **27**, 2-3, pp. 207–228.
Gourieroux, C., Jasiak, J., and Le Fol, G. (1999). Intra-day market activity, *Journal of Financial Markets* **2**, 3, pp. 193–226.
Gourieroux, C. and Le Fol, G. (1997). Volatilités et mesures du risque, *Journal de la Société Française de Statistique* **138**, 4, pp. 7–32.
Hansen, L. P. and Sargent, T. J. (2007). *Robustness* (Princeton University Press, Princeton, New Jersey).
Hill, B. M. (1975). A simple general approach to inference about the tail of a distribution, *Annals of Statistics* **3**, 5, pp. 1163–1174.
Hosking, J. R. M. and Wallis, J. R. (1987). Parameter and quantile estimation for the generalized pareto distribution, *Technometrics* **29**, 3, pp. 339–349.
Ibragimov, R. (2009). Portfolio diversification and value at risk under thick-tailedness, *Quantitative Finance* **9**, 5, pp. 565–580.
Jacod, J. and Shiryaev, A. N. (2002). *Limit Theorems for Stochastic Processes* (Springer, 2nd edition).

Janicki, A. and Weron, A. (1993). *Simulation and Chaotic Behavior of α-Stable Stochastic Processes* (Marcel Dekker, New York).

Kanter, M. (1975). Stable densities under change of scale and total variation inequalities, *Annals of Probability* **3**, 4, pp. 697–707.

Keynes, J. M. (1911). The principal averages and the laws of error which lead to them, *Journal of the Royal Statistical Society* **74**, 3, pp. 322–331.

Keynes, J. M. (1983). Letter to F. C. Scott, February 6, 1942, in Donald Moggridge (ed.), *The Collected Writings of John Maynard Keynes*, Vol. 12 (Cambridge University Press, New York), pp. 81–83.

Kimball, M. (1990). Precautionary saving in the small and in the large, *Econometrica* **58**, 1, pp. 53–73.

Kimball, M. (1992). Precautionary motives for holding assets, in Peter Newman, Murray Milgate, and John Eatwell (eds.), *The New Palgrave Dictionary of Money and Finance* (Stockton Press, New York), pp. 158–161.

Koponen, I. (1995). Analytic approach to the problem of convergence of truncated Lévy flights towards the Gaussian stochastic process, *Physical Review E* **52**, 1, pp. 1197–1199.

Korn, R. and Korn, E. (2001). *Option Pricing and Portfolio Optimization: Modern Methods of Financial Mathematics* (American Mathematical Society, Providence, Rhode Island).

Kotz, S., Kozubowski, T., and Podgorski, K. (2001). *The Laplace Distribution and Generalizations: a Revisit with Applications to Communications, Economics, Engineering, and Finance* (Birkhäuser, Boston).

Kou, S. G. (2002). A jump diffusion model for option pricing, *Management Science* **48**, 8, pp. 1086–1101.

Kraus, A. and Litzenberger, R. H. (1976). Skewness preference and the valuation of risky assets, *Journal of Finance* **31**, 4, pp. 1085–1100.

Laplace (de), P. S. (1774). Mémoire sur la probabilité des causes par les événements, *Mémoires de Mathématiques et de Physique, Présentés à l'Académie Royale des Sciences par Divers Savans et Lûs dans ses Assemblées*.

Laplace (de), P. S. (1778). Mémoire sur les probabilités, *Mémoires de l'Académie Royale des Sciences de Paris*.

Le Courtois, O. (2011). On portfolio optimization with Lévy processes, *EM Lyon working paper*.

Le Courtois, O. (2012). On prudence and temperance. Illustration with monoperiodic portfolio optimization, *EM Lyon working paper*.

Le Courtois, O. and Walter, C. (2011). A study on Value-at-Risk and Lévy processes, *EM Lyon working paper*.

Lévy, P. (1924). Théorie des erreurs. La loi de gauss et les lois exceptionnelles, *Bulletin de la Société Mathématique de France* **52**, pp. 49–85.

Lévy-Véhel, J. and Walter, C. (2002). *Les Marchés Fractals. Efficience, Ruptures et Tendances sur les Marchés Financiers* (PUF, Paris).

Madan, D. B., Carr, P., and Chang, E. C. (1998). The Variance Gamma process and option pricing, *European Finance Review* **2**, 1, pp. 79–105.

Madan, D. B. and Milne, F. (1991). Option pricing with VG martingale components, *Mathematical Finance* **1**, 4, pp. 39–55.
Madan, D. B. and Seneta, E. (1990). The Variance Gamma (V.G.) model for share market returns, *Journal of Business* **63**, 4, pp. 511–524.
Maenhout, P. (2004). Robust portfolio rules and asset pricing, *Review of Financial Studies* **17**, 4, pp. 951–983.
Majri, M. (2005). Une alternative au choix du seuil des extrêmes dans le cadre d'une modélisation par la loi GPD, *Working paper*.
Mandelbrot, B. (1963). The variation of certain speculative prices, *Journal of Business* **36**, 4, pp. 394–419.
Mandelbrot, B. (1997). *Fractals and Scaling in Finance. Discontinuity, Concentration, Risk* (Springer, New York).
Mandelbrot, B. and Taylor, H. (1967). On the distribution of stock price differences, *Operations Research* **15**, 6, pp. 1057–1062.
Markowitz, H. M. (1952). Portfolio selection, *Journal of Finance* **7**, 1, pp. 77–91.
Markowitz, H. M. (1992). Discours d'introduction à la journée d'étude organisée par la société de statistique de Paris sur le thème des marchés financiers et de la gestion de portefeuille le 26 mars 1992, *Journal de la Société de Statistique de Paris* **133**, 4, pp. 13–33.
Marshall, A. W. and Olkin, I. (1979). *Inequalities: Theory of Majorization and its Applications* (Academic Press, New York).
Merton, R. C. (1969). Lifetime portfolio selection under uncertainty: the continuous-time case, *Review of Economics and Statistics* **51**, 3, pp. 247–257.
Merton, R. C. (1971). Optimum consumption and portfolio rules in a continuous-time model, *Journal of Economic Theory* **3**, 4, pp. 373–413.
Merton, R. C. (1976). Option pricing when underlying stock returns are discontinuous, *Journal of Financial Economics* **3**, 1–2, pp. 125–144.
Merton, R. C. (1992). *Continuous-Time Finance* (Blackwell, Malden, Massachusetts).
Mitton, T. and Vorkink, K. (2007). Equilibrium underdiversification and the preference for skewness, *Review of Financial Studies* **20**, 4, pp. 1255–1288.
Müller, U. A., Dacorogna, M. M., Dave, R. D., Olsen, R. B., Pictet, O. V., and von Weizscker, J. (1997). Volatilities of different time resolutions. analyzing the dynamics of market components, *Journal of Empirical Finance* **4**, 2–3, pp. 213–239.
Müller, U. A., Dacorogna, M. M., Olsen, R. B., Pictet, O. V., Schwarz, M., and Morgenegg, C. (1990). Statistical study of foreign exchange rates, empirical evidence of a price change scaling law, and intraday analysis, *Journal of Banking and Finance* **14**, 6, pp. 1189–1208.
Nunno (di), G. and Oksendal, B. (2011). *Advanced Mathematical Methods for Finance* (Springer-Verlag, Berlin, Heidelberg).
Oksendal, B. and Sulem, A. (2010). *Applied Stochastic Control of Jump Diffusions* (Springer-Verlag, Berlin, Heidelberg, 2nd edition).
Pickands, J. (1975). Statistical inference using extreme order statistics, *Annals of Statistics* **3**, 1, pp. 119–131.

Pratt, J. W. (1964). Risk aversion in the small and in the large, *Econometrica* **32**, 1–2, pp. 122–136.
Prause, K. (1999). *The generalized hyperbolic model: estimation, financial derivatives, and risk measures*, Ph.D. thesis, University of Freiburg.
Press, S. (1967). A compound events model for security prices, *Journal of Business* **40**, 3, pp. 317–335.
Quetelet, L. A. (1835). *Sur l'Homme et le Développement de ses Facultés, ou Essai de Physique Sociale, 2 volumes* (Bachelier, Paris).
Quittard-Pinon, F. (2003). *Marchés des Capitaux et Théorie Financière* (Economica, Paris, 3rd edition).
Regnault, J. (1863). *Calcul des Chances et Philosophie de la Bourse* (Mallet-Bachelier et Castel, Paris).
Rubinstein, M. E. (1973). The fundamental theorem of parameter-preference security valuation, *Journal of Financial and Quantitative Analysis* **8**, 1, pp. 61–69.
Samorodnitsky, G. and Taqqu, M. S. (1994). *Stable Non-Gaussian Random Processes* (Chapman and Hall).
Samuelson, P. A. (1965). Proof that properly anticipated prices fluctuate randomly, *Industrial Management Review* **6**, 2, pp. 41–49.
Satō, K.-I. (1999). *Lévy Processes and Infinitely Divisible Distributions* (Cambridge University Press, Cambridge).
Schoutens, W. (2003). *Lévy Processes in Finance: Pricing Financial Derivatives* (Wiley, Chichester).
SMABTP (2004a). Éléments de réflexion sur la stratégie financière, *Working paper*.
SMABTP (2004b). Méthode de gestion SMABTP, *Working paper*.
Uppal, R. and Wang, T. (2003). Model misspecification and underdiversification, *Journal of Finance* **58**, 6, pp. 2465–2481.
Van Nieuwerburgh, S. and Veldkamp, L. (2007). Information acquisition and under-diversification, *Review of Economic Studies* **77**, 2, pp. 779–805.
Vasicek, O. (1977). An equilibrium characterisation of the term structure, *Journal of Financial Economics* **5**, 2, pp. 177–188.
Walter, C. (1996). Une histoire du concept d'efficience sur les marchés financiers, *Annales. Histoire, Sciences Sociales* **51**, 4, pp. 873–905.
Walter, C. (1999). Aux origines de la mesure de performance des fonds d'investissement: les travaux d'Alfred Cowles, *Histoire & Mesure* **14**, 1-2, pp. 163–197.
Walter, C. (2004a). Le modèle linéaire dans la gestion des portefeuilles: une perspective historique, *Cahiers du Centre d'Analyse et de Mathématiques Sociales* **163**, 1, pp. 1–28.
Walter, C. (2004b). Performance concentration, *AFFI conference proceedings*.
Walter, C. (2005). La gestion indicielle et la théorie des moyennes, *Revue d'Économie Financière* **79**, 2, pp. 113–136.
Walter, C. (2011). Performation et surveillance du système financier, *Revue d'Économie Financière* **101**, 1, pp. 105–116.

Wang, S., Young, V., and Panjer, H. (1997). Axiomatic characterization of insurance prices, *Insurance: Mathematics and Economics* **21**, 2, pp. 173–183.

Williams, J. B. (1938). *The Theory of Investment Value* (Harvard University Press, Cambridge, Massachusetts).

Wilson, O. (1923). First and second law of errors, *Journal of the American Statistical Association* **18**, 143, pp. 841–852.

Index

activity, 63
approach
 largest values, 181
 perturbative, 296
aversion to tail thickness, 262

certainty equivalent, 255
change of scale, 26
coherent risk measure, 228
comonotonic additivity, 229
concentration measure, 333
concentration of results, 202
conditional tail expectation (CTE), 232
conditional value-at-risk (CVaR), 232
continuity, 230
convolution, 28
curve
 cumulative frequency, 219
 Lorenz, 203

distribution
 Cauchy, 91
 Fréchet, 184
 gamma, 120
 generalized hyperbolic, 142
 generalized inverse Gaussian, 144
 generalized Laplace, 131
 generalized Pareto, 195
 Gumbel, 184
 hyperbolic, 136
 infinitely divisible, 54
 inverse Gaussian, 92
 Jenkinson–von Mises, 186
 mixture, 116
 normal inverse Gaussian, 142
 Poisson, 55
 self-similar, 79
 stable, 78
 Weibull, 184

equation
 Bellman, 326
 Hamilton–Jacobi–Bellman, 305
 partial differential, 316
 partial integro-differential, 322

factor 3 (Basel agreement), 251
Fisher–Tippett theorem, 184
Fourier transform, 242
function
 Bessel, 141
 Schur-convex, 337

gain, 10

hierarchy of large values, 216

infinitesimal operator, 313
invariance by translation, 228
invariance in law, 229

Lévy measure, 62
 stable, 82

Lévy–Khintchine formula, 60
  Laplace, 115
  stable, 81, 86
Laplace motion, 117
law of 80/20, 202
laws
  of errors, 106
  of maxima, 182

maxima of Lévy processes, 190
maximum domain of attraction, 187
mean excess loss, 232
mean loss beyond VaR, 231
mean-reverting clock, 174
model
  Carr and Wu, 100
  CGMY, 102
  Koponen, 102
  Mandelbrot, 100
model risk, 235
moment, 19, 67
monotonicity, 228

optimization program
  dynamic, 305
  monoperiodic, 277

Pearson–Fisher coefficient, 19, 35
performance concentration, 208
positive homogeneity, 228
process
  auto-regressive, 174
  CGMY, 102, 171
  CIR, 175
  compensated compound Poisson, 59
  compensated Poisson, 57
  compound Poisson, 57
  gamma, 124
  ICIR, 177
  Lévy, 60
  Laplace, 134
  Poisson, 56
  time change, 177
  time-changed Laplace, 173
  variance gamma, 117, 170

prudence, 261
psychological value, 254

random measure
  compensated Poisson, 310
  Poisson, 309
return
  continuous, 10
  simple, 10
risk aversion
  absolute, 260
  relative, 261
risk premium, 256

simulation
  Box–Muller, 97
  Cauchy distribution, 96
  centered reduced stable distribution, 98
  Chambers–Mallows–Stuck, 98
  exponential distribution, 96
social time, 149
sources of performance, 15
strategy
  investment, 12
  self-financed, 16
sub-additivity, 229
subordination, 153
subordinator, 155

tail conditional expectation (TCE), 232, 245
taste for asymmetry, 261
temperance, 262
test
  continuity, 45
  discontinuity, 39
  finiteness of activity, 47
  Jarque-Bera, 37
  Kolmogorov, 37
  normality, 34
threshold exceedances, 194
time change, 148

uncertainty, 297, 325
utility function

CARA, 269
CRRA, 273
DARA, 268
exponential, 271
HARA, 270
IARA, 268

logarithmic, 271
power, 271

value-at-risk (VaR), 231, 243
variation, 64

Printed in the United States
By Bookmasters